REFERENCE COORDINATE SYSTEMS FOR EARTH DYNAMICS

ASTROPHYSICS AND SPACE SCIENCE LIBRARY

A SERIES OF BOOKS ON THE RECENT DEVELOPMENTS
OF SPACE SCIENCE AND OF GENERAL GEOPHYSICS AND ASTROPHYSICS
PUBLISHED IN CONNECTION WITH THE JOURNAL
SPACE SCIENCE REVIEWS

VOLUME 86
PROCEEDINGS

REFERENCE COORDINATE SYSTEMS FOR EARTH DYNAMICS

PROCEEDINGS OF THE 56TH COLLOQUIUM
OF THE INTERNATIONAL ASTRONOMICAL UNION HELD IN
WARSAW, POLAND, SEPTEMBER 8–12, 1980

Edited by

E. M. GAPOSCHKIN

Smithsonian Astrophysical Observatory, Cambridge, MA, U.S.A.

and

B. KOŁACZEK

Polish Academy of Sciences, Warsaw, Poland

Springer-Science+Business Media, B.V.

Library of Congress Cataloging in Publication Data
Main entry under title:

Reference coordinate systems for earth dynamics.

    (Astrophysics and space science library ; v. 86. Proceedings)
    Includes index.
    1. Coordinates—Congresses. 2. Geodynamics—
Congresses. I. Gaposchkin, E. M. II. Kołaczek, Barbara.
III. International Astronomical Union. IV. Series: Astrophysics
and space science library ; v. 86. V. Series: Astrophysics and
space science library. Proceedings.
QB147.R43   526.6   81–5066
                                 AACR2
ISBN 978-94-009-8458-5   ISBN 978-94-009-8456-1 (eBook)
DOI 10.1007/978-94-009-8456-1

TABLE OF CONTENTS

PREFACE

The IAU Colloquium No. 56, the Second IAU Colloquium, "On Reference Coordinate Systems for Earth Dynamics," co-sponsored by the COSPAR and the International Association of Geodesy of IUGG was held in Warsaw, Poland, on September 8-12, 1980.

The Colloquium was organized by the Space Research Centre of the Polish Academy of Sciences and the Smithsonian Institution Astrophysical Observatory. It was sponsored by the Committee of Astronomy, the Committee of Geodesy, and the Committee of Space Research of the Polish Academy of Sciences. The first Colloquium devoted to this subject was held in Toruń, Poland, in 1974. The Scientific Organizing Committee consisted of:

Cochairmen

Dr. E. M. Gaposchkin	USA
Dr. B. Kołaczek	Poland

Members of the Program Committee

Prof. J. Kovalevsky	France
Prof. I. I. Mueller	USA
Prof. M. Rochester	Canada

Members

Dr. M. Bursa	Czechoslovakia
Dr. H. K. Eichhorn	USA
Prof. W. Fricke	FRG
Dr. E. Hog	Denmark
Dr. Y. Kozai	Japan
Dr. Y. S. Yatskiv	USSR

The Local Organizing Committee consisted of:

Dr. B. Kołaczek, Chairman
Dr. W. Pachelski, Secretary
Dr. W. Dobaczewska
Dr. J. Kryński
Dr. G. Sitarski
Prof. J. Smak
Dr. J. B. Zieliński
Mr. W. Zarnowiecki

E. M. Gaposchkin and B. Kołaczek (eds.), Reference Coordinate Systems for Earth Dynamics, ix–x.
Copyright © 1981 by D. Reidel Publishing Company.

The duties of chairmen of the sessions were performed by Dr. J.
Zieliński, Dr. P. Bender, Prof. W. Fricke, Dr. B. Kołaczek,
Prof. K. Lambeck, Prof. H. Moritz, Prof. G. Veis, Dr. J. D.
Mulholland, Prof. M. C. Rochester, Dr. C. Murray, and Dr. E. M.
Gaposchkin. In addition, the occasion of the Colloquium was used
to have a meeting of the IAU Working Group on Nutation organized
by Dr. K. Seidelmann and a meeting of IAG Commission VIII headed
by Prof. I. I. Mueller. The volume contains the texts or ab-
stracts of papers presented at the Colloquium. An analysis and
synthesis of what was learned and the consensus of the meeting on
progress made and future prospects written by Prof. J. Kovalevsky
and Prof. I. I. Mueller is also included.

This Colloquium was brought about by the unprecedented increase in accuracy of metric measurements from the ground and
space, achieved in recent years, especially since the first
Colloquium, and in the future, as well as significant progress of
theory and model development. Geodesists and geophysicists are
re-examining the principles and methods used for establishing
terrestrial reference frames in the light of new understanding of
the mobile Earth. Satellite and solar system dynamics require
more careful definition. Galactic and extragalactic stellar
reference frames are being redefined. All of these reassessments
are fundamental, and none is independent of the others. This
Colloquium brought together a unique combination of geophysicists,
geodesists, dynamicists, astronomers, and astrometricists to discuss their common problems with reference frames and reference
systems, each from his own point of view. The discussions were
penetrating and on many points a consensus was actually obtained.

The editors want to thank Ms. J. Horn and Ms. Korniewicz for
assistance in preparing the proceedings, Ms. P. Looney and Mr. S.
Terry for excellent typing of manuscripts, the program committee
for reviewing all the papers, the referees at the meeting who
helped to insure a high standard of the proceedings, and Y. Bock
for compiling the index. However, any errors in compiling these
proceedings are the responsibility of the editors.

The organizers are very grateful for the financial support
of the Polish Academy of Sciences (PAS), the Polish Astronomical
Society, the Smithsonian Institution, IAU, COSPAR, and to the
Copernicus Astronomical Centre of the PAS for making the meeting
facilities available.

LIST OF PARTICIPANTS

AARDOOM, L.	An Apeldoorn, The Netherlands
ADAM, J.	Budapest, Hungary
ANDERSON, A. J.	Hällby, Uppsala, Sweden
BARAN, W.	Olsztyn, Poland
BEM, J.	Wrocław, Poland
BENDER, P. L.	Boulder, Colorado, USA
BIENIEWSKA, H.	Warszawa, Poland
BORKOWSKI, K.	Warszawa, Poland
BOUCHER, C.	St. Mande, France
BRUMBERG, V. A.	Leningrad, USSR
BRZEZIŃSKI, A.	Warszawa, Poland
BUJAKOWSKI, K.	Krakow, Poland
BUTKIEWICZ, E.	Borowiec, Poland
CALAME, O.	Grasse, France
CANNON, W. H.	Toronto, Canada
CHANDA, S. R.	Sriharikota, India
CHOJNICKI, T.	Warszawa, Poland
CZARNECKI, K.	Warszawa, Poland
DEBARBAT, S.	Paris, France
DOBACZEWSKA, W.	Warszawa, Poland
DOBRZYCKA, M.	Warszawa
DOMARADZKI, S.	Warszawa
DOMIŃSKI, I.	Borowiec
DREWES, H.	München, Federal Republic of Germany
EMENIKE, E. N.	Enugu, Nigeria
FEISSEL, M.	Paris, France
FRICKE, W.	Heidelberg, Federal Republic of Germany
FROESCHLÉ, M.	Grasse, France
GAJDEROWICZ, I.	Olsztyn, Poland
GALAS, R.	Warszawa, Poland
GAMBIS, D.	Paris, France
GAPOSCHKIN, E. M.	Cambridge, Massachusetts, USA
GENDT, G.	Potsdam, German Democratic Republic
GOAD, C. C.	Rockville, Maryland, USA
GÓRAL, W.	Kraków, Poland
GRZEDZIELSKI, S.	Warszawa, Poland
GUINOT, B.	Paris, France
HALMOS, F.	Sopron, Hungary
HEMENWAY, P. D.	Austin, Texas, USA
HEMMLEB, G.	Potsdam, German Democratic Republic
HURNIK, H.	Poznań, Poland
IWANOWSKA, W.	Toruń, Poland
JAKŚ, W.	Borowiec, Poland
JASECKI, M.	Olsztyn, Poland

KACZOROWSKI, M.	Warszawa, Poland
KALMAR, J.	Sopron, Hungary
KAMELA, Cz.	Warszawa, Poland
KANIUTH, K.	München, Federal Republic of Germany
KELM, R.	München, Federal Republic of Germany
KLOKOCNIK, J.	Ondrejov, Czechoslovakia
KOŁACZEK, B.	Warszawa, Poland
KOVALEVSKY, J.	Grasse, France
KOWALCZYK, Z.	Warszawa, Poland
KOZAI, Y.	Tokyo, Japan
KOZIEL, K.	Wisla, Poland
KREMERS, H.	West Berlin
KRYSZKIEWICZ, E.	Poznań, Poland
KURZYŃSKA, E.	Poznań, Poland
LACHAPELLE, G.	Ottawa, Canada
LAMBECK, K.	Canberra, Australia
LAMPARSKI, J.	Olsztyn, Poland
LASKOWSKI, P.	Warszawa
LATKA, J.	Warszawa, Poland
LEHMANN, M.	Borowiec, Poland
LELGEMANN, D.	Frankfurt, Federal Republic of Germany
LIESKE, J. H.	Pasadena, California, USA
LYSZKOWICZ, A.	Olsztyn, Poland
MADSEN, F.	Charlottenlund, Denmark
MAJEWSKI, E.	Warszawa, Poland
MALKOWSKI, M.	Warszawa, Poland
MANABE, S.	Mizusawa, Japan
MARCINIAK, J.	Borowiec, Poland
MARCUS, E.	Bucharest, Romania
McCARTHY, D. D.	Washington, D. C., USA
MEYER, C.	Grasse, France
MICHAEL, W. H.	Hampton, Virginia, USA
MICHALSKI, W.	Warszawa, Poland
MIETELSKI, J.	Kraków, Poland
MILANI, A.	Pisa, Italy
MIRONOV, N. T.	Kiev, USSR
MORITZ, H.	Graz, Austria
MORRISON, L. V.	Hailsham, England
MUELLER, I. I.	Columbus, Ohio, USA
MULHOLLAND, J. D.	Austin, Texas, USA
MURRAY, C. A.	Hailsham, England
NASKRECKI, W.	Poznań, Poland
NIEMEIER, W.	Hannover, Federal Republic of Germany
NOBILI, A. M.	Pisa, Italy
OPALSKI, W.	Warszawa, Poland
OSZCZAK, S.	Olsztyn, Poland
OTERMA, L.	Turku, Finland
PACHELSKI, W.	Warszawa, Poland
PAPO, H. B.	Haifa, Israel
PASZOTTA, S.	Olsztyn, Poland
PELLINEN, L. P.	Moscow, USSR
PINES, S.	New York, New York, USA

POMA, A.   Cagliari, Italy
RICHTER, B.   Stuttgart, Federal Republic of Germany
ROBERTSON, D. S.   Rockville, Maryland, USA
ROCHESTER, M. G.   Newfoundland, Canada
ROGOWSKI, J.   Warszawa, Poland
RUTKOWSKA, M.   Warszawa, Poland
SACERDOTE, F.   Pisa, Italy
SCHILLAK, S.   Borowiec, Poland
SEEBER, G.   Hannover, Federal Republic of Germany
SEEGER, H.   Bonn, Federal Republic of Germany
SEIDELMANN, P. K.   Washington, D. C., USA
SHELUS, P. J.   Austin, Texas, USA
SHKODROV, V. G.   Sofia, Bulgaria
SITARSKI, G.   Warszawa, Poland
ŚLEDZIŃSKI, J.   Warszawa, Poland
SMITH, M. L.   Boulder, Colorado, USA
SOOP, E. M.   Darmstadt, Federal Republic of Germany
STANGE, L.   Dresden, German Democratic Republic
STEINERT, G.   Dresden, German Democratic Republic
STOFFELS, K. G.   West Berlin
STRUGALSKA, E.   Warszawa, Poland
ŚWIATEK, K.   Olsztyn, Poland
ŚWIERKOWSKA, S.   Poznań, Poland
SZALIŃSKA, E.   Warszawa, Poland
SZPUNAR, W.   Warszawa, Poland
TEICHRT, B.   West Berlin
VAN GELDER, B. H. W.   Delft, Netherlands
VARGA, M.   Budapest, Hungary
VEIS, G.   Athens, Greece
WALTER, H. G.   Heidelberg, Federal Republic of Germany
WILKINS, G. A.   Hailsham, England
WNUK, E.   Poznań, Poland
WYTRZYSZCZAK, I.   Poznań, Poland
XIAO, N.   Paris, France
YOKOYAMA, K.   Mizusawa, Japan
ZABEK, Z.   Warszawa, Poland
ZARNOWIECKI, W.   Warszawa, Poland
ZARNOWSKA, J.   Warszawa, Poland
ZHONGOLOVITCH, I. D.   Leningrad, USSR
ZIELIŃSKI, J. B.   Warszawa, Poland
ZIOLKOWSKI, K.   Warszawa, Poland

REFERENCE COORDINATE SYSTEMS FOR EARTH DYNAMICS: A PREVIEW

Ivan I. Mueller
Dept. of Geodetic Science, Ohio State University
Columbus, Ohio 43210 USA

ABSTRACT. A common requirement for all geodynamic investigations is a well-defined coordinate system attached to the earth in some
prescribed way, as well as a well-defined inertial coordinate system in
which the motions of the terrestrial system can be monitored. This
paper deals with the problems encountered when establishing such coordinate systems and the transformations between them. In addition, problems related to the modeling of the deformable earth are discussed.

1. INTRODUCTION

Geodynamics has become the subject of intensive international research during the last decade, involving plate tectonics, both on the
intra-plate and inter-plate scale, i.e., the study of crustal movements,
and the study of earth rotation and of other dynamic phenomena such as
the tides. Interrelated are efforts improving our knowledge of the gravity and magnetic fields of the earth. A common requirement for all
these investigations is the necessity of a well-defined coordinate system (or systems) to which all relevant observations can be referred and
in which theories or models for the dynamic behavior of the earth can be
formulated. In view of the unprecedented progress in the ability of
geodetic observational systems to measure crustal movements and the rotation of the earth, as well as in the theory and model development,
there is a great need for the definition, practical realization, and
international acceptance of suitable coordinate system(s) to facilitate
such work. Manifestation of this interest has been the numerous specialized symposia organized during the past decade or so, such as those
held in Stresa [Markowitz and Guinot, 1968], Morioka [Melchior and Yumi,
1972; Yumi, 1971], Torun [Kołaczek and Weiffenbach, 1974], Columbus
[Mueller, 1975b and 1978], Kiev [Fedorov, Smith and Bender, 1980] and
San Fernando [McCarthy and Pilkington, 1979]. There seems to be general
agreement that only two basic coordinate systems are needed: a Conventional Inertial System (CIS), which in some "prescribed way" is attached
to extragalactic celestial radio sources, to serve as a reference for
the motion of a Conventional Terrestrial System (CTS), which moves and

1

rotates in some average sense with the earth and is also attached in
some "prescribed way" to a number of dedicated observatories operating
on the earth's surface. In the latter, the geometry and dynamic behavior of the earth would be described in the relative sense, while in the
former the movements of our planetary system (including the earth) and
our galaxy could be monitored in the absolute sense. There also seems
to be a need for certain interim systems to facilitate theoretical calculations in geodesy, astronomy, and geophysics as well as to aid the
possible traditional decomposition of the transformations between the
frames of the two basic systems. This scheme is shown in the figure
below. The Earth Model block represents the current best knowledge of
the geometry and dynamic behavior of the earth, partially deduced from
the measurements made at the Dedicated Observatories. This model is
continuously improving as more data of increasing accuracy becomes available, and it includes both the local (L) and global (G) phenomena which
have theoretical foundations based on physical reality and are mathematically describable. In the final and ideal situation, which may be
achieved only after several iterations over an extended period of time,
the global part of the model should be identical to the connection between the CIS and CTS frames. Departures (v) from the model (L') observed at the observatories (j) or at other stations (i) are of course
most important since they represent new information based on which the
model can be improved, after observational random and systematic errors
have been taken into proper consideration. The model could eventually
include the solid earth as well as the oceans and the atmosphere.

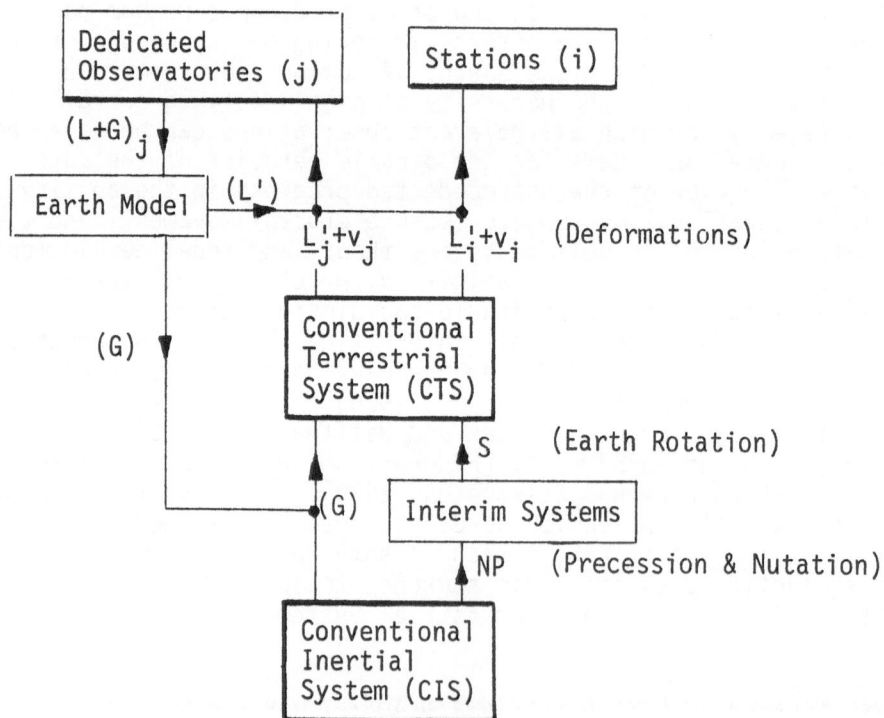

As we will see later, there already seems to be understanding on how the two basic reference systems should be established; certain operational details need to be worked out and an international agreement is necessary. There are, however, a number of more or less open questions which will have to be discussed further. These include the type of interim systems needed and their connections to both CIS and CTS, the type(s) of observatories, their number and distribution, whether all instruments need to be permanently located there or only installed at suitable regular intervals to repeat the measurements; how far the model development should go so as not to become impractical and unmanageable; and how independent observations should be referenced to the CTS, i.e., what kind of services need to be established and by whom.

2. CONVENTIONAL INERTIAL SYSTEMS (CIS) OF REFERENCE

2.1 Basic Considerations

The first law of Newton is as follows: "Every body persists in its state of rest or uniform motion in a straight line unless it is compelled to change that state by forces impressed on it" [Newton, 1686]. It should be obvious that the above *law of inertia* cannot hold in any arbitrary reference frame so that only certain specific reference frames are acceptable. In classical mechanics, reference frames in which the above law is valid are called *inertial frames*. Such "privileged" frames move through space with a constant translational velocity but without rotational motion. Another privileged frame in classical mechanics is the *quasi-inertial*, which also moves without rotational motion, but its origin may have acceleration. Such a frame would be, for example, a non-rotating geocentric Cartesian coordinate system whose origin due to the earth orbit around the sun would move with a non-constant velocity vector. Inertial reference frames thus are either at rest or are in a state of uniform rectilinear motion with respect to *absolute space*, a concept also mentioned by Newton and visualized as being observationally defined by the stars of invariable positions, a dogma in his time.

The refinement of classical mechanics through the theory of relativity requires changes in the above concepts. The theory of special relativity allows for privileged systems, such as the inertial frame but in the *space-time continuum* instead of the absolute space [Moritz, 1967]. Transformation between inertial frames in the theory of special relativity are through the so-called Lorentz transformations, which leave all physical equations, including Newton's laws of motion, and the speed of light invariant. The special theory of relativity holds only in the absence of a gravitational field.

In the theory of general relativity, Einstein defined the inertial frames as "freely falling coordinate systems" in accordance with the local gravitational field which arises from all matter of the universe. Thus the inertial frames lose their privileged status. Concerning the existence of inertial frames in the extended portions of the space-time

continuum, Einstein [1956] states that
   "there are finite regions, where, with respect to a suitably
   chosen space of reference, material particles move freely
   without acceleration, and in which the laws of special relativity hold with remarkable accuracy."
In other words, one can state [Weinberg, 1972] that
   "At every space-time point in an arbitrary gravitational field,
   it is possible to choose a locally inertial coordinate system
   such that, within sufficiently small region of the point in
   question, the laws of nature take the same form as in unaccel-
   erated Cartesian Coordinate system in the absence of gravitation."
(i.e., as in the theory of special relativity). Our sphere of interest,
the area of the solar system, where the center of mass of the earth-moon
system is "falling" in an elliptic orbit around the sun, in a relatively
weak gravitational field, seems to qualify as such a "small region."
Thus we may assume that inertial or quasi-inertial frames of reference
exist, and any violation of principles when using classical mechanics
can be taken into account with small corrections appropriately applied
to the observations and by an appropriate "coordinate" time reference.
The effects of special relativity for a system moving with the earth
around the sun are in the order of 10^{-8}, while those of general relativity are 10^{-9} [Moritz, 1979]. Since 10^{-8} on the earth's surface corresponds to about 6 cm, corrections at least for special relativity effects
are needed when striving for such accuracies. Other than this, the prob
lem, in the conceptual sense, need not be considered further.

2.2 Inertial Systems in Practice

   2.21 Extragalactic Radio Source System. This system is attached to
radio sources which generally either are quasi-stellar objects (quasars)
or galactic nuclei. Very long baseline interferometers rotating with the
earth determine the declinations of these sources with respect to the
instantaneous rotation axis of the earth, as well as their right ascension differences with respect to a selected source (3C273, NRAO 140,
Persei (Algol), etc.). In addition, the observations also determine
changes in the earth rotation vector with respect to a selected initial
state, the baseline itself, and certain instrumental (clock) corrections
The frame of the Radio Source-CIS can be defined by the adopted true or
mean coordinates of appropriately selected sources referred to some
standard epoch. The mean coordinates naturally will depend on the model
of the transformation from the true frame of date to the adopted mean
standard. If, however, the reduction procedure is correct (see more on
this later), there are no known reasons for non-radial relative motions
of the sources, i.e., for the rotation of the frame. Thus, such a frame
could be considered inertial or at least quasi-inertial. The equatorial
system of coordinates may be retained for convenience, but the frame
could be attached to the sources in any other arbitrary way should this
be necessary.

   As far as the accuracy of the Radio Source-CIS is concerned, the
question has meaning only in the sense of the formal precisions of the

source positions in the catalogue. At the Torun meeting, this number
was 0.1 [Moran, 1974]; now it is at most 0".01 [Purcell et al., 1980].
It is hoped that within a few years the precision should reach 0".001
(5×10^{-9}). The problem on this level is that the densification of such
a catalogue will be very difficult, since only a relatively few well-
defined point-like radio sources have been observed. Others have structures such that identification of the center of the radiation with such
accuracy may not be possible. This situation may change when the astrometric satellites (see below) are launched.

2.22 <u>Stellar System</u>. This system will be attached to stars in the
FK5 catalogue, i.e., the adopted right ascensions and declinations of
the FK5 stars will define the equator and the equinox and thus the frame
of the Stellar-CIS. The FK5, to be effective in 1984, will be the fifth
fundamental catalogue in a series which began with the FC in 1879
[Fricke and Gliese, 1978]. In the fundamental catalogues the equator is
determined from zenith distance (or distance difference) observations of
the stars themselves, but the equinox determination also necessitates
measurements of the sun or other members of the planetary system. It
was always tacitly assumed that coordinate systems attached to the fundamental catalogues were quasi-inertial. However, as more and more observations became available for proper motions and on the various members of the planetary systems, certain small rotations were discovered,
which require changes in the positions of the fundamental equator and
equinox, in the proper motions and in the precessional constant (all in-
tricately interwoven) when one fundamental catalogue replaces the other.
This slow and painstaking process should lead to a quasi-inertial system
eventually. We hope that the FK5 will be such a system.

When the FK4 was compiled, a small definitive correction to the
declination of FK3 was applied, but there seemed to be no need to change
the position of the equinox or the precessional constant [Fricke, 1974].
The FK5 will be a considerably different and improved catalogue. The
main changes with respect to the FK4, regarding the issue of the coordinate systems, are as follows [Fricke, 1979a]: (1) New value of general
precession in longitude adopted by the IAU in 1976 will be used (more on
this later). (2) The centennial proper motions in right ascension will
be increased by 0.086/century (this number is provisional) to eliminate
the motion of the FK4 equinox with respect to the dynamical equinox
(the FK4 right ascensions are decreasing with time due to an error in the
FK4 proper motions, see below). (3) Rotation of the FK4 equinox at 1950
by the amount of 0.040 (also a provisional value)so that the FK5 and the
dynamic equinoxes will be identical (the FK4 right ascensions at 1950
are too small). (4) Elimination of inhomogeneities of the FK4 system by
means of absolute and quasi-absolute observations. (5) Determination of
individual correction to positions and proper motions of FK4 stars. (6)
Addition of new fundamental stars to extend the visual magnitude from
7.5 to about 9.2. More than 1500 new stars are to be added.

It should be mentioned that the above improvements are possible because of the availability and/or reanalysis of observations of the sun

(1900-1970), of lunar occultations (1820-1977), of Mars (1941-1971), of minor planets (1850-1977), and the JPL DE-108 Ephemeris based on optical or radar observations of the sun, planets and some space probes (Mariner 9, Viking). All in all the number of these observations exceeds 350,000. In addition, more than 150 catalogues of star observations have become available since the completion of the FK4 [Fricke, 1979b].

One should also take note here of the FK5sup catalogue, which will contain the FK5 coordinates of a few extragalactic radio sources with radio and optical positions and thus provide the connections between the Stellar-CIS and the Radio Source-CIS, though with somewhat limited accuracy ($\sim0\overset{''}{.}1$). Improvement of this particular problem is expected from the Space Telescope [Van Altena, 1978] which could increase the number of radio stars, observable by VLBI, in the FK5 to about 50. Such missions (e.g., Hipparcos) could also contribute to the determination of the fundamental equator and equinox with increased accuracies, by observations of the minor planets. This, of course, would mean improved ties with the planetary-CIS (discussed below) which nowadays is based on the observations mentioned in connection with the establishment of the FK5 equator and equinox. The astrometric satellite Hipparcos is described to be able to measure relative positions of some 100,000 stars to a precision of $0\overset{''}{.}0015$ and annual proper motions to $0\overset{''}{.}002$ over a lifetime of 2.5 years [Barbieri and Bernacca, 1979]. A second mission ten years later could improve this figure by a factor of 5. This compares well indeed with the precision of ground based observations of $0\overset{''}{.}04$ at best, requiring something like 50 years to obtain proper motions of comparable precision ($0\overset{''}{.}002$).

As far as the accuracy of the FK5-CIS is concerned, the question again is meaningful only in the sense of how precise the star positions in the FK5 will be. It is hoped that in the worst regions this will not be worse than $0\overset{''}{.}02$ in position and $0\overset{''}{.}0015$ in the annual proper motion. There should be better regions, of course.

2.23 <u>Dynamical Systems</u>. The dynamics expressed in the equations of motion define a number of non-rotating planes which could be the basis of reference frames. Considering the observable planes that could be the basis of such a Dynamic-CIS, there are the planetary (including the earth-moon barycenter) orbital planes, the equator, the lunar orbital plane, and the orbital planes of certain high flying, thus only slightly perturbed, artificial earth satellites (e.g., Lageos or GPS). Since all of these planes have relative rotations, it is possible to derive a mean plane for a given epoch from an observable apparent plane, or a non-observable invariant plane could be adopted [Duncombe et al., 1974]. At this point, the definition of the origin of the system becomes important also, because relativisitc effects necessitate the distinction between proper and coordinate times. In the radio-source or stellar quasi-inertial systems, the question of origin can be settled through appropriate corrections for aberration and parallax, etc., but here it is also necessary that a uniform and unambiguous time scale referenced to a non-rotating frame of specified origin be established

(coordinate time). The practical implications of a global coordinate
time scale is not treated here, but the problem should not be ignored
(cf. [Ashby and Allan, 1978]). In more practical (observational) terms
one can distinguish between Planetary, Lunar and (artificial) Satellite
CIS's, each frame defined, in theory, by two of the above-mentioned
planes, and in practice, by the available ephemerides.

In the case of the *planetary systems*, the defining planes are the
equator and the ecliptic, their intersection being the line of the equinoxes. In practical terms the frame of the Planetary-CIS is defined by
the ephemerides of the centers of masses of the planets, including the
barycenter of the earth-moon system. The ephemerides, such as the JPL
DE-108 mentioned earlier, are based on observations of the sun, the
planets, possibly space probes. Since most modern ephemerides are computed through the numerical integration of the orbital equations of motion, the degree of satisfaction that can be obtained depends only on
the completeness of the modeling, including the astronomical constants,
the determination of the starting conditions and, of course, on the type,
accuracy and distribution of the observed data. In this sense each planetary ephemeris defines its own reference frame. These should agree with
each other within the observational accuracies. Connection between the
Planetary-CIS's and the Stellar-CIS's is through the determination of
the equinox and the equator, as explained earlier.

In the case of the *lunar system*, the main references are the orbital
plane of the moon and the equator of the earth. In practice the Lunar-
CIS frame is again defined by the lunar ephemeris, which nowadays is most
accurately determined from lunar laser observations made from the surface
of the earth to reflectors deposited on the lunar surface. For this reason, the adequacy of the definition also depends on how well the lunar
rotation (librations) can be computed. Since the most frequently used
lunar ephemerides are generally calculated through numerical integration,
the above dependence on modeling (especially on the effect of tidal dissipation in the earth), and on initial conditions, apply here also. The
identity of the coordinate frame, such defined, may be compared to the
other frames to certain accuracies. Lunar occultation of stars, or the
earlier Markowitz moon-camera photography, provide a connection to the
Stellar-CIS; differential VLBI observations between radio sources deposited on the moon and the extragalactic ones would tie to the Radio Source-
CIS. The connection to the Planetary-CIS is through solar eclipse observations, and also through the planetary ephemeris used when calculating
the lunar ephemeris. There are also some other looser connections stem-
ming from the orientation of the earth when its non-spherical gravitational effects on the lunar motions are taken into consideration. Present observations reveal a residual rotation (or accelerations) in the
order of a few seconds of arc per century squared. This seems to be the
present stability (i.e., the accuracy) of this quasi-inertial frame. It
is unlikely that without stronger connections to a frame of better stability, this rotation can be eliminated. As it is, the accuracy of this
CIS should compare favorably with that defined by the FK5 but only over
a period of, say, a decade [Kovalevsky, 1979].

In the case of *satellite systems*, the problem is compounded by additional modeling problems related to the force field in which the satellite moves and by the fact that nowadays there are no direct connections to other frames of reference. Modern satellite tracking techniques (laser, Doppler, etc.) all basically observe ranges or range differences and contain no directional information. The main reference planes, the orbital plane of the satellite and the equator, intersect along the line of nodes, the initial orientation of which therefore must be defined more or less arbitrarily. In the "old days" of satellite geodesy, when satellites were observed photographically in the background of stars, this direction could be determined with respect to the FK4, though not much better than a few tenths of a second of arc. The accumulation of errors in describing the motion of the node with respect to a selected zero point, even for the most suitable high flying and small heavy spherical satellites (Lageos), prevents a Satellite-CIS from being accurate over a long period of time, say beyond several months. In any case, in observational terms such a frame would be defined by the satellite ephemeris made available to the users by organizations which provide for the continuous tracking of the satellite in question. A current example would be the Precise Ephemeris of the U.S. Navy Navigational Satellite (Transit) System. As far as the connections to other systems are concerned, the only accurate possibility seems to be indirectly through the tracking stations. If two observational systems occupy the same station, one observing the satellite, the other, say, the radio sources, either simultaneously or after a short time interval (during which the movement of the station can be modeled), the connection between the satellite and radio source frames can be established. In fact, the now classical disparity between the JPL and SAO frames came to light just through such an arrangement, when the SAO longitudes determined from satellite camera tracking (thus in the FK4 frame) differed by those determined by JPL space probe tracking (in the planetary frame) by an amount (about $0\overset{''}{.}7$ in the early 1970's) consistent with the FK4 equinox motion with respect to the dynamical equinox, mentioned earlier. Only through such continuously maintained connections can the lifetime of a Satellite-CIS be extended, thus its accuracy increased

2.3 Conclusions

From the above discussion, the following conclusions can be drawn:

1. The most accurate, long-term CIS will be the one attached to extragalactic radio sources. It is accessible through VLBI observations Other systems can be accurately connected to it by station collocation or the Space Telescope.

2. The CIS attached to the FK5 is somewhat less accurate. Direct access to it is through optical star observations, which by nature are generally less accurate than VLBI observations. Its main value is in defining the fundamental mean system of coordinates and thereby providing a direction (the FK5 equinox) for the time (UT1) definition, and for the possible orientation of the Radio Source-CIS. The latter

function, however, stems from more of a traditional requirement and not from theoretical needs.

3. Of the Dynamical-CIS's, the accuracy of the planetary system should be equivalent to the FK5. The lunar and satellite systems by themselves are suitable for medium-term to short-term work only. Their stability can be extended by connections to the Radio Source-CIS through accurate and continuous observations at collocated stations. Ties between the radio source and the planetary systems may also be available through the proposed Very Large Array (VLA) observations of minor planets. Solar eclipse observations provide a connection between the lunar and planetary systems.

It is an unavoidable conclusion that for geodetic and geodynamic applications the most useful CIS is the one attached to the extragalactic radio sources, observable by VLBI, whose orientation is defined through ties to the FK5. The origin of the system can be chosen at will at the center of mass of the earth, of the solar system or elsewhere depending on the application or on operational convenience.

3. CONVENTIONAL TERRESTRIAL SYSTEMS (CTS) OF REFERENCE

As mentioned in the Introduction, the CTS is in some "prescribed way" attached to observatories located on the surface of the earth. The connection between the CTS and CIS frames by tradition (to be preserved) is through the rotations [Mueller, 1969]

$$[\underline{CTS}] = SNP [\underline{CIS}]$$

where P is the matrix of rotation for precession, N for nutation (to be discussed in Section 4), and S for earth rotation (including polar motion). Polar motion thus is defined as the angular separation of the third (Z) axis of the CTS and the axis of the earth for which the nutation (N) is computed (e.g., instantaneous rotation axis, Celestial Ephemeris Pole, Tisserand mean axis of the mantle (see Section 4)).

3.1 Current Situation

At present the internationally accepted Woolard series of nutation (the IAU 1979 series becomes effective only with the 1984 ephemerides) is computed for the instantaneous rotation axis of the rigid earth, and the Z axis of the CTS is the Conventional International Origin (CIO), defined by the adopted astronomic latitudes of the five International Latitude Service (ILS) stations, located approximately on the 39°08' parallel. These are assumed to be motionless relative to each other, and without variations in their respective verticals (plumb lines) relative to the earth. Thus, conceptually, polar motion should be determined from latitude observations only at these ILS stations. This has been done for over 80 years, and the results are the best available *long-term* polar motions, properly, but not very accuractely, determined.

The first axis of the CTS is defined by the assigned astronomic longitudes of time observatories (around 50) participating in the work of the Bureau International de l'Heure (BIH).

Due to the fact that in most geodetic and astronomical applications accurate shorter-term variations of polar motion are needed, which are not available with sufficient accuracy from the ILS observations, polar motion is also determined from latitude and/or time observations at a larger number of observatories participating in the work of the International Polar Motion Service (IPMS), as well as of the BIH. In the resulting calculations the earlier definition of the CIO cannot be maintained. The common denominator being the Woolard series of nutation, observationally the Z axis of the CTS is defined by the coordinates of the pole as published by the IPMS or by the BIH. Thus it is legitimate to speak of IPMS and BIH poles of the CTS (in addition to the CIO). The situation recently has become even more complicated because Doppler and laser satellite tracking, VLBI observations, and lunar laser ranging also can determine *variations* in the earth rotation vector (including polar motion), some of which are incorporated in the BIH computations. Further confusion arises due to the fact that the BIH has two systems: the BIH 1968 and the BIH 1979, the latter due to the incorporation of certain annual and semiannual variations of polar motion determined from the comparisons of astronomical (optical) results with those from Doppler and lunar laser observations [Feissel, 1980].

Though naturally every effort is made to keep the IPMS and BIH poles of the CTS as close as possible to the CIO, the situation cannot be considered satisfactory from the point of view of the geodynamic accuracy requirement of a few parts in 10^9. The current accuracy of the pole position is estimated to be $0\overset{\prime\prime}{.}01$, and that of the UT1, 1 ms ($\sim 5 \times 10^{-8}$) for five-day averages [Guinot, 1978]. These figures, of course, do not include biases from the definition problems mentioned.

3.2 The Future CTS

There seems to be general agreement that the new CTS frame conceptually be defined similarly to the CIO-BIH system [Bender and Goad, 1979; Guinot, 1979; Kovalevsky, 1979; Mueller, 1975a], i.e., it should be attached to observatories located on the surface of the earth. The main difference in concept is that these can no longer be assumed motionless with respect to each other. Also they must be equipped with advanced geodetic instrumentation like VLBI or lasers, which are no longer referenced to the local plumblines. Thus the new transformation formula may have the form

$$[\underline{OBS}]_j = \underline{L}'_j + [\underline{CTS}]_j + \underline{v}_j = \underline{L}'_j + SNP\,[\underline{CIS}]_j + \underline{v}_j \quad ,$$

where $\underline{L}'_j$ is the vector of the "j" observatory's movement on the deformable earth with respect to the CTS, computed from suitable models (see the figure and Section 4); NP, the nutation and precession matrices computed with the new 1976 IAU constants and the 1979 IAU series of nutation (provided the latter is not going to be changed; see Section 4); and S, the rotation matrix between the CTS and the true frame for which the nutation is computed. Variations in S can be determined by a future international service (like the BIH) by comparing repeatedly observed observatory coordinates ($[\underline{OBS}]_j$), corrected for the modelable deformations ($-\underline{L}'_j$), and by minimizing the residuals ($\underline{v}_j$) in the least squares sense. This rotation can either be determined from the residuals of the Cartesian coordinates (e.g., [Moritz, 1979]) or, for possible better sensitivity, since the rotation is least sensitive to variations in height, only from those of the horizontal coordinates (geodetic latitude and longitude) (e.g., [Bender and Goad, 1979]). It is unlikely that the rotation will continue to be determined (as presently) from astronomical coordinates, i.e., from the direction of the vertical, for the reason of inadequate observational accuracy.

As far as the origin of the CTS is concerned, it could be centered at the center of mass of the earth, and its motion with respect to the stations can be monitored either through observations to satellites or the moon, or, probably more sensitively, from continuous global gravity observations at properly selected observatories [Mather et al., 1977]. For the former method a translational term may easily be incorporated in the above transformation equation.

Since the above method or some variation thereof provides only changes in S and in the translation, the new CTS needs to be initialized in a way to provide continuity. This could be done through the CIO or the IPMS or BIH poles, and the BIH zero meridian, at the selected initial epoch, uncertainties in their definition mentioned earlier mercifully ignored.

It is probably not useless to point out that if such a system is established, the most important information for the users will be the transformation parameters, but for the scientist new knowledge about the behavior of the earth will come from the analysis of the residuals after the adjustment.

It is hoped that the IAU and IUGG will make practical recommendations on the establishment of such or a very similar Conventional Terrestrial System, including the necessary plans for supporting observatories and services. One of the recommendations ought to be that due to the fact that the ultimate goal is the determination of the total transformation between the CTS and CIS, the future service must publish not only the parameters of the S matrix determined from the repeated comparisons (the situation at present), but also the models and parameters in L' as well as in NP, i.e., the parameters defining the whole *system*.

4. MODELING THE DEFORMABLE EARTH

In this section we will try to highlight the modeling problems associated with the components of transformation between the CIS and CTS mentioned in Section 3.

4.1 Precession (P)

At the XVIth General Assembly in Grenoble in 1976, the IAU adopted a new speed of general precession in longitude of 5029″.0966 per Julian century at the epoch J2000.0 (JED 2451545.0). This value when referred to the beginning of the Besselian year B1900.0 is 5026″.767 per tropical century, which may be compared to the previously adopted (and presently still used) value of 5025″.64 per tropical century at B1900.0. The change was calculated by Fricke [1977] from proper motions of stars in the systems GC, FK3, N30, and FK4. From the results, the correction of +1″.10 per century to Newcomb's luni-solar precession in longitude was recommended. This value combined with a correction to Newcomb's planetary precession, due to the improved 1976 IAU values of planetary masses, resulted in the above new precessional constant. Expressions to compute the effect of precession from one epoch to another were developed by Lieske et al. [1977]; and the usual equatorial parameters, z, θ, ζ_0, to be used in the precession matrix [Mueller, 1969],

$$P \;=\; R_3(-z)\, R_2(\theta)\, R_3(-\zeta_0) \quad,$$

to and from the epoch J2000 were computed by Lieske [1979]. The above matrix allows the currently best transformation between the CIS (say, the FK5 at J2000.0) and an interim "Mean Equator and Equinox Frame" of some date.

4.2 Nutation (N)

The nutation story is much more complex. First of all, the nutation matrix is [Mueller, 1969]

$$N \;=\; R_1(-\varepsilon -\Delta\varepsilon)\, R_3(-\Delta\psi)\, R_1(\varepsilon) \quad,$$

where ε is the obliquity of the ecliptic, $\Delta\varepsilon$ is the nutation in obliquity, and $\Delta\psi$ the nutation in longitude, computed from a certain theory of nutation. This matrix allows transformation from the aforementioned interim mean frame of date to the (also) interim true frame of the same date. This part is clear and without controversy. The complexities are in the agreement reached (or still to be reached) on the theory of nutation when computing the above parameters. Kinoshita et al. [1979] give an historical review:

"In astronomical ephemerides, nutation has been computed until now by the formulae which were given by Woolard (1963). The coefficients of the formulae are calculated assuming that the Earth is rigid. However, it has been found in recent analyses of observations ... that some coefficients of actual nutations are in better agreement with values calculated by the non-rigid Earth theory.

"Moreoever, Woolard (1953 gave the nutation of the axis of rotation. Therefore, a small and nearly diurnal variation appears in the latitude and time observations, which is the so-called dynamical variation of latitude and time, or Oppolzer terms. In the global reduction of latitude and time observations, such as polar motion or time services, the Oppolzer terms have been until now removed from the data at each station (cf. BIH Rapport Annuel 1977, pA-3) or counted out as a part of the non-polar common z and τ-terms (IPMS Annual Report 1974, p. 11). On the other hand, Atkinson (1973) pointed out that if the (forced) nutation of the axis of figure is calculated instead of rotation axis, such a complicated treatment becomes unnecessary.

"Considering these situations, the IAU investigated the treatment of nutations, together with the system of astronomical constants which should be used in new ephemerides, and set up the 'Working Group of IAU Commission 4, on Precession, Planetary Ephemeris, Units, and Time-Scales'. The results by the Working Group are given in the report of Joint Meeting of Commissions 4, 8, and 31, in Grenoble, 1976 (Duncombe et al. 1976). In the report, the proposal by Atkinson is adopted, and the formula for computing the (forced) nutation of figure axis is shown clearly and in detail, by using the equation-numbers given by Woolard (1953). However, the amendments of coefficients taking account of the non-rigidity of the Earth have not been adopted. In regard to this problem, it was noted that there should be a possibility of making further amendments in Kiev Symposium ... .

"At the IAU Symposium No. 78 in Kiev in 1977, the problem with the non-rigid values of nutation was discussed, and a series of new values were recommended which seemed to be based on Molodenskij's non-rigid theory. In the Symposium, however, it was recommended that the axis for which the nutation should be computed was the axis of rotation. This recommendation reversed the resolution given at Grenoble.

"In accordance with the resolution at the Kiev Symposium, an 'IAU Working Group on Nutation under Commission 4' was set up and is investigating these two problems, in order to prepare a fully documented proposal for the next IAU General Assembly in Montreal in 1979. In the second draft of the Working Group circulated on Nov. 16, 1978, the following conclusions are reported: (1) as for the axis to be referred, the Grenoble resolution is still valid, and (2) as for the coefficients of nutation series, the value in which the non-rigidity of the Earth is taken into account should be adopted as a working standard of astronomical observations. In the draft, a table of nutation series is given, and the numerical values in the table are based on the rigid theory by Kinoshita (1977), with use of IAU (1976) System of Astronomical Constants, and are modified by Molodenskij's non-rigid theory (Molodenskij 1961)."

As we understand it, the Kinoshita theory above is for the nutation of the axis of maximum moment of inertia of the "mean shape of the elastic mantle" (briefly, "mean axis of figure of the mantle"). To add to the

history, after the above-quoted Working Group Report was circulated, a
new proposal was made by J.M. Wahr and M.L. Smith of CIRES that it
would be preferable to adapt the non-rigid earth results of Wahr [1979]
for the earth model 1066A developed by Gilbert and Dziewonski [1975].
This model is a rotating, elliptically stratified linearly elastic and
oceanless earth with a fluid outer core and a solid inner core. The
nutations are computed for the "Tisserand mean figure axis of the surface," which is also a mean mantle fixed axis [Wahr, 1979]. The IAU in
Montreal in 1979 considered both proposals and opted for the Kinoshita
et al. [1979] series. A few months later in December, 1979, the IUGG
in Canberra, in Resolution No. 9 addressed to the IAU, requested recon-
sideration in favor of the Wahr model. This is where the matter stands
now.

It should be pointed out that regardless of the fact that in geodetic or geodynamic applications we are only concerned with the total
transformation SNP, it is of scientific importance to understand clearly the definition of the interim true equator and equinox frame of date,
more specifically, the exact definition and the desirability (from the
observability point of view) of the axis for which the nutation is computed.

In order to simplify the discussion, let us start with the rigid
model. The motion of each of the axes, i.e., the axis of figure (F)
(maximum moment of inertia), of the angular momentum (H), and the instantaneous rotation axis (I) are described by differential equations. If
we want to refer to one of these axes we have to consider the complete
solution of the differential equations, i.e., the free solution and the
forced solution components. Confusion can arise if one refers to only
one solution component (forced or free), but still calls it axis of figure, instantaneous rotation axis, etc. It is mandatory to point out
which solution component one refers to. Neglecting to do so has been
the reason for the by now classical confusing controversy about the Atkinson papers, though Atkinson [1975, p.381] clearly states:
  "Accordingly, when we speak of computing the nutations for either
  axis, we mean here computing the _forced motion_ only, excluding the
  appropriate fraction of the non-computable Chandlerian wobble."
Unfortunately, he, and others as well, then continue to use the term
"axis of figure" sometimes in the sense of the axis of maximum moment of
inertia and at other times in the sense of the forced motion of the axis
of figure.

A remark concerning the "Eulerian pole of rotation" (E_0) as given
by Woolard seems in order also. Quoting once again Atkinson [1976]:
  "The wording of the resolution on nutation, and the notes on it,
  which have been circulated by the Working Group, avoid all explicit mention of the axis of figure, even though they specify that the
  coefficients which Woolard gives for that axis shall be inserted,
  and they refer to the "Eulerian pole of rotation" although this
  cannot ever, in principle, coincide with the celestial pole and
  really has no more direct connection with the observations than

is shown for it in [his] Fig. 2, i.e., none at all."
The difference between the Eulerian pole of rotation (E_0) and the pole
which Atkinson talks about is due to a homogeneous solution component.
(E_0) is obtained from the complete solution of (I) by subtracting the
periodic diurnal body-fixed motions of (I).

Consequently, the point E_0 has no periodic motion with respect to
the crust, but it does have such a motion in space which is exactly the
free nutation. Although this spatial motion is conceptually insignificant considering the observation technique (fundamental observations at
both culminations), one gets another point, which is called the (true)
Celestial Pole (C) in [Leick and Mueller, 1979], by subtracting the
forced body-fixed motions of (H) from the complete nutation set of (H).
The thus obtained axis (C) has no periodic diurnal spatial motion because the homogeneous solution of the angular momentum (H) is constant
(zero). Equivalently, one can say that the nutations of (C) correspond
to the forced solution of the axis of figure (rigid case, of course).
This is the pole which Atkinson talks about and which is called (mistak-
enly) the "mean axis of figure." There is no doubt that this is the
point to which the astronomical observations as well as lunar laser ranging refer, and the nutation should be adopted for this point. As for
terminology, the IAU in 1979 named this (C) pole appropriately the *Celestial Ephemeris Pole* because its motion characteristics, i.e., no periodic
diurnal motion relative to crust or space, have always been associated
with the concept of the celestial pole. It would be preferred that the
word "figure" be dropped entirely for several reasons. First, one intuitively associates the axis of figure with the one for which the moment
of inertia is maximum. This is true for the (C) only if the free solution (Chandler) is zero. But this is, generally, not the case. Second,
the conceptual definition of (C) can easily be extended to elastic models
or models with liquid core (the IAU 1979 case). Moreover, in order to
emphasize that the observations take place on the earth surface, it would
be useful to denote the actual pole accessible to the fundamental observation techniques by another designation, e.g., (CO), similarly to UTO.
The "O" would indicate that the nutations of this pole can in principle
be determined only from observations because of the lack of a perfect
earth model. Any nutation set based on a model is only an approximation
to the nutations of the (CO). In this sense the rigid earth nutations
of (I), (H) or (F) are all equivalent. Each of these nutations defines
its own pole which has a diurnal motion around the (CO). The purpose of
the measuring efforts is to find the corrections to the adopted set of
nutations in order to get those of the (CO), the only pole which is observable.

Some have suggested the term "zero excitation figure axis" for what
is called above the (CO). The term "zero excitation" would not reduce
the confusion. The spatial motion of this axis is computed by adding
Atkinson's terms to Woolard's series, but this is equivalent to the
forced motion of the axis of figure (rigid case). The observed motion
of the (CO) relative to the crust only appears as a motion of zero excitation (free motion) at the first sight. Since the conceptual observation

time of one position determination is one day, the observed position of the (CO) will always include effects due to oceans, atmospheric mass redistribution, etc., i.e., the geophysical nutations. These motions are better known as the annual polar motion and the sub-harmonics. Therefore, the zero-excitation pole is not directly observable. On the other hand, the concept of the (CO) can still be used in this case since it is by definition the pole which has no periodic diurnal motions relative to the crust or to space.

There is also the common offset of both the rotation axis and the (CO) caused by the tidal deformation [McClure, 1973]. This is an offset of (I) and (CO) relative to (H) for the perfectly elastic model as compared with the rigid model. We have to remember, again, that the observations refer to the (CO). Therefore, any nutation correction which is derived from observations (based on an adopted set of nutations) will automatically give the corrections to the (CO). Consequently, there is no need for a special consideration of this possible separation, at least not for those harmonic motions whose amplitudes are derived from observations. In fact, the analysis of the observed fortnightly term seems to contradict somewhat the predicted amplitude for the perfectly elastic model.

From the above discussion, it also seems clear that ideas advocating the adoption of nutations for the axis of angular momentum violate the concept of observability. It is true that the direction of (H) in space is the same for the rigid, elastic, or any other reasonable earth model. But this property is not of much interest to the astronomer or geodesist who tries to determine the orientation of the earth. It is *conceptually simpler* to refer to an axis which is observable.

Returning now to the problem of the IAU 1979 adopted set of nutations, there seems to be little difference whether the Kinoshita series is retained or the Wahr set is adopted. Using more and more realistic earth models is certainly appealing. On the other hand, severely model-dependent developments are liable to change as models improve. A more important point is that whichever series is adopted, it should be for the Celestial Ephemeris Pole (C), which (again) has no periodic diurnal motion relative to the *crust* (not the mantle!) or the CIS.

4.3 Earth Rotation (S)

The two components of the S matrix [Mueller, 1969],

$$S = R_2(-x_p) \, R_1(-y_p) \, R_3(\phi) \quad ,$$

are the rotational angle of the first (X) axis of the CTS with respect to the first axis of the interim true equator and equinox frame of date, measured in the equator of the Celestial Ephemeris Pole (or whatever is defined in the N matrix), also known as Apparent Sidereal Time (ϕ), and the polar motion coordinates (x_p, y_p) referred to the same pole and the Z axis of the CTS.

In this connection it should be mentioned that some authors prefer a different "true"frame, which would have no rotation about the Z axis [Guinot, 1979; Murray, 1979; Kinoshita et al., 1979]. It is in such an interim frame where, for example, a nutational theory can be conveniently developed, or satellite orbits calculated [Kozai, 1974]. Such a frame can be obtained from the CIS by a modified NP transformation, where

$$N = R_1(-\Delta\epsilon\cos M + \Delta\psi\sin\epsilon\sin M)\ R_2(\Delta\psi\sin\epsilon\cos M + \Delta\epsilon\sin M), \text{ and}$$

$$P = R_3(-z + M)\ R_2(\theta)\ R_3(-\zeta_0),$$

where M is the precession in right ascension.

In this case the rotation of CTS about the Z axis (ϕ) is the Apparent Sidereal Time from which the general precession and nutation in right ascension are removed. What is left, thus, is the rotational angle of the X axis of the CTS directly with respect to that of the CIS. Such a definition of the sidereal angle would, of course, necessitate the redefinition of UT1, a possibility for controversy. It should be noted also, that the above transformation is independent of the ecliptic, a preference of many astronomers.

Here there is not very much modeling that can be considered really useful. Of course, the rotation rate of the earth could be modeled as constant and possibly in the UTC scale. This would then mean that observed departures could immediately be referenced to that scale, a current practice. If one really wanted to go overboard, polar motion could also be modeled with the Chandlerian cycle of, say, 428 days and a circular movement of radius 0."15, centered at the Z axis of the CTS. More complex models may be developed (e.g., Markowitz, 1976, 1979], but since there are no valid physical concepts yet for the excitation of the amplitude of the Chandler motion, such modeling would not serve much purpose.

4.4 Deformations (L')

The deformations which reasonably can be modeled at the present state of the art are those due to the tidal phenomena and to tectonic plate movements.

4.41 <u>Tidal Deformations</u>. Tides are generated by the same forces which cause nutation; thus models developed for the latter should be useful for the former. One would think that for earth tides it may not be necessary to use the theories based on the very sophisticated earth models: the amplitude of the phenomena being only around 30 cm, an accuracy of 3% should be adequate for centimeter work. This should be compared, for example, with the accuracy of the Wahr nutation model claimed to be at the 0.3% level. However, the tides and nutations differ in one important respect. The nutations hardly depend upon the elasticity and are affected only slightly by the liquid core (this is one reason why modern theories such as those of Wahr and Kinoshita give only slightly different results). Thus, except perhaps for the largest terms, one can depend upon theory when dealing with nutation. The tides, on the other hand, depend intimately upon the internal properties of the earth, and one must use tidal theories with caution [Newton, 1974]. Additional problems are handling the transformation of the potential into physical displacements and on the calculations of regional (ocean loading) or local tidal deformations.

As far as the transformation of the tidal potential into displacement is concerned, the traditional way to do this is through the Love numbers for the solid effect and through "load" numbers for ocean loading. These numbers, however, are spherical approximations which, for the purely elastic earth, are global constants. For more sophistication elliptic terms can be added, but they will change the results by 1-2% only. A liquid core model produces resonance effects, which will result in a frequency dependency. The actual numbers representative for a given location can be determined only through in situ observations, such as gravity, tilt, deflections, which are all sensitive to certain Love number combinations and frequencies. Difficulties in this regard include the frequency dependence of the Love number. For example, the Love number h for radial (vertical) displacement can be determined locally from combined gravity and tilt meter observations by the analysis of the O_1 tidal component, but the real radial motion of geodetic interest is influenced by the M_2 and other semidiurnal tidal components.

Tidal loading effects have recently been very successfully computed by Goad [1979] using the 1° square Schwiderski [1978] M2 ocean tide model. Global results show agreement with gravimetrically observed deformation on the 0.5 μgal (5×10^{-10}) level. From this it would seem that with good quality ocean tide models and with proper attention to the frequency dependence, this problem is manageable.

Suitable equations for displacement, gravity change, deflection change, tilt and strain calculations due to tides may be found in [Melchior, 1978; Vanicek, 1980] and in [Wahr, 1979] for the elliptic case.

As a conclusion one can reasonably state that the global and regional station movements due to tides can be estimated today within centimeters. Local effects, however, can be sizable and unpredictable, and therefore they are best determined from in situ observations. Thus most of the tidal effect in fact can and should be removed from the observations.

4.42 <u>Plate Tectonic Mass Transfer</u>. The concept that the earth lithosphere is made up of a relatively small number of plates which are in motion with respect to each other is the central theme of global plate tectonics. The theory implies the transfer of masses as the plates move with velocities determined from geologic evidence (see, e.g., [Solomon and Sleep, 1974; Kaula, 1975; or Minster and Jordan, 1978]). Material rises from the asthenosphere and cools to generate new oceanic lithosphere, and the lithospheric slabs descend to displace asthenospheric material (see, e.g., [Chapple and Tullis, 1977]). A good example of how such a theory can be used to estimate the vertical motions of observatories located on the lithosphere (in terms of changes in geoid undulations) is given in [Larden, 1980], based on specific models constructed in [Mather and Larden, 1978]. The results indicate that changes in the geoid can reach 150 mm/century. Horizontal displacements can be estimated from the plate velocity models mentioned directly with certain possible amendments [Bender, 1974].

4.43 <u>Other Deformations</u>. If one wants to carry the modeling further, it is possible to estimate seasonal deformations due to variations in air mass and groundwater storage, for which global data sets are available [Van Hylckama, 1956; Stolz and Larden, 1979; Larden, 1980]. A more esoteric effect would be the expansion of the earth (e.g., [Dicke, 1969; Newton, 1968]). The rate of possible expansion is estimated to be 10 - 100 mm/century.

One could continue with other modeling possibilities, but there is a real question on the usefulness of modeling phenomena of this level of magnitudes and uncertainties. As a general philosophy, one could accept the criteria that modeling should be attempted only if reliable and global data is available related to the phenomena in question, and if the magnitudes reach the centimeter per year level or so.

One last item which should be brought up is the fact that the issue of referencing observations and/or geodynamic phenomena is not exhausted by the establishment of reference frames of the Cartesian types discussed in this paper. An outstanding issue is still the geoid as a reference surface. Though it is true that three-dimensional advanced geodetic observational techniques do not need the geoid as a reference, there are still others, such as spirit leveling, which are used in the determination of crustal deformations in the local scale. In addition, the geoid is needed to reference gravity observations on a global scale (one should remember that a 1 cm error in the geoid corresponds to a 3 μgal error in the gravity reduction, which is (or soon will be) the accuracy of modern gravimeters). Further, in connection with the use of satellite altimetry for the determination of the departures of sea surface topography from the equipotential geoid (a topic of great oceanographic interest), there is a requirement for a geoid of at least 10 cm accuracy. The determination of such a geoid globally, or even over large areas, is a very difficult problem, which, however, is not the subject of the present paper.

ACKNOWLEDGEMENT. This work was supported by NASA Grant No. NSG 5265 and by Grant No. 4810/79 of the Deutsche Forschungsgemeinschaft (DFG) while the author visited at the Geodetic Institute, Stuttgart University, during June-July, 1980.

REFERENCES

Ashby, N. and Allan, D.W.: 1979, *Radio Science* <u>14</u>, 649.
Atkinson, R.d'E.: 1973, *Astron. J.* <u>78</u>, 147.
Atkinson, R.d'E.: 1975, *Monthly Notices Roy. Astron. Soc.* <u>71</u>, 381.
Atkinson, R.d'E.: 1976, 'On the Earth's Axes of Rotation and Figure',

pres. at XVI General Assembly of IUGG, Grenoble.

Barbieri, C. and Bernacca, P.L.(eds.): 1979, *European Satellite Astrometry*, Ist. di Astronomia, Univ. di Padova, Italy.

Bender, P.L.: 1974, *see* Kołaczek and Weiffenbach (eds.), 85.

Bender, P. and Goad, C.: 1979, *The Use of Artificial Satellites for Geodesy and Geodynamics, Vol. II*, National Tech. Univ. Athens.

Chapple, W.M. and Tullis, T.E.: 1977, *J. Geophys. Res.* 82, 1967.

Dicke, R.H.: 1969, *J. Geophys. Res.* 74, 5895.

Duncombe, R.L., Seidelmann, P.K. and Van Flandern, T.C.: 1974, *see* Kołaczek and Weiffenbach, 223.

Duncombe, R.L., Fricke, W., Seidelmann, P.K. and Wilkins, G.A.: 1976, *Trans. IAU* XVIB, 52.

Einstein, A.: 1956, *The Meaning of Relativity*, Princeton Univ. Press, Princeton, New Jersey.

Fedorov, E.P., Smith, M.L. and Bender, P.L. (eds.): 1980, *IAU Symp.* 79.

Feissel, M.: 1980, *Bull. Geodes.* 54, 81.

Fricke, W.: 1974, *see* Kołaczek and Weiffenbach, 201.

Fricke, W.: 1977, *Veröffentlichungen Astron. Rechen-Inst. Heidelberg* 28, Verl. G. Braun, Karlsruhe.

Fricke, W. and Gliese, W.: 1978, *see* Prochazka and Tucker, 421.

Fricke, W.: 1979a, 'Progress Rept. on Preparation of FK5', pres. at Commission IV, IAU XVII General Assembly, Montreal.

Fricke, W.: 1979b, *see* Barbieri and Bernacca, 175.

Gilbert, F. and Dziewonski, A.M.: 1975, *Phil. Trans. R. Soc. London* A278, 187.

Goad, C.C.: 1979, 'Gravimetric Tidal Loading Computed from Integrated Green's Functions', NOAA Tech. Memorandum NOS NGS 22, NOS/NOAA, Rockville, Md.

Guinot, B.: 1978, *see* Mueller, 1978, 13.

Guinot, B.: 1979, *see* McCarthy and Pilkington, 7.

Kaula, W.M.: 1975, *J. Geophys. Res.* 80, 244.

Kinoshita, H.: 1977, *Celes. Mechan.* 15, 227.

Kinoshita, H., Nakajima, K., Kubo, Y., Nakagawa, I., Sasao, T. and Yokoyama, K.: 1979, *Publ. Int. Lat. Obs. of Mizusawa* XII, 71.

Kołaczek, B. and Weiffenbach, G. (eds.): 1974, *On Reference Coordinate Systems for Earth Dynamics*, IAU Colloq. 26, Smithsonian Astrophys. Obs. Cambridge, Mass.

Kovalevsky, J.: 1979, *see* McCarthy and Pilkington, 151.

Kozai, Y.: 1974, *see* Kołaczek and Weiffenbach, 235.

Larden, D.R.: 1980, 'Some Geophysical Effects on Geodetic Levelling Networks', *Proc. 2nd International Symp. on Problems Related to the Redefinition of North American Vertical Geodetic Networks*, Canadian Inst. o. Surveying, Ottawa.

Leick, A. and Mueller, I.I.: 1979, *manuscripta geodaetica* 4, 149.

Lieske, J.H., Lederle, T., Fricke, W. and Morando, B.: 1977, *Astron. Astrophys.* 58, 1.

Lieske, J.H.: 1979, *Astron. Astrophys.* 73, 282.

Markowitz, Wm. and Guinot, B. (eds.): 1968, *IAU Symp.* 32, Reidel.

Markowitz, Wm.: 1976, 'Comparison of ILS, IPMS, BIH and Doppler Polar Motions with Theoretical', Rep. to IAU Comm. 19 and 31, IAU General Assembly, Grenoble.

Markowitz, Wm.: 1979, 'Independent Polar Motions, Optical and Doppler; Chandler Uncertainties', Rep. to IAU Comm. 19 and 31, IAU General Assembly, Montreal.

Mather, R.S., Masters, E.G. and Coleman, R.: 1977, *Uniserv G* 26, Univ. of New South Wales, Sidney, Australia.

Mather, R.S. and Larden, D.R.: 1978, *Uniserv G* 29, 11, Univ. of New South Wales, Sidney, Australia.

McCarthy, D.D. and Pilkington, J.D.H. (eds.): 1979, *IAU Symp.* 82, Reidel.

McClure, P.: 1973, 'Diurnal Polar Motion', GSFC Rep. X-592-73-259, Goddard Space Flight Center, Greenbelt, Md.

Melchior, P. and Yumi, S. (eds.): 1972, *IAU Symp.* 48, Reidel.

Melchior, P.: 1978, *The Tides of the Planet Earth*, Pergamon Press, Oxford.

Minster, J.B. and Jordan, T.H.: 1978, *J. Geophys. Res.* 83, 5331.

Molodenskij, M.S.: 1961, *Comm. Obs. R. Belgique, 188 S. Geoph.* 58, 25.

Moran, J.M.: 1974, *see* Kolaczek and Weiffenbach, 269.

Moritz, H.: 1967, *Dept. of Geod. Sci. Rep.* 92, Ohio State Univ., Columbus.

Moritz, H.: 1979, *Dept. of Geod. Sci. Rep.* 294, Ohio State Univ., Columbus.

Mueller, I.I.: 1969, *Spherical and Practical Astronomy As Applied to Geodesy*, Ungar Publ. Co., New York.

Mueller, I.I.: 1975a, *Geophys. Surveys* 2, 243.

Mueller, I.I. (ed.): 1975b, *Dept. of Geod. Sci. Rep.* 231, Ohio State Univ., Columbus.

Mueller, I.I. (ed.): 1978, *Dept. of Geod. Sci. Rep.* 280, Ohio State Univ., Columbus.

Murray, C.A.: 1979, *see* McCarthy and Pilkington, 165.

Newton, I.: 1686, *Philosophiae Naturalis Principia Mathematica*, Univ. of California Press, 1966.

Newton, R.R.: 1968, *J. Geophys. Res.* 73, 3765.

Newton, R.R.: 1974, *see* Kolaczek and Weiffenbach, 181.

Prochazka, F.V. and Tucker, R.H. (eds.): 1978, *Modern Astrometry, IAU Colloq.* 48, Univ. Obs. Vienna.

Purcell, G.H., Jr., Fanselow, J.L., Thomas, J.B., Cohen, E.J., Rogstad, D.H., Sovers, O.J., Skjerve, L.J. and Spitzmesser, D.J.: 1980, *Radio Interferometry Techniques for Geodesy*, p. 165, NASA Conference Publ. 2115, NASA Scientific & Tech. Information Office, Washington, D.C.

Schwiderski, E.M.: 1978, 'Global Ocean Tides, Part 1: A Detailed Hydrodynamical Interpolation Model', US Naval Surface Weapons Center TR-3866, Dahlgren, Va.

Solomon, S.C. and Sleep, N.H.: 1974, *J. Geophys. Res.* 79. 2557.

Stolz, A. and Larden, D.R.: 1979, *J. Geophys. Res.* 84, 6185.

Van Altena, W.: 1978, *see* Prochazka and Tucker, 561.

Van Hylckama, T.E.A.: 1956, *Climatology* 9, 59.

Vanicek, P.: 1980, 'Tidal Corrections to Geodetic Quantities', NOAA Tech. Rep. NOS 83 NGS 14, NOS/NOAA, Rockville, Md.

Wahr, J.M.: 1979, *The Tidal Motions of a Rotating, Elliptical, Elastic and Oceanless Earth*, PhD diss., Dept. of Physics, Univ. of Colorado, Boulder.

Weinberg, S.: 1972, *Gravitation and Cosmology: Principles and Applications of the General Theory of Relativity*, Wiley & Sons, New York.

Woolard, E.W.: 1953, *Astronomical Papers Prepared for the Use of the American Ephemeris and Nautical Almanac*, XV, Part I, US Govt. Printing

Office, Washington, D.C.

Yumi, S. (ed.): 1971, *Extra Collection of Papers Contributed to the IAU Symposium No. 48, "Rotation of the Earth"*, International Latitude Obs., Mizusawa, Japan.

ESTABLISHMENT OF TERRESTRIAL REFERENCE FRAMES BY NEW OBSERVATIONAL TECHNIQUES

Peter L. Bender[*]
Joint Institute for Laboratory Astrophysics,
National Bureau of Standards and University of Colorado,
Boulder, Colorado 80309 USA

ABSTRACT

The use of space techniques for establishing transcontinental and intercontinental distances is progressing very rapidly. We can think of the set of station locations used in either LAGEOS ranging or VLBI measurements as forming the vertices of a polyhedron. After correcting for tectonic plate motions using an adopted model, we expect the geometry of the polyhedron to be fairly stable over periods of the order of a year. However, after some period of time, a new set of station coordinates will be required because of improved data, unexpected station motions, etc. Methods for maintaining agreement with the previous set of station coordinates in some average sense are discussed in this paper. Some of the contributions expected from other new measurement methods also are described.

INTRODUCTION

This article is addressed to the question of how we can use new observational techniques to establish worldwide terrestrial reference frames. For simplicity, the main emphasis will be on geometrical reference frames that can be established by laser distance measurements to the LAGEOS satellite and by very long baseline radio interferometry (VLBI). These two techniques are likely to be the most important ones in the 1980s for establishing basic worldwide networks of geometrical reference points. However, it should be noted that much larger numbers of points in and around seismic zones are likely to be determined with similar accuracy by observing the signals from the NAVSTAR Global Positioning System satellites. These points will have to be tied to the basic worldwide reference frames in some way.

---

[*]Staff Member, Quantum Physics Division, National Bureau of Standards.

E. M. Gaposchkin and B. Kołaczek (eds.), Reference Coordinate Systems for Earth Dynamics, 23–36.
Copyright © 1981 by D. Reidel Publishing Company.

One of the basic reasons for establishing accurate worldwide
geometrical reference frames is connected with the definition of the
earth's rotation. The angular motion of such frames with respect to
a conventional celestial reference frame is likely to be used in the
future to define UT1 and polar motion. Any inadequacies in the reference frames thus are likely to show up as limitations on the accuracy
of the UT1 and polar motion determinations. In particular, this means
that the coordinates for the different stations which make observations used in determining UT1 and polar motion should be as consistent
as possible. Otherwise changes in the distribution of data between
different stations, either at different times of the year or over
shorter periods, will lead to errors in the results.

Because of the above requirement, it appears desirable to adopt a
model for the motions of the various tectonic plates for use in determining the reference system. This question has been discussed previously (Bender, 1974), and may be considered further in other papers
in this volume. For example, any of the four absolute plate motion
models referred to in Table 8 of Minster and Jordan (1978) might be
a reasonable choice until a substantial amount of new information on
plate motions from space techniques becomes available. The important
point is that some of the LAGEOS ranging and VLBI stations either are
or will be on the relatively high velocity Pacific and Indian plates.
Without a plate motion model, the coordinates of such stations will
become inconsistent with respect to other stations quite quickly.

While there definitely is a possibility that present rates of
plate motions will differ from the long-term average rates determined
from the geological record and other information (Bender, 1974,1978),
this would require a substantial change in the picture most geophysicists have of tectonic plate motions. It appears difficult to see how
the present rates of motion out in the centers of the major plates can
be much different from the long-term average rates unless there is a
layer in the asthenosphere with much lower viscosity than is expected
from studies of post-glacial rebound. This is because the main driving forces for plate motions at ridges and rises and the viscous
stresses on the bottoms of the plates are likely to be quite stable
in time. For values of the order of 10^{20} poise for the viscosity
and 100 km for the thickness of the low-viscosity part of the asthenosphere, the transient effect of even a great earthquake at the front
of the plate would not propagate out to the center of the plate during
typical recurrence times for great earthquakes. Thus it seems best
for geodesists to regard the null hypothesis which we wish to test as
being that the present rates of motion in the centers of plates are
equal to the long-term average rates.

One other question which comes up frequently concerning models
of plate motions is based on the fact that only relative motions of
the different plates are actually observed. Fortunately, studies by
Solomon and Sleep (1974), Kaula (1975), and Minster and Jordan (1978)

have shown that quite different geophysical assumptions about plate motions give fairly similar absolute plate motion velocities with respect to the bulk of the material in the mantle. The differences appear to be only about 1 cm per year. While the choice is necessarily somewhat arbitrary, it appears to make little practical difference which of the suggested geophysical constraints are used to determine the absolute plate velocities.

Another requirement on the terrestrial reference systems used in determining UT1 and polar motion is that they should be as close as possible to the systems that might be used at later times for more detailed analyses of worldwide crustal movements. In such crustal movement studies, scientists undoubtedly will establish new reference systems that are optimized for long-term stability. However, substantial differences from the systems used for determining UT1 and polar motion initially would lead to large revisions in these quantities. Also, stability of the defined reference frames will help in making it possible to detect anomalous motions of individual stations at an early date.

The next section will describe the worldwide position measurements by LAGEOS ranging and VLBI which are likely to be carried out in the 1980's. Part of the material in this section is taken from another article which was prepared recently (Bender, 1980). This will be followed by a discussion of possible procedures for maintaining terrestrial reference systems. Finally, the expected contributions from other new observational techniques will be described briefly.

WORLDWIDE POSITION MEASUREMENTS

The basic approach used in the space techniques is to measure the distance or difference in distance from points on the ground to extraterrestrial reference points such as satellites or astronomical radio sources. The accuracy that appears to be achievable in such distance measurements is roughly 0.3 to 3 cm, and depends on many factors. For laser distance measurements, the inaccuracy is mainly due to uncertainty in the integrated atmospheric density along the line of sight and to the inadequacy of most present procedures for measuring the round-trip travel time. For radio measurements, the main problem is the uncertainty in the correction for the integrated amount of water vapor along the path. The effect of the ionosphere also has to be considered, but this can be corrected for accurately by comparing results of measurements made at two substantially different frequencies.

The LAGEOS satellite (Smith et al., 1979a,b; Smith and Dunn, 1980) is in a nearly circular orbit with 110° inclination and 5900 km altitude. It is spherically symmetric and has a high density to minimize perturbations due to radiation pressure and atmospheric drag. About 15 ground stations currently are making range measurements to

LAGEOS, which probably is at least as many as are needed to maintain
good coverage of the orbit. However, changes in the locations of some
of the ground station would be quite helpful in improving the southern
hemisphere coverage and in making use of sites with better weather,
where the gaps in the data obtained would be considerably shorter.

The basic approach is to make range measurements to LAGEOS whenever possible from roughly 10 to 15 fixed or semi-fixed stations, and
to determine the orbit and variations in the Earth's rotation from the
data (Smith et al., 1979a). Arc lengths of a week or so may be used
in such fits. At suitable intervals, solutions involving a considerably larger amount of data can be carried out in order to obtain corrections to the station coordinates (Smith et al., 1979b), to some
of the gravitational harmonic coefficients, to the ocean tide models
(Eaves et al., 1979; Smith and Dunn, 1980), and to a few other parameters. Simulations (Bender and Goad, 1979) have indicated that such
solutions can give accuracies of 4 or 5 cm for intercontinental distances, even without additional new information about the Earth's
gravity field from other sources beyond that contained in NASA's GEM-
10 gravity field model and the LAGEOS ranging data itself. Information on changes in the station positions should be considerably more
accurate.

The main improvement required in LAGEOS ranging is in the basic
measurement accuracy. At present many of the stations have 10 cm accuracy for the average residual over 100 returns, and the precision is
considerably better. A few stations have 2 to 4 cm accuracy. What is
needed is to upgrade roughly 10 of the stations to 1 cm basic measurement accuracy for 2 or 3 minute intervals as rapidly as possible.

For very long baseline radio interferometry, a number of fixed
stations already are involved in accurate measurements for geodynamics
studies. These include five in the U.S., the Onsala Observatory in
Sweden, the Effelsberg Observatory in the Federal Republic of Germany,
and the NASA Deep Space Network stations in Spain and Australia. With
the introduction of the new Mark III ground systems at a number of the
sites, the accuracy of the results is expected to be limited mainly by
the uncertainty in the tropospheric propagation velocity due to water
vapor.

A network of three stations in the U.S. to monitor the Earth's
rotation at least several times per week is being set up by the
National Geodetic Survey (Carter et al., 1979; Carter, 1980). Hopefully, this will become part of a worldwide system in the future. A
number of countries have expressed interest in taking part in such a
program. Encouraging results already are being obtained by several
groups (Fanselow et al., 1979; Robertson et al., 1980). The total
number of stations required to monitor the Earth's rotation is smaller
than for LAGEOS because measurements can be made even during cloudy
weather or moderate rain. Thus about eight stations may be sufficient

for this purpose, with some additional ones in the southern hemisphere needed part of the time to serve as reference points for crustal movement measurements.

The main area where improvements will be needed concerns the determination of the water vapor correction. The most promising approach is to use water vapor radiometers to measure the emission from water molecules along the line of sight. The observed power in a H_2O molecular emission line can be combined with an estimate of the average atmospheric temperature to give the integrated water vapor content of the atmosphere. Results obtained with several somewhat different types of radiometers have been reported by Guiraud et al. (1979), Moran and Rosen (1980), and Resch and Claflin (1980). The most extensive results so far are those of Guiraud et al. (1979), but observations were reported only for vertical paths. Seven additional radiometers have been designed and assembled recently by the Jet Propulsion Laboratory for use in VLBI measurements (Resch and Claflin, 1980).

There is a good theoretical basis for expecting that 1 cm accuracy can be achieved with water vapor radiometers, even at elevation angles as low as 20° (Westwater, 1978; Wu, 1979). But direct measurements of radiometer performance for low elevation angles and under varying atmospheric conditions still are needed (Resch and Claflin, 1980), in support of both VLBI and Global Positioning System measurement programs.

Measurements of baseline accuracy and reproducibility now are available for a number of VLBI baselines. Results obtained with a 1.24 km baseline over a period of 15 months show rms variations of 3 mm, 5 mm, and 7 mm respectively in the baseline length, azimuth, and elevation (Rogers et al., 1978). The mean results agree to 6 mm or better in each coordinate with the values obtained by careful ground surveying (Carter et al., 1980). Measurements at different times with a 9 m mobile station at two sites separated by 42 km gave a baseline length which agreed with ground survey measurements to 6±10 cm (Niell et al., 1979). And measurements over a 4000 km baseline between the Haystack Observatory in Massachusetts and the Owens Valley Radio Observatory in California have given results with a 4 cm rms reproducibility over a two-year period (Robertson et al., 1979). Recently, the first measurements of intercontinental distances with sub-decimeter reproducibility have been reported (Herring et al., 1980). It is encouraging that the above results were obtained even without the use of water vapor radiometers.

It appears likely that there will be a total of 20 to 25 fixed LAGEOS ranging and VLBI stations by the mid-1980's which will be accurate enough to contribute substantially to determining worldwide reference frames. In addition, the positions of a considerably larger number of sites probably will have been determined by high-mobility stations using both of the techniques (Silverberg, 1978; Niell et al.,

1979). Such stations are expected to take from two days to one or two weeks to determine their locations. One question which undoubtedly will arise is how to make use of the results from such stations for determining reference frames. At first glance, it might seem desirable to use such sites only to determine secondary reference systems. On the other hand, information from sites that are visited more than once at intervals of 1 or 2 years will be valuable in determining where relative motions are occurring within the plates. It may be desirable to incorporate such information into the definition and maintenance of the primary reference frames in some way.

Another important question concerns whether it is desirable to keep the sets of coordinates for LAGEOS ranging stations and VLBI stations separate, or whether to combine them in a joint set of coordinates. There can be a large number of ties between the two sets of station locations which are made by the high-mobility stations. If the sets of coordinates are not combined, a six-parameter transformation between the two networks still would be determined. This transformation could be used in combining the UT1 and polar motion results, as well as for putting the coordinates of the VLBI stations on a geocentric basis. The argument against using a combined set of coordinates is that any distortions introduced by one technique would influence the accuracy of the UT1 and polar motion results from the other technique.

It should be noted that lunar laser range measurements also are likely to contribute to worldwide position measurements, UT1, and polar motion. However, almost all lunar laser ranging stations are likely to be located at sites where LAGEOS range measurements also are made. For this reason, it does not seem necessary to treat lunar laser ranging separately in considering terrestrial reference frames.

It may be worthwhile to comment briefly at this point on the fact that many people are reluctant to use the term accuracy in discussing the results of geophysical measurements. The statement sometimes is made that accuracy can be established only by comparison with results obtained with a more accurate technique. This is incorrect, since the preparation and publication of careful error budgets which include allowances for systematic errors has been found to be very useful for determining the accuracy of measurements in other research areas such as primary frequency standards (Bender, 1974). The statement as given also is self-contradictory, which can be seen as follows. If one technique were believed to be the most accurate one, there would be no way to determine its accuracy. Thus there would be no way to determine that it was more accurate than the method believed to be the next most accurate. It therefore could not be used to determine the accuracy of that method, and so on down the line.

PROCEDURES FOR MAINTAINING TERRESTRIAL REFERENCE SYSTEMS

To be as specific as possible, we will assume that a decision is made at some particular time to adopt certain sets of coordinates for the LAGEOS ranging and VLBI stations and a plate motion model. These would be used for determining the earth's rotation and the positions of high-mobility stations until evidence began to accumulate that some of the adopted coordinates were inconsistent. At that point there would be justification for modifying the sets of coordinates, and possibly the plate motion model. For convenience, we assume that only the coordinates will be adjusted. Three possible approaches for carrying out the adjustment in such a way that the average orientation of the coordinate system is maintained are discussed below.

One approach which appears attractive is to carry out a free network adjustment using some suitable block of data. The station coordinates would be adjusted jointly with solutions for UT1 and polar motion as functions of time. Constraints are needed in order to remove the ambiguity between changes in all the station coordinates due to a rotation about some axis and fixed offsets in UT1 and polar motion. For LAGEOS ranging, one possibility that has been suggested is to minimize the weighted sum of squares of the horizontal position changes for the stations under rotations about three orthogonal axes (Bender and Goad, 1979). This leads to the following constraint equations:

$$\sum_i w_i (x_i \sin \phi_i \cos \lambda_i - y_i \sin \lambda_i) = 0$$

$$\sum_i w_i (x_i \sin \phi_i \sin \lambda_i + y_i \cos \lambda_i) = 0$$

$$\sum_i w_i (x_i \cos \phi_i) = 0 \quad .$$

Here x_i and y_i are the changes in the station coordinates in the eastern and northern directions respectively, ϕ_i and λ_i are the latitude and longitude, and w_i is the weight for a particular station. For VLBI stations the corresponding approach would be to minimize the weighted sum of squares of three-dimensional position changes under both rotations and translations.

Another possible approach would be to use principal value decomposition in order to reduce the number of degrees of freedom being solved for by three. I am not familiar with the literature concerning the use of such methods in geodesy, and hope that another paper in this volume will cover the subject better. The generalized inverse method and the Bjerhammar inverse method presumably are closely related. If the station coordinates were the only parameters being solved for, these methods probably would give essentially the same results as the free network adjustment approach. However, since UT1,

polar motion, and possibly orbit parameters are included in the solutions, it seems likely that the size of the corrections needed in
these quantities might influence the results, since the covariance
matrix wouldn't be diagonal. For this reason, my guess is that the
principal value decomposition type of approach is less desirable than
a free network adjustment method.

A third possible approach can be called sequential adjustment.
In this approach the values of UT1 and polar motion derived from the
available data using the initial set of coordinates are kept fixed.
An appropriate block of data is then reanalyzed using the previously
determined values of UT1 and polar motion, and a new set of station
coordinates are derived. The solution should be well determined in
this case, and no constraints will be needed. This approach is similar to the one used by the BIH for maintaining the origin of longitude.

The main advantage of the sequential adjustment approach is that
it is simple to carry out and minimizes any discontinuity in UT1 and
polar motion at the time that the new set of coordinates are adopted.
This is because the new coordinates are required to be as consistent
as possible with the old UT1 and polar motion values during the adjustment period. If the time interval between the end of the data set
that is fit and the switch to the new station coordinates is fairly
short, there is little time for discrepancies in UT1 and polar motion
to build up.

The main advantage of the free network adjustment method is that
it does not appear to introduce any unnecessary errors in the resulting station coordinates. Thus the fluctuations in comparing sets of
station coordinates obtained at different epochs would be minimized.
However, there probably will be somewhat larger offsets in UT1 and
polar motion when the switch to the new coordinates is made.

The choice between the free network adjustment approach and the
sequential adjustment approach may be a difficult one. My own preference would be for the free network approach, since I think that the
scientific benefits of keeping the reference system for position measurements with the high-mobility stations as stable as possible are
substantial. Also, the resulting offsets in UT1 and polar motion
are likely to be quite small. However, both approaches appear to be
viable.

Another important question involves the extent to which the
expected geophysical stability of sites is taken into account in
weighting the results. There are strong arguments in favor of having
a LAGEOS ranging station at the European Southern Observatory in
Chile, for example. The weather is good enough so that observations
might be obtained on nearly 300 days per year. Thus the efficiency of
such a station for determining UT1 and polar motion is extremely high,
and it would be an excellent reference site for high-mobility station

measurements. On the other hand, we know little about the possible
frequency of irregular crustal movements in this area, even well away
from the time of great earthquakes.

One way of handling the site stability problem is to carry out
solutions in which different weightings of data are used for stations
in different types of areas. Another approach is to use high-mobility
stations to tie questionable sites directly to stable locations on the
same continent, possibly as often as several times per year. A third
approach is to use GPS receivers or high-mobility stations to tie in
the sites with respect to the surrounding areas. The entire question
of site stability and of weighting data from different sites is a complicated enough one so that recommendations may be needed from some
group that is familiar with both the geophysical factors involved and
the sources of systematic measurement errors.

EXPECTED CONTRIBUTIONS FROM OTHER NEW TECHNIQUES

As mentioned earlier, it seems likely that measurements using
signals from the NAVSTAR Global Positioning System satellites will be
used extensively in the 1980's to study crustal movements in seismic
areas and for other geodetic purposes. This topic is discussed elsewhere in this volume by Goad, and thus will be mentioned only briefly
here. Measurement times of about 2 hours or less per site are expected. For moderate baseline lengths, measurement errors due to
other sources apparently can be made small compared with the uncertainties in the water vapor corrections determined by water vapor
radiometers. The GPS receiver and a water vapor radiometer could be
mounted in a light truck, with only a driver needed, and measurements
made with respect to a receiver and radiometer at a fixed site in the
area. For differential measurements over moderate baselines, the
water vapor correction accuracy appears likely to be about 1 cm, even
at low elevation angles of about 20°.

Recently, additional studies of the GPS accuracy capability have
been carried out by Larden and Bender (1980). A modified form of
"worst case error analysis" was used in order to obtain as reliable
accuracy estimates as possible. This approach was needed because of
the probability that the water vapor correction errors would be far
from random from measurement to measurement, and would vary with a
similar time scale to that of the satellite motions. Despite the type
of analysis used, expected accuracies of 1 to 2 cm for the baseline
components were found when 1 cm errors were assumed for the water
vapor corrections.

Another technique that is likely to make important contributions
to maintaining terrestrial reference frames during the 1980's is absolute gravity measurements. Substantial progress has been made in the
past decade in developing absolute gravimeters. An instrument located

in Paris has been operating for some time with a reported accuracy of
a few microgals (Sakuma, 1974a,b), and another similar instrument has
been built in Japan. Transportable absolute gravimeters were first
used to provide some of the reference points for the 1971 International
Gravity Network. More recent transportable instruments have been developed in western Europe by joint Italian and French efforts (Cannizzo et
al., 1978), in the U.S. (Hammond and Iliff, 1978), and in the U.S.S.R.
(Boulanger, 1979). The accuracy achieved is believed to be about 10
microgals, although the discrepancies found at a few sites have been
considerably larger. Work also is proceeding on a more portable absolute gravimeter which has an accuracy goal of 3 microgals (Faller et
al., 1979).

    If the sources of occasional discrepancies in absolute measure-
ments with present transportable instruments are found and eliminated,
as seems likely, and particularly if higher accuracy can be achieved,
then absolute gravity measurements can provide valuable checks on
vertical movements. While there are some situations where elevation
changes can occur without variations in gravity, this probably is
rare. The change in some important cases is roughly two to three
microgals per centimeter of uplift. If both gravity changes and elevation changes can be measured, the combination provides an additional
constraint on the physical mechanism responsible for the variations.
Thus absolute gravity measurements at many of the fixed sites for
worldwide terrestrial reference networks are highly desirable.

    The one major restriction on the use of gravity measurements
concerns local hydrology. Sites for gravity measurements have to be
chosen carefully, since effects such as the withdrawal of water from
aquifers can change the results. However, the choice of adequate
sites on crystalline rock outcrops appears to be feasible in many
non-sedimentary areas. Use should be made of absolute gravity measurements at such sites whenever they exist close to the sites used
for geometrical position measurements.

    It has been suggested that absolute gravity measurements also
should be used to monitor variations in the position of the center of
mass of the entire earth-oceans-atmosphere system with respect to
points on the crust. However, the expected amplitudes of the seasonal
and longer-term center-of-mass motions are quite small, as indicated
by studies such as those of Stolz and Larden (1979) and Larden (1980).
In view of the hydrology problem and the difficulties of the measurements, it seems best during the 1980's to use the combination of LAGEOS
range measurements and absolute gravity measurements at crystalline
rock sites to look for evidence of center-of-mass motion, rather than
using gravity measurements alone.

    So far the discussion has been limited to terrestrial geometric
reference frames. However, space techniques also are likely to contribute to tying together the vertical control networks on different

continents. This question was discussed recently by Colombo (1980), and results were given on what could be achieved with worldwide gravity field models complete to degree 20. Leveling, position, and gravity measurements over 5° or 10° caps on two continents were assumed in order to relate the equipotential surfaces of the worldwide model to the local undulations of the real surfaces.

If one extrapolates to the type of gravity field model expected from a GRAVSAT mission, considerably improved results could be expected even with ground measurements over 1° or 2° caps. Looking even further into the future, superconducting gravity gradiometers (Paik et al., 1978; Paik, 1980) or possibly laser interferometry measurements in orbit (Bender et al., 1980) might some day improve the worldwide gravity model resolution down to 40 to 50 km. Ground measurements might then be needed over only perhaps 0.5 to 1° caps in order to be able to identify points of equal gravitational potential on different continents.

CONCLUDING REMARKS

De facto terrestrial reference systems for geodynamics studies which include adopted plate motion models may well come into use quite soon for analyzing both LAGEOS ranging data and VLBI data. It will be helpful if the practical choices involved in establishing and maintaining such systems are discussed clearly at IAU Colloquium No. 56. Hopefully, this will reduce the number of changes needed later in the procedures for maintaining the systems.

Acknowledgments. Conversations with many different people have been helpful in clarifying the ideas presented in this paper. Discussions of new measurement techniques with my colleagues at JILA and with E. C. Silverberg, C. C. Goad, and C. C. Counselman III have been particularly valuable. The work has been supported partly by the NASA Office of Space and Terrestrial Applications under Contract S-65000-B and partly by the National Bureau of Standards.

REFERENCES

Bender, P.L.: 1974, Reference coordinate system requirements for geodynamics, in IAU Colloquium No. 26, On Reference Coordinate Systems for Earth Dynamics, 26-31 August, 1974, Torun, Poland (B. Kolaczek and G. Weiffenbach, eds., Polish Academy of Science, Warsaw), pp. 85-92.
Bender, P.L.: 1978, Some U.S. views on scientific opportunities in ocean and earth dynamics, in Space Oceanography, Navigation and Geodynamics, Proceedings, European Space Agency Workshop, Schloss Elmau, Germany (S. Hieber and T. C. Guyenne, eds., ESA SP-137, Paris), pp. 19-23.

Bender, P.L.: 1980, Improved methods for measuring present crustal movements, in Dynamics of Plate Interiors, Final Report of W.G.7 of the Inter-Union Commission on Geodynamics (R.I. Walcott, A.W. Bally, P.L. Bender, T.R. McGetchin and W.R. Peltier, eds., American Geophysical Union, Geodynamics Series, Washington, D.C.), in press.

Bender, P.L. and Goad, C.C.: 1979, Probable LAGEOS contributions to a worldwide geodynamics control network, in The Use of Artificial Satellites for Geodesy and Geodynamics, Vol. II (G. Veis and E. Livieratos, eds., National Technical University, Athens), pp. 145-161.

Bender, P.L., Bertotti, B. and Faller, J.E.: 1980, Laser inter-ferometer method for mapping ultra-short-wavelength structure in the earth's gravitational field, in preparation.

Boulanger, J.D.: 1979, Certain results of absolute gravity determina-tions by the instrument of the U.S.S.R. Academy of Sciences, in Publication Dedicated to T. J. Kukkamäki on the Occasion of his 70th Anniversary (J. Kakkuri, ed., Pub. No. 89 of the Finnish Geodetic Institute, Helsinki), pp. 20-26.

Cannizzo, L., Cerutti, G. and Marson, I.: 1978, Absolute-gravity mea-surements in Europe, Il Nuovo Cimento, 1C, pp. 39-85.

Carter, W.E.: 1980, Project POLARIS: A status report, in Radio Inter-ferometry Techniques for Geodesy (NASA Conf. Pub. 2115, Wash., D.C.), pp. 455-460.

Carter, W.E., Robertson, D.S. and Abell, M.D.: 1979, An improved po-lar motion and Earth rotation monitoring service using radio in-terferometry, in IAU Symp. No. 82, Time and the Earth's Rotation (D.D. McCarthy and J.D.H. Pilkington, eds., D. Reidel, Dordrecht, Holland), pp. 191-197.

Carter, W.E., Rogers, A.E.E., Counselman III, C.C. and Shapiro, I.I.: 1980, Comparison of geodetic and radio interferometric measure-ments of the Haystack-Westford baseline vector, J. Geophys. Res. 85, pp. 2685-2687.

Colombo, O.L.: 1980, A World Vertical Network (Report No. 296, Dept. of Geodetic Science, Ohio State University, Columbus).

Eaves, R.J., Schutz, B.E. and Tapley, B.D.: 1979, Ocean tide perturba-tions on the LAGEOS orbit (abst.), Trans. AGU 60, p. 808.

Faller, J.E., Rinker, R.L. and Zumberge, M.A.: 1979, Plans for the development of a portable absolute gravimeter with a few parts in 10^9 accuracy, Tectonophys. 52, pp.107-116.

Fanselow, J.L., Thomas, J.B., Cohen, E.J., MacDoran, P.F., Melbourne, W.G., Mulhall, B.D., Purcell, G.H., Rogstad, D.H., Skjerve, L.J., Spitzmesser, D.J., Urech, J. and Nicholson, G.: 1979, Determina-tion of UT1 and polar motion by the Deep Space Network using very long baseline interferometry, in IAU Symp. No. 82, Time and the Earth's Rotation, loc. cit., pp. 199-209.

Guiraud, F.O., Howard, J. and Hogg, D.C.: 1979, A dual-channel micro-wave radiometer for measurement of precipitable water vapor and liquid, IEEE Trans. on Geoscience Elect. GE-17, pp. 129-136.

Hammond, J.A. and Iliff, R.L.: 1978, The AFGL absolute gravity program, in Applications of Geodesy to Geodynamics (I.I. Mueller, ed., Report No. 280, Dept. of Geodetic Science, Ohio State University, Columbus), pp. 245-254.

Herring, T.A., Corey, B.E., Counselman III, C.C., Shapiro, I.I., Rönnäng, B.O., Rydbeck, O.E.H., Clark, T.A., Coates, R.J., Ma, C., Ryan, J.W., Vandenberg, N.R., Hinteregger, H.F., Knight, C.A., Rogers, A.E.E., Whitney, A.R., Robertson, D.S. and Schupler, B.R.: 1980, Geodesy by radio interferometry: Intercontinental distance determinations with subdecimeter precision, J. Geophys. Res., submitted.

Kaula, W.M.: 1975, Absolute plate motions by boundary velocity minimization, J. Geophys. Res. 80, pp. 244-248.

Larden, D.R.: 1980, Monitoring the Earth's Rotation by Lunar Laser Ranging (Thesis, Univ. of New South Wales, Sydney), in preparation.

Larden, D.R. and Bender, P.L.: 1980, Expected accuracy of geodetic baseline determinations using the GPS reconstructed carrier phase method, Bull. Geod., in preparation.

Minster, J.B. and Jordan, T.H.: 1978, Present-day plate motions, J. Geophys. Res. 83, pp. 5331-5354.

Moran, J.M. and Rosen, B.R.: 1980, The estimation of the propagation delay through the troposphere from microwave radiometer data, Radio Science, submitted.

Niell, A.E., Ong, K.M., MacDoran, P.F., Resch, G.M., Morabito, D.D., Claflin, E.S. and Dracup, J.F.: 1979, Comparison of a radio interferometric differential baseline measurement with conventional geodesy, Tectonophys. 52, pp. 49-58.

Paik, H.J.: 1980, Superconducting tensor gravity gradiometer for satellite geodesy and inertial navigation, J. Astr. Sci., submitted.

Paik, H.J., Mapoles, E.R. and Wang, K.Y.: 1978, Superconducting gravity gradiometers, in Proc. Conf. on Future Trends in Superconductive Electronics, Charlottesville, Virginia (Amer. Inst. of Phys. Conf. Proc. No. 44, B.S. Deaver, Jr., C.M. Falco, J.M. Harris, and S.A. Wolff, eds., New York), pp. 166-170.

Resch, G.M. and Claflin, E.S.: 1980, Microwave radiometry as a tool to calibrate tropospheric water-vapor delay, in Radio Interferometry Techniques for Geodesy, loc. cit., pp. 377-384.

Robertson, D.S., Carter, W.E., Corey, B.E., Counselman III, C.C., Shapiro, I.I., Wittels, J.J., Hinteregger, H.F., Knight, C.A., Rogers, A.E.E., Whitney, A.R., Ryan, J.W., Clark, T.A., Coates, R.J., Ma, C. and Moran, J.M.: 1979, Recent results of radio interferometric determinations of a transcontinental baseline, polar motion, and earth rotation, in IAU Symp. No. 82, Time and the Earth's Rotation, loc. cit., pp. 217-224.

Robertson, D.S., Clark, T.A., Coates, R.J., Ma, C., Ryan, J.W., Corey, B.E., Counselman III, C.C., King, R.W., Shapiro, I.I., Hinteregger, H.F., Knight, C.A., Rogers, A.E.E., Whitney, A.R., Pigg, J.C. and Schupler, B.R.: 1980, Polar motion and UT1: Comparison of VLBI, lunar laser, satellite laser, satellite Doppler, and conventional astrometric determinations, in Radio Interferometry Techniques for Geodesy, loc. cit., pp. 33-44.

Rogers, A.E.E., Knight, C.A., Hinteregger, H.F., Whitney, A.R., Counselman III, C.C., Shapiro, I.I., Gourevitch, S.A. and Clark, T.A.: 1978, Geodesy by radio interferometry: Determination of a 1.24-km baseline vector with ~5 mm repeatibility, J. Geophys. Res. 83, pp. 325-334.

Sakuma, A.: 1974a, Report on absolute measurement of gravity, in Proceedings of the International Symposium on the Earth's Gravitational Field and Secular Variations in Position (Dept. of Geodesy, Univ. of New South Wales, Sydney), pp. 674-684.

Sakuma, A.: 1974b, Report on absolute measurements of gravity, in Bull. d'Information No. 35, Bureau Gravimetrique International, pp. I-39-I-42.

Silverberg, E.C.: 1978, Mobile satellite ranging, in Applications of Geodesy to Geodynamics, loc. cit., pp. 41-46.

Smith, D.E. and Dunn, P.J.: 1980, Long term evolution of the LAGEOS orbit, Geophys. Res. Lett. 7, pp. 437-440.

Smith, D.E., Kolenkiewicz, R., Dunn, P.J. and Torrence, M.: 1979a, Determination of polar motion and Earth rotation from laser tracking of satellites, in IAU Symp. No. 82, Time and the Earth's Rotation, loc. cit., pp. 247-255.

Smith, D.E., Kolenkiewicz, R., Dunn, P.J. and Torrence, M.: 1979b, Determination of station coordinates from LAGEOS, in The Use of Artificial Satellites for Geodesy and Geodynamics, Vol. II, loc. cit., pp. 162-172.

Solomon, S.C. and Sleep, N.H.: 1974, Some simple physical models for absolute plate motions, J. Geophys. Res. 79, pp. 2557-2567.

Stolz, A. and Larden, D.R.: 1979, Seasonal displacement and deformation of the earth by the atmosphere, J. Geophys. Res. 84, pp. 6185-6194.

Westwater, E.R.: 1978, The accuracy of water vapor and cloud liquid determination by dual-frequency ground-based microwave radiometry, Radio Science 13, pp. 677-685.

Wu, S.-C.: 1979, Optimum frequencies of a passive microwave radiometer for tropospheric pathlength correction, IEEE Trans. Ant. Prop. AP-27, pp. 233-239.

IDEAL REFERENCE FRAMES, CONCEPTS AND INTERRELATIONSHIPS

George Veis
National Technical University, Athens

ABSTRACT. Many problems of geodynamics depend on spatial relationship
of points and their temporal variations. To solve these problems it is
convenient, but not necessary, to use a reference frame. To use a
reference frame, a scheme is needed by which the coordinates of any
point expressed in this frame could be obtained. Coordinates are hardly
ever measured directly. Instead, they are computed from measurements
of other quantities within the framework of a theory that relates the
measured quantities with the coordinates. Such a scheme will be called
a reference system. Reference systems have been realized using simple
theories and reducing the measurements with the best available, but
not always complete, theories. This geometric (static) method has been
used to a great extend to define astronomic reference systems (star
catalogues) and geodetic reference systems (geodetic datums). With
space techniques, a method can be used based on dynamic principles. A
space object moving according to a certain theory (assumed to be known)
defines in a time dependent way the representative points. A reference
system of this type is the WGS 72.

Most of the problems connected to Astronomy, Geodesy, and Geodynamics are related with the spatial relationship of points, and the
temporal variation of their position. In principle these problems can
be formulated by expressing the position of each point by a variety of
methods but it becomes much easier if the positions of all points are
expressed with the same simple scheme. This is accomplished by using
a reference frame in which the positions are expressed with coordinates.
In what will follow we assume that euclidean space is a sufficiently
good approximation and that any deviations from it can be taken care
of, by appropriate small corrections.

In a one-dimensional space a point can be positioned if its distance from an arbitrarily selected point, used as origin, is given.
Here we make the assumption that we have defined a scale for measuring
distances. Moving in two and three dimensional spaces we need,
respectively, to measure two or three distances from two or three
selected points as origins.

E. M. Gaposchkin and B. Kołaczek (eds.), Reference Coordinate Systems for Earth Dynamics, 37–41.
Copyright © 1981 by D. Reidel Publishing Company.

Defining positions with distances from selected points is not
convenient because the formulation of our problems becomes rather complicated. Instead we choose to define positions by giving the distances
from two lines or three planes, for a two or three dimensional space
respectively. In order to make it even simpler we also choose the
lines or planes to intersect at right angles. Three mutually perpendicular vectors of unit length define a triad. We will call such a
triad a reference frame and the distances from the three unit vectors
cartesian or rectangular coordinates. Cartesian coordinates are a
commonly used form of coordinates but other forms, such as spherical,
cylindrical, ellipsoid or geoidal are also used. Spherical coordinates
are used in astronomy and ellipsoidal in geodesy (ellipsoidal coordinates also require to define that the parameters of the reference
ellipsoid be defined). Assuming that these different forms of coordinates are referred to the same reference frame (or triad) their transformation is a trivial problem.

A three-dimensional reference frame can be arbitrarily located and
oriented in space giving six degrees of freedom, three for the location
and three for the orientation. There is no way by which this arbitrariness can be removed without additional information. However if the
relative position and orientation of two reference frames are known
(which means six parameters if they have the same scale) the conversion
of coordinates from one frame to another is a simple matrix operation.

The freedom in choosing the origin and the orientation of the
reference frame, as well as the form of the coordinates, could allow
the selection of a coordinate system to optimize its use. As a rule
this is accomplished if: a) observables are easily and simply expressed
in this coordinate system, b) needed computations are easily and simply
performed, and c) coordinates can readily be used for different applications. If a coordinate system is to be used only for the solution
of a theoretical problem, the selection of an optimum system is not so
difficult as a rule.

A coordinate system however is not used only for the solution of
theoretical problems. There are many applications not only of geodesy
and astronomy but also of geophysics, surveying, navigation, engineering etc. that require the use of a coordinate system. What this means
is, to be able to find, in a certain reference frame, the coordinates
of a physical point, and to be able to find the point in space with
given coordinates. If the points are in motion, the coordinates should
refer to a certain epoch.

The implementation of the selected coordinate system is a complicated operation. Except for very special cases, coordinates are not
measured directly as distances from axes materialized on the ground.
Instead they are computed from other measured quantities mathematically
related with the coordinates.

This mathematical relation is based on a certain theory and physical laws which may also include some parameters. In addition the measurements may have to be corrected before they are used. An example is the measurement of the distance between two points with a microwave measuring instrument. This distance is related with a simple formula with the rectangular coordinates of the two points. The measured distance however has to be corrected for refraction before being used in the formula. The use of a theory will not affect the arbitrariness of the origin and orientation of the reference frame, unless some outside positional information is introduced with the theory.

A scheme including the necessary measurements, theories, and computations, by which coordinates in a certain reference frame can be computed, defines a reference system. The internal consistancy of such a system will depend on the accuracy of the measurements and their corrections, on the completeness of the theories and the correctness of the constants, and on the precision of the computations. Coordinates derived from two different reference systems will not agree if the measurements, the theories and the computations are not consistent. In order to relate two reference systems it is necessary to find the relations between the two theories (and constants), the two sets of observations (and their corrections) and the two computations used for their definition.

The Reference Systems that have been implemented up to now in classical astronomy and geodesy (star catalogues and geodetic datums) were based on simple (euclidian) geometry. The measurements used were primarily angles with very few distances. Only during the last 20 years the number of distance measurements has been increased, with the introduction of new instruments such as EDM, laser etc.

In astronomy the coordinates of stars were determined using mainly Meridian instruments. These observations were reduced using a certain theory for precession and nutation to a common epoch and corrected for polar motion. The astronomic coordinates of the instruments were also unknown. With this method only a few thousand stars (fundamental stars) were determined. This fundamental network was densified by relative measurements on photographic plates and extended to some hundred thousands of stars. Fundamental stars are expected to have an uncertainty of $\pm$ 0''02 but the remaining stars may be more than 10 times worse.

In geodesy the coordinates of geodetic points (triangulation points) were determined by establishing a geodetic network of which angles and distances were measured. By adjusting this network (assuming it to have been normally projected on a reference ellipsoid) the coordinates of all the points of the network were computed. The origin and the orientation of the network was not arbitrarily set, but adopted from astronomic observations as a good approximation. So geodetic datums are, practically speaking, derived from free networks.

Within a geodetic datum the internal relative accuracy is anywhere between 10^{-6} and 10^{-5}, depending on the quality of the observations, the accuracy of the applied corrections (mainly refraction) and
the completeness of the reductions (mainly from geoid to ellipsoid).
Datums computed before 1950, without the use of computers, were
adjusted with approximate methods and this is another source of error.
A characteristic of a geodetic datum is that the positional accuracy
depends on the location of the triangulation point in the network and
that these errors propagate with distance. However, for most practical
applications, such as cartography and engineering, the relative accuracy is sufficient.

If we exclude distortions in the networks, two geodetic reference
systems (datums) should be related by translation and rotation (six
parameters). The relative rotation should be rather small (of the
order of 1" or less), since both datums were oriented with astronomic
observations. On the other hand, since old networks have used
standards of length of low accuracy for their scale, two datums may
differ in scale by several parts per million. As a result, for the
relation of two geodetic systems a scale factor also has to be introduced, bringing the number of parameters to seven.

The realization of these static reference systems is materialized
by assigning nominal coordinates, using the above mentioned techniques,
to a number of selected stars or triangulation points. It is the
catalogue of these selected points (stars or triangulation points) and
their coordinates that really defines the reference system. In order
to calculate the coordinates of a new point in the same reference system, similar measurements are made between some of the selected points
and the new one and coordinates are obtained using the same theory "de
proche en proche".

With simultaneous or quasi-simultaneous observations to artificial
satellites, which are used simply as moving beacons, three-dimensional
geodetic networks (free nets) have been established, adjusted and
calculated. The relative positions for stations separated by intercontinental distances have been determined. These nets have given
positional accuracies of about ± 5 m.

Instead of using a static approach to determine nominal coordinates
for selected points, we could use a moving object whose motion, expressed in some reference frame, is precisely known. Assuming we can
time the observations to the needed accuracy and freeze the moving
object, the problem becomes equivalent with the static geometric one.
Artificial Satellites, (and the Moon) provide, at least in principle,
such a moving object, since their moon is governed by the laws of
celestial mechanics. The theory is there but in practice the problem
is much more complicated to account for the great number of parameters
whose values must be known (mainly in order to express the gravity
field of the earth). These parameters can be calculated as auxiliary
unknowns by successive approximations. Results obtained with this

Dynamic Satellite Geodesy method are rather promising, especially with some special geodetic satellites like Lageos. This method has two advantages. First, it gives positions expressed in a dynamically defined inertial reference system (for periods where the secular perturbations may be ignored), and second it can give the best approximation to a geocentric system. Such a dynamic reference system becomes much more complicated and complex to define, since it includes a very complicated theory (and calculations) as well as a great number of parameters. Such a reference system is the WGS 72, especially when associated with the Transit System, which gives a very high degree of internal consistency. We should expect the GPS to be even better.

When comparing or combining dynamic systems we have to be careful, since the theories and constants used for the determination of the orbits, even the computer programs, may be different. To a first approximation, however, the dynamic systems are also related to each other by a translation, rotation and scale, and thus also with the static ones.

If we consider that the points on the earth's crust undergo secular, periodic and abrupt displacements in a pseudosystematic and random way, the implementation of a static geodetic reference system providing fixed nominal coordinates to a number of points becomes utopia. In this case we either use the minimum number of arbitrary points (6 parameters) or we use very many points assuming that the motion of the mean will be zero. The use of a dynamic reference system may be in principle a better solution, since it will eventually relate the positions to an almost inertial system. Such a system will be improved continuously be successive approximations. For better results and faster conversion, it is essential that an integrated large scale satellite system be used, and as such the GPS looks to be the most promising system.

RELATIVISTIC EFFECTS IN REFERENCE FRAMES

H. Moritz

Institute of Physical Geodesy
Technical University
Graz, Austria

ABSTRACT

The impact of relativistic theories of space, time and gravitation on the problem of reference systems is reviewed.

First, the concept of inertial systems is discussed from the point of view of the special and the general theory of relativity. Then, relativistic corrections of Doppler, laser and VLBI, and similar effects are reviewed; they are usually on the order of 10^{-8} . Finally, the problem of a possible variation of the gravitational constant G (on the order of 10^{-11}/year) is outlined; such a variation does not occur in special and general relativity, but is implied by certain generalized field theories which are less commonly accepted.

1. INTRODUCTION

We all know that the special theory of relativity is a refinement of classical mechanics for the case that we are dealing with very high velocities, and that the general theory of relativity provides a refinement of the Newtonian theory of gravitation, relevant for very strong gravitational fields such as the fields of black holes, and for problems of cosmology. For the gravitational field of the Earth and for satellite motion in this field, as well as for terrestrial reference systems, classical mechanics is sufficient; relativistic effects are negligible or can be taken into account by very small corrections.

E. M. Gaposchkin and B. Kołaczek (eds.), Reference Coordinate Systems for Earth Dynamics, 43–58.
Copyright © 1981 by D. Reidel Publishing Company.

This well-known fact will, of course, be confirmed by the present paper, but conceptually the situation is not always obvious at first sight. The following example will certainly strike us as paradoxical: The most accurate means for practically establishing an inertial system is VLBI using quasars; however, quasars are a typical phenomenon of an expanding universe which must be described by general relativity and for which, therefore, rigorously no inertial systems exist!

We shall come back to this paradox later on. It already indicates that an understanding of the basic principles of reference systems from a relativistic point of view is of conceptual significance.

The present paper attempts a review of the impact of relativity on the problem of reference systems. We shall first discuss the concept of inertial system from the point of view of both the special and the general theory of relativity, then give a review of relativistic corrections and similar effects by which relativistic geometry and mechanics differ from the classical situation, and finally discuss the problem of a possible variation of the gravitational constant G which, however, goes beyond Einsteinian relativity.

There are a number of excellent textbooks on the theory of relativity. The most elegant presentation is perhaps (Synge, 1960, 1972), the most comprehensive and modern text is certainly (Misner et al., 1973), and a very readable and useful recent book is (Ohanian, 1976). An excellent review article on applications to space science is (Dicke and Peebles, 1965). In a previous work (Moritz, 1967), the author has treated in some detail the question of inertial systems from a relativistic standpoint, especially with a view to separation of gravitation and inertia which is not usually considered in standard textbooks (except Synge, 1960). The present paper partly follows (Moritz, 1979).

Like all great and deep theories, Einstein's theory admits of various, often controversial, interpretations. It has even been argued that the name, general relativity, is not entirely appropriate since the essence of this theory is not the general "relativity" of all reference systems but rather the fact that the theory provides a mathematical description of "absolute" curved space-time. This point of view seems to be rather widely accepted at present (e.g. Fock, 1959 ; Synge, 1960 ; Misner et al., 1973 ; Ohanian, 1976); it is also favored in the present article.

2. INERTIAL SYSTEMS AND RELATIVITY

Inertial systems in special relativity. In the special theory of relativity, inertial systems play a basic role as privileged coordinate systems in space-time: in such a system, the four-dimensional line element has the simple form

$$ds^2 = dx^2 + dy^2 + dz^2 - c^2dt^2 = dx_1^2 + dx_2^2 + dx_3^2 + dx_4^2 . \quad (1)$$

Here $x = x_1$, $y = x_2$, and $z = x_3$ denote rectangular coordinates in space, t designates the time, and c denotes the constant velocity of light in a vacuum; we have to put $x_4 = ict$, where $i^2 = -1$. As in classical mechanics, a reference system moving with constant velocity with respect to an inertial system, is again an inertial system.

Transformations between inertial systems are such as to leave the line element (1) invariant (unchanged); the set of such "Lorentz transformations" form a group, the Lorentz group, which describes the symmetry of the space-time of special relativity.

No inertial systems in general relativity. The special theory of relativity holds only in the absence of a gravitational field. Gravitational fields are treated by the general theory of relativity. Here the line element has the form

$$ds^2 = \sum_{\alpha=1}^{4} \sum_{\beta=1}^{4} g_{\alpha\beta} dx^\alpha dx^\beta = g_{\alpha\beta} dx^\alpha dx^\beta \quad (2)$$

where x^α denotes coordinates x^1, x^2, x^3, x^4 in space-time, which will in general be curvilinear rather than rectangular. The $g_{\alpha\beta}$ are functions of these coordinates. Indices such as α and β run from 1 to 4 ; lower indices are called covariant, and upper indices, contravariant. The Einstein summation convention, which will be used in this section, prescribes summation with respect to any index that occurs in both an upper and a lower position, as shown in eq. (2). The coordinates x^α now have upper indices because the differentials dx^α form a "contravariant vector".

The line element (2) relates to (1) in much the same way as a line element on a curved surface,

$$ds^2 = Edu^2 + 2Fdudv + Gdv^2, \quad (3)$$

relates to a line element in the plane,

$$ds^2 = dx^2 + dy^2. \quad (4)$$

Here, u, v are curvilinear coordinates and E, F, G form the "metric tensor"

$$\begin{bmatrix} E & F \\ F & G \end{bmatrix} = \begin{bmatrix} g_{11} & g_{12} \\ g_{12} & g_{22} \end{bmatrix}. \tag{5}$$

In space-time, the metric tensor $[g_{\alpha\beta}]$ is a 4x4 matrix, so that there is full analogy between the general forms (2) and (3) on the one hand, and between the "inertial forms" (1) and (4) on the other hand. In a way, the general theory of a relativity is nothing else but an extension of the theory of two-dimensional surfaces to four-dimensional space-time.

This analogy will help understand an important point. On a curved surface one can introduce coordinates which, in an infinitesimal neighborhood of a point, give a line element

$$ds^2 = du^2 + dv^2 \tag{6}$$

which has the same form as the plane element (4). Geometrically, this means that, in a small neighborhood of this point, the surface is approximated by its tangent plane. However, it will not be possible, in general, to introduce coordinates in such a way that the "inertial form" (6) holds on the whole surface (or even in a finite part of it).

Transferred to four dimensions, this reasoning shows that, in a curved space-time, it will be possible to introduce coordinates which correspond to an inertial system in an infinitesimal neighborhood of a point; but it is not possible to introduce an inertial system that is valid for the whole space-time.

In this sense, there are no inertial systems in general relativity. All possible coordinate systems are, in principle, equivalent; there are no privileged systems. This is Einstein's Principle of General Covariance, or General Relativity.

Another important principle in this theory is the Principle of Equivalence, according to which gravitational and inertial forces (such as the centrifugal or Coriolis force) are basically identical: both are effects of a deviation of the coordinate system of line element (2) from an inertial system of line element (1). Thus gravitation is interpreted geometrically as an effect of the curvature of space-time.

Both the Principle of Equivalence and the Principle of General Covariance have played a fundamental heuristic role in Einstein's considerations leading to his theory of gravitation around 1915 because these principles provide a natural transition from the flat space-time of special relativity to the curved space-time of general relativity. Einstein's heuristic procedure is still the

best way for understanding this theory; hence it is strongly
emphasized in almost every textbook on general relativity.

The relativistic treatment of reference systems, however,
requires some subtler distinctions which show that, after all,
privileged systems can be introduced which serve as practically
satisfactory approximations to inertial systems, both on a local
and on a global level.

Local inertial systems. Just as a curved surface can be
approximated locally by a tangent plane, so curved space-time can
be approximated, in the neighborhood of a certain point, by a tangent "plane" space-time in which an inertial system can be introduced. Thus, in a certain "small" region, inertial systems are
possible even in general relativity. Since our space-time is only
very slightly curved, the gravitational field in the solar system
being very weak, the "small" region just mentioned certainly covers
the solar system and even extends well beyond. According to
Eddington (1924, pp.99) a local inertial system will deviate from
a global system by about 2 seconds of arc in a century.

Global nearly-inertial systems. The application of the relativistic theory of gravitation to the region of our solar system
requires boundary conditions at infinity: with increasing distance from the attracting masses the effect of gravitation vanishes,
and the curved space-time becomes flat at infinity. This fact permits the introduction of uniquely defined privileged systems, the
harmonic coordinate systems. These systems rigorously refer to
curved space-time. At infinity they reduce to inertial systems of
form (1) , and within the solar system they approximate inertial
systems practically very well.

In this sense, the harmonic coordinates form a privileged
coordinate system, which is a natural generalization of an inertial
system to curved space-time. This has been particularly emphasized
by Fock (1959); see also (Weinberg, 1972, p. 162).

Quasi-inertial systems and Fermi propagation. Let us introduce the concept of quasi-inertial systems. They are three-dimensional cartesian systems whose origin is moving arbitrarily but
whose axes remain always parallel; a physical realization is by
means of axes whose direction is stabilized by means of gyroscopes.
The underlying principle is that the axis of a freely spinning
gyroscope maintains its direction even if its frame is accelerated
or rotated; furthermore, the axis is unaffected by gravity.

Quasi-inertial systems differ from inertial systems in the
strict sense by the fact that they can be in nonuniform (accelerated) motion with respect to each other, as long as the coordinate
axes remain parallel. Inertial systems are always in uniform motion,

that is, they move with a constant velocity vector with respect to each other. A geocentric system of which the axes have a constant direction in space is an example of a quasi-inertial system: the origin (the geocenter) moves along an ellipse around the sun, rather than along a straight line with constant velocity.

This concept of a quasi-inertial frame can be defined also in general relativity. The relevant concept is <u>Fermi propagation</u>, or <u>Fermi-Walker transport</u>, which is considered in detail and used extensively in (Synge, 1960). It is also treated in (Misner et al., 1973, p. 170), but hardly elsewhere in standard textbooks. Therefore we shall briefly consider it here, following (Moritz, 1967).

The equation of Fermi-Walker transport may be written

$$\frac{\delta \lambda^{\alpha}}{\delta s} = \lambda_{\beta} \left(\frac{\delta u^{\beta}}{\delta s} u^{\alpha} - \frac{\delta u^{\alpha}}{\delta s} u^{\beta} \right) \tag{7}$$

(Synge, 1960, p. 13). Here λ^{α} (or λ_{β}) are the contravariant (or covariant) components of the vector undergoing Fermi propagation, related by

$$\lambda_{\alpha} = g_{\alpha\beta} \lambda^{\beta} . \tag{8}$$

The vector u^{α} is the four-velocity

$$u^{\alpha} = \frac{dx^{\alpha}}{ds} , \tag{9}$$

the unit vector of the tangent to the world line of the particle to which the vector λ_{α} is attached. The symbol δ denotes covariant differentiation:

$$\frac{\delta \lambda^{\alpha}}{\delta s} = \frac{d\lambda^{\alpha}}{ds} + \Gamma^{\alpha}_{\beta\gamma} \lambda^{\beta} u^{\gamma} , \tag{10}$$

where $\Gamma^{\alpha}_{\beta\gamma}$ are the Christoffel symbols, and analogously for $\delta u^{\alpha}/\delta s$.

In our case, the vector λ^{α} represents the spin axis of the gyroscope. It lies in the instantaneous three-dimensional space of the spinning particle and is therefore orthogonal to u^{α} :

$$u^{\alpha} \lambda_{\alpha} = 0 . \tag{11}$$

Hence   (7)   reduces to

$$\frac{\delta\lambda^\alpha}{\delta s} = \lambda_\beta \frac{\delta u^\beta}{\delta s} u^\alpha .$$ (12)

This equation holds for Fermi-Walker transport of a space-like vector satisfying (11) . It expresses the fact that the change $\delta\lambda^\alpha/\delta s$ has the direction of u^α and has no component in the instantaneous three-space of the observer. Thus the change λ^α is purely in time: the vector λ^α remains unchanged in space, it is transported parallelly. This shows that Fermi propagation is related to spatial parallelism.

Consider now a system of three mutually orthogonal space-like vectors λ^α , each of which is represented by the axis of a freely spinning gyroscope. In this way the axes of a rectangular xyz system which is transported parallelly in space, may be realized physically.

It can be shown (Moritz, 1967, p. 47) that the change $\delta\lambda^\alpha/\delta s$ is small of order c^{-2} , c being the velocity of light. To this accuracy, the direction of Fermi-propagated axes remains constant in space; it furthermore is practically unaffected by the gravitational field.

This shows that gyroscopically stabilized "quasi-inertial systems" are possible even in the context of general relativity.

Separation of gravitation and inertia. After this discussion of "privileged" coordinate systems which seem to contradict the Principle of General Covariance, let us now briefly remark on the separation of gravitational and inertial forces, which seems to violate the Principle of Equivalence. This question is related to the problem of reference systems only indirectly; it has been dealt with rather fully in two reports (Moritz, 1967, 1971).

The Principle of Equivalence states that, because of the identity of gravitational and inertial mass (shown experimentally by R. Eötvös around 1900 to an accuracy of 5×10^{-9} !) the resultant of gravitational and inertial forces acting at one point cannot be separated into a gravitational and an inertial part; both are equivalent and cannot be distinguished.

Matters are different if we consider, not only one point, but a region in space, which may be arbitrarily small. In the theory of surfaces, the Gaussian curvature K provides a criterion for distinguishing a curved surface from a plane, depending on whether K is nonzero or zero. The generalization of the Gaussian curvature to four dimensions is the Riemannian curvature tensor

$R_{\alpha\beta\gamma\delta}$; again, space-time is flat if $R_{\alpha\beta\gamma\delta} = 0$ and curved otherwise. Now, curvature of space-time $R_{\alpha\beta\gamma\delta}$ is an objective criterion for the presence of a genuine gravitational field, so that, according to (Synge, 1960, p. 109), we may write

$$R_{\alpha\beta\gamma\delta} = \text{gravitational field .} \tag{13}$$

The Riemann curvature thus provides a criterion for the presence of a gravitational field, but not yet a means for the separation of gravitational and inertial effects. In flat space-time, inertial forces have an objective significance since they are due to the deviation of the observer's coordinate system from an inertial system. Similarly in a weak gravitational field, a separation of gravitation and inertia is feasible if we succeed in introducing a privileged coordinate system similar to an inertial system. In this way, the separation of gravitation and inertia is intimately connected with the question of an "almost" inertial reference system, such as the harmonic system mentioned above.

We finally point out that in such a system there is approximately (Moritz, 1967, p. 43)

$$c^2 R_{i4j4} = \frac{\partial^2 V}{\partial x_i \partial x_j} \tag{14}$$

where i and j are spatial indices running from 1 to 3 . Thus, second-order gradients of the potential V are purely gravitational. In (Moritz, 1971) we have shown that using a combination of accelerometers, measuring first-order gradients, and gradiometers, measuring second-order gradients, a separation of the gravitational signal from inertial disturbances can be effected even with first-order gradients, that is, in the gravitational force.

Cosmological questions. For a homogeneous and isotropic universe, the line element (2) has the form (Bondi, 1960, p. 102)

$$ds^2 = dt^2 - [R(t)]^2 \frac{dx^2 + dy^2 + dz^2}{[1 + (k/4)(x^2 + y^2 + z^2)]^2} . \tag{15}$$

Here $R(t)$ is a time-dependent scale factor by means of which the expansion of the universe can be described. The constant k may have the values +1, 0, or -1 . For $k = 0$, space is Euclidean; for $k = 1$, space has constant positive curvature, and for $k = -1$, constant negative curvature. The space-time described by (15) is called the Robertson-Walker model. (For $k = 0$ and $R = c^{-1}$, (15) reduces to (1) , apart from the irrelevant factor $(-c^2)$.)

This model appears well suited to describe mathematically the large-scale space-time structure of the universe, apart, of

course, from "local" gravitational irregularities such as caused
by our solar system. On the basis of present observational data
it is not possible to decide clearly whether k is positive, negative or zero, although there is some indication that space may
have negative curvature (cf. Ohanian, 1976, p. 416).

At any rate, Robertson-Walker space-time will not in general
be the flat space-time of special relativity (1) . Thus, strictly speaking, inertial systems in the usual sense will not exist.
This leads us to the paradox already mentioned in the introduction:
The most accurate means of practically establishing an inertial
system is VLBI using quasars; however, quasars are a typical
phenomenon of an expanding universe which is described by the
curved space-time (15) <u>for which no inertial system exists!</u>

This paradox, however, is a theoretical curiosity rather
than a fact of particular significance. Indeed, as we have seen
above, all our practical inertial systems are nonrigorous in the
sense of general relativity but still perfectly useful. For the
region of our galaxy, we may easily consider space-time to be
essentially flat, apart from local gravitational irregularities.
The same holds <u>a forteriori</u> for our solar system. Furthermore, it
is possible to study cosmology within the frame of special relativity and even of classical mechanics (Bondi, 1960, Chapters XI
and IX).

3. RELATIVISTIC CORRECTIONS

The mathematical description of geometry and gravitational
field around the Earth (geodesy, geodynamics, satellite dynamics)
and in the solar system (celestial mechanics, classical astronomy)
uses Euclidean geometry and classical mechanics. Such a description
is valid to an accuracy of about 1 part in 10^8 . For higher
precisions, the special and general theories of relativity must be
taken into account. This is best done by applying small "relativistic corrections" to the classical formulas.

<u>Post-Newtonian approximation</u>. Let us formulate the equa-
tions of general relativity in an approximate form which is sufficiently accurate for one purpose and, at the same time, comes
close to classical potential theory. Such "nearly-Newtonian gravitational fields" or post-Newtonian approximations" to Einstein's
theory are treated in almost every text on relativity; cf. (Misner
et al., 1973, pp. 445, 1068) and (Boucher, 1979).

For this case the general line element reduces to

$$ds^2 = (1 + 2 \frac{V}{c^2})(dx^2 + dy^2 + dz^2) - (1 - 2 \frac{V}{c^2}) c^2 dt^2 . \quad (16)$$

Here x, y, z are rectangular spatial coordinates as usual, and
t is a time coordinate; it will be called <u>coordinate time</u>. The
symbol V denotes the classical Newtonian gravitational potential,
defined in the "geodetic" way (everywhere positive and tending to
zero at infinity; physicists frequently use the opposite sign)
and c is the velocity of light as usual. This equation is linear
in V/c^2 ; higher powers are consistently neglected in this theory.

What is the order of magnitude of V/c^2 ? The gravity potential W at the surface of the earth is approximately

$$W = 6.3 \times 10^7 m^2 s^{-2} ;$$

cf. the value U_o given in (Heiskanen and Moritz, 1967, p. 80);
for the present purpose, the gravitational potential V and the
gravity potential W (including the centrifugal force)   are
nearly equal. Then

$$\frac{V}{c^2} \doteq \frac{W}{c^2} \doteq \frac{6.3 \times 10^7 m^2 s^{-2}}{(3 \times 10^8 ms^{-1})^2} \doteq 0.7 \times 10^{-9} . \quad (17)$$

Thus, V/c^2 is a dimensionless quantity of order 10^{-9} at the
earth's surface (and smaller at higher elevations). If we neglect
this small quantity, then the line element (16) reduces to the
simple line element (1)) of special relativity.

<u>Time</u>. Since time can be measured by means of atomic clocks
far more accurately (to order 10^{-13} or better) than any other
relevant quantity, relativistic effects show here quite well and
must be taken into account. <u>Atomic time</u> which an atomic clock
measures, has the character of a <u>proper time</u> and will be denoted
by τ . The element of proper time, dτ , is proportional to the
element ds , given by (16) , of the world line of the atomic
clock:

$$ds = icd\tau , \qquad\qquad d\tau = ds/ic . \qquad (18)$$

Hence, for a clock at rest (dx = dy = dx = 0) ,

$$d\tau = (1 - 2 \frac{V}{c^2})^{\frac{1}{2}} dt \doteq (1 - \frac{V}{c^2}) dt . \qquad (19)$$

If the atomic clock is fixed to the rotating Earth, then the gravitational potential V in (19) must be replaced by the gravity potential W which is the sum of V and of the centrifugal potential. Thus atomic clocks depend on the <u>potential</u> in a similar way as the old pendulum clocks depend on <u>gravity</u>, through incomparably less. Just as gravity, or better gravity differences, can be measured by means of pendulums, so the potential, or better potential differences, can, in principle, be measured by atomic clocks. (It would, however, be premature to hope for a new geodetic instrument measuring potential differences in this way: 1 cm in elevation would correspond to 10^{-18} in time!)

Of such nature is the experiment   by Pound and Rebka described in (Misner et al., 1973, p. 1057) and (Ohanian, 1976, p.212), which measures the gravitational redshift of γ-rays using the Mössbauer effect and hence the potential difference. (Redshift occurs if the "clock" represented by the emitting source is slower.)

Related phenomena are the time delay of radar echoes from Mercury, Venus and Mars due to their gravitational fields as measured by Shapiro and others (Ohanian, 1976, p. 128), and time dilation experiments measuring the redshift of different spectral lines of the sun and other stars. (Ohanian, 1976,p.214). We shall consider such an effect below when discussing laser distance measurements.

Another question is the relation between Atomic Time   (AT) and Ephemeris Time   (ET). Conceptually, AT is the time of quantum theory, and ET is the time of mechanics (classical or relativistic). If general relativity is correct, then AT = ET. On the other hand, (Duncombe et al., 1974, p. 232) state that empirical observations tend to indicate that these two time scales are not equivalent. As an explanation they suggest that the gravitational constant G decreases at the rate of about 10^{-11} per year. We shall consider the question of temporal variability of G in sec. 4. For the time being, however, we shall limit ourselves to general relativity in the Einsteinian sense, for which G is constant.

As a practical consequence we note that eq. (19)   can be used <u>to reduce atomic time</u> τ <u>to coordinate time</u> t . A more general expression is the well known formula

$$\frac{d\tau}{dt} = 1 - \frac{v^2}{2c^2} - \frac{V}{c^2} , \qquad (20)$$

which is an immediate consequence of (16)   and   (18) , putting $dx^2 + dy^2 + dz^2 = v^2 dt^2$ and neglecting the term Vv^2/c^4 as being of higher order. Here v is the velocity of the clock in the basic system xyzt .

The term v^2/c^2 is a special-relativistic correction, and V/c^2 represents a general-relativistic contribution. The orders of magnitude of these corrections are as follows:

$$\frac{v^2}{c^2} \doteq 10^{-8} \quad \text{for the Earth's orbital speed (30 km/sec)}$$

$$\frac{v^2}{c^2} \doteq 10^{-12} \quad \text{for the Earth's rotational speed (0.46 km/sec at the equator)}$$

$$\frac{V}{c^2} \doteq 10^{-8} \quad \text{for the Sun's gravitational potential at the Earth's orbit}$$

$$\frac{V}{c^2} \doteq 10^{-9} \quad \text{for the geopotential at the Earth's surface; cf. also (17).}$$

The reduction from atomic time to coordinate time is thus given by

$$t = \int (1 + \frac{v^2}{2c^2} + \frac{V}{c^2}) \, d\tau \ . \tag{21}$$

This reduction permits us to get a uniform time scale which is not affected by motion and by gravitational irregularities (Thomas, 1975).

Length. The present definition of the meter in terms of a certain multiple of the orange line of krypton will probably be given up in the near future. It will be redefined in terms of the atomic second and the velocity of light, of which the present accepted value is

$$c = (299\ 792\ 458 \pm 1.2) \text{ms}^{-1} \tag{22}$$

(Moritz, 1975). This indirect definition of length will be more accurate.

In fact, since c is accurate to about 4 parts in 10^9, the new definition of length will be as accurate (time being defined with superior precision). Relativistic effects are below this level, so that the influence of these effects on length will be negligible still for some time.

Doppler measurement. Doppler shift (changes in frequency) and aberration (changes of direction) of light or of another electromagnetic wave are fundamental phenomena in special relativity and are treated in almost all textbooks on relativity (for an astronomical presentation cf. Schneider, 1979, ch. 11). Relativistic Doppler shift and aberration differ from their classical counterparts by the factor $(1-v^2/c^2)$, which is very close to unity.

If λ is the wave length as omitted by the source, λ' is the received wave length, and $\Delta\lambda = \lambda' - \lambda$, then

$$\frac{\Delta\lambda}{\lambda} = (1 + \frac{v_r}{c}) (1 - \frac{v^2}{c^2})^{-\frac{1}{2}} - 1 \doteq \frac{v_r}{c} + \frac{v^2}{2c^2} . \qquad (23)$$

Here v_r is the radial component of the velocity vector $\underline{v}$ of the source with respect to the observer, and v is the norm of $\underline{v}$. Eq. (23) differs from the classical Doppler shift v_r/c by the "second order Doppler correction" $v^2/2c^2$, which is a special-relativistic effect. The latter is present even if the velocity vector $\underline{v}$ has no radial component v_r. Therefore $v^2/2c^2$ is also called "transversal Doppler effect"; it is due to the apparent retardation of the moving clock (the source).

If one considers also the effect of gravitation on the clock frequency by (20), then (23) becomes

$$\frac{\Delta\lambda}{\lambda} = \frac{v_r}{c} + \frac{v^2}{2c^2} + \frac{V_S - V_R}{c^2} , \qquad (24)$$

where $V_S - V_R$ is the potential difference between sender (source, clock) S and receiver R.

If the sender is in a satellite and the receiver is at the Earth's surface, then (24) applies to geodetic Doppler observations. The second-order corrections (second and last term on the right-hand side) cancel partly since $V_R > V_S$. For normal satellite heights, the general-relativistic correction is smaller by an order of magnitude; however, if the satellite height reaches half of the Earth's radius, then the two second-order corrections cancel completely (Weinberg, 1972, p. 84). For a satellite height of 1000 km, the second-order Doppler correction gives $\Delta\lambda/\lambda \doteq v^2/2c^2 \doteq 3 \times 10^{-10}$. It is also worth noting that in satellite Doppler positioning, second-order Doppler effects (from both special and general relativity) can be treated as constant frequency bias (Blais, 1977; Boucher, 1976, 1978).

Laser distance measurements. The velocity of light in the presence of a gravitational field is not c but $v = c(1 - 2V/c^2)$; this is an immediate consequence of (16) on putting $ds = 0$ for light and $dx^2 + dy^2 + dz^2 = v^2dt^2$. Thus the distance s computed by multiplying the travel time by c must be diminished by

$$\delta s = \frac{2}{c} \int_P^S V dt \ .$$

Here P is the laser station and S is the reflecting satellite. To a sufficient accuracy we may put $V = GM/r$ and integrate simply from $r = R$ to $r = R + H$, where R is the Earth's radius and H is the satellite elevation, putting $dt = dr/c$. Thus

$$\delta s = \frac{2GM}{c^2} \int_R^{R+H} \frac{dr}{r} = \frac{2GM}{c^2} \ln \frac{R+h}{R} \ , \tag{25}$$

which reaches 1 mm for $H = 1000$ km and 4 cm for the moon.(For lunar laser ranging cf. Mulholland, 1977; Stolz, 1979.)

Very-long- baseline interferometry. Here it is customary to reduce observed atomic time to coordinate time with respect to an inertial system with origin at the center of the solar system. The reduction formula, obtained by an appropriate evaluation of (21), has the principal term (Thomas, 1975; Robertson, 1975b)

$$\Delta t = - \frac{1}{c^2} \underline{v} \cdot \underline{x} \ , \tag{26}$$

where $\underline{v}$ is the Earth's orbital speed and $\underline{x}$ is the clock's geocentric position vector. This term has a daily period and an amplitude of about 1.5 μs ; it can also be explained as a classical aberration effect. For a detailed discussion see (Thomas, 1975). Aberration effects in VLBI are considered from the standpoint of special relativity in (Robertson, 1975a).

Deflection of light. Light rays can be regarded as straight except under unusual circumstances. Classical is the deviation of a light ray grazing the sun during an occultation. Modern results concerning this phenomenon and concerning analogous deflections of radio waves are given in (Ohanian, 1976, pp. 124-125); the order of magnitude is 1 - 2 seconds of arc.

Gyroscopic effects. Above we have seen that gyroscopes undergoing Fermi-Walker transport behave very much as in classical mechanics. Small relativistic effects ("geodetic precession") are described in (Ohanian, 1976, pp. 292-298).

Influence on planetary motion. The classical example is the precession of the perihelion of the orbit of the planet Mercury (about 40" per century). There are also periodic relativistic effects in earth-moon separation on the order of 1 m , which can be measured by lunar lasar ranging (Misner et al., 1973, p. 1048).

(For relativistic effects on satellite orbits, cf. Rubincam,1977.)

4. IS THE GRAVITATIONAL CONSTANT CONSTANT?

As an explanation of certain astronomical and geophysical phenomena, it has been suggested (Dicke, 1964; Dicke and Peebles, 1965; Duncombe et al., 1974) that the gravitational constant G is decreasing by a few parts in 10^{11} per year. The evidence is not clear, however; it seems to be difficult to separate a true change of G from other systematic influences (Ohanian, 1976, p. 188; Stephenson, 1978). The evidence on which the conclusions of Dicke and Peebles (1965) are based, is now superseded by recent data on the secular variation of the earth's rotation, summarized in (Lambeck, 1980, pp. 299-319). An experimental bound, $|\dot{G}/G| < 4\times10^{-10}$ has been obtained by Shapiro et al. (1971) by analyzing radar-echo time delays between Earth and Mercury (see also Williams et al.,1978).

In Einstein's general theory of relativity, G is constant. A changing G requires a different theory; such theories have been proposed by Jordan, Brans, and Dicke (Misner et al., 1973, p.1070) and Treder (1977). Since Einstein's theory is of incomparable simplicity and perfection, most physicists would be willing to give it up only in the presence of very solid empirical evidence. For the purposes of reference systems and time scales it thus appears permissible at present to take a conservative attitude and remain within the frame of Einstein's theory.

REFERENCES

Blais, J.A.R.: 1977, *Inertia, Inertial Reference Systems and Physical Geodesy,* Am. Geophys. Union Fall Meeting, San Francisco, Calif.
Bondi, H.: 1960, *Cosmology,* Cambridge Univ. Press.
Boucher, C.: 1976, *Modelisation des mesures Doppler effectuées par le récepteur JMR-1 sur le système Transit,* No. 26748, Inst. Géogr. Nation., Paris.
Boucher, C.: 1978, *Relativistic Correction to Satellite Doppler Observation,* Meeting of IAG SSG 4.45, Mathematical Structure of the Gravity Field, Lagonissi (Greece).
Boucher, C.: 1979, *The Relativistic Representation of the Gravity Field,* IUGG/IAG General Assembly, Canberra.
Dicke, R.H.: 1964, in De Witt and De Witt (eds.), *Relativity, Groups and Topology,* Gordon and Breach, New York, pp.163-313.
Dicke, R.H., Peebles, P.J.: 1965, Space Sci.Rev. 4, pp. 419-460.
Duncombe, R.L., Seidelmann, P.K., Van Flandern, T.C.: 1974, IAU Coll. 26, Torún, pp. 223-233.
Eddington, A.S.: 1924, *The Mathematical Theory of Relativity,* Cambridge University Press.

Fock, V.: 1959, *The Theory of Space Time and Gravitation*, Pergamon Press, London.

Heiskanen, W.A., Moritz, H.: 1967, *Physical Geodesy*, W.H. Freeman, San Francisco.

Lambeck, K.: 1980, *The Earth's Variable Rotation*, Cambridge University Press.

Misner, C.W., Thorne, K.S., Wheeler, J.A.: 1973, *Gravitation*, W.H. Freeman, San Francisco.

Moritz, H.: 1967, Report No. 92, Dept. of Geodet. Sci., Ohio State Univ.

Moritz, H.: 1971, Report No. 165, Dept. of Geodet. Sci, Ohio State Univ.

Moritz, H.: 1975, Bull. Géod. 118, pp. 398-408.

Moritz, H.: 1979, Report No. 294, Dept. of Geodet. Sci, Ohio State Univ.

Mulholland, J.D. (ed.): 1977, *Scientific Applications of Lunar Laser Ranging*, D. Reidel, Dordrecht, Holland (articles by J.D. Mulholland and J.G. Williams).

Ohanian, H.C.: 1976, *Gravitation and Spacetime*, W.W. Norton, New York.

Robertson, D.S.: 1975a, Nature 257, pp. 467-468.

Robertson, D.S.: 1975b, Report X-922-77-228, Goddard Space Flight Center, Greenbelt, Md.

Rubincam, D.P.: 1977, Celestial Mechanics 15, pp. 21-33.

Schneider, M.: 1979, *Himmelsmechanik*, Bibliographisches Institut, Zürich.

Shapiro, I.I., Smith, W.B., Ash, M.B., Ingalls, R.P., Pettengill, G.H.: 1971, Phys. Rev. Lett. 26, pp. 27-30.

Stephenson, F.R.: 1978, in P. Brosche and J. Sündermann (eds)., *Tidal Friction and the Earth's Rotation*, Springer, Berlin, pp. 5-21.

Stolz, A.: 1979, Deutsche Geodätische Kommission, A,90, München.

Synge, J.L.: 1960, *Relativity: The General Theory*, North Holland Publ. Co., Amsterdam.

Synge, J.L.: 1972, *Relativity: The Special Theory*, North Holland Publ. Co., Amsterdam.

Thomas, J.B.: 1975, Astron. J. 80, pp. 405-411.

Treder, H.J.: 1977, in *Third International Symposium, Geodesy and Physics of the Earth*, Veröffentlichungen des Zentralinstituts für Physik der Erde, 52(1), pp.119-127, Potsdam.

Weinberg, S.: 1972, *Gravitation and Cosmology*, Wiley, New York.

Williams, J.G., Sinclair, W.S., and Yoder, C.F.: 1978, Geophys. Res. Lett. 5, pp. 943-946.

KINEMATIC AND DYNAMIC REFERENCE FRAMES

E. M. Gaposchkin
Smithsonian Astrophysical Observatory
Cambridge, Massachusetts

ABSTRACT

The fundamental properties of kinematic and dynamic reference frames are defined, first as an abstract concept and second in a practical (although idealized) thought experiment. A four dimensional space-time description in coordinate free notation illustrates the properties, limitations, and relationships between kinematic and dynamic reference frames. Kinematic reference frames can be defined quite rigorously. Dynamic reference frames cannot be defined so well, but are nonetheless very useful. In practice a combination has been generally adopted. Presently we can materialize purely kinematic terrestrial and celestial reference frames.

INTRODUCTION

Ultimately our observations are applied to the study of the physical world. We identify or label observations using the term "coordinates". This labeling is an orderly way to classify events and objects. We can also extend this function in discussing interactions among "labels" (coordinates), and tend to give the coordinates physical significance. As long as we completely understand what we are doing there is nothing wrong with this practice. However, we must be very careful that we correctly interpret the physical significance of coordinates, and not study some phenomenon of a reference frame that we have invented.

For example, in studying earth satellites, one can use the earth's equator as a reference. The equations of motion will take a form that depends on the theoretical definition used to describe the equator's motion in space. Depending on the choice of definition, satellite perturbations will arise. In the Kozai and Kinoshita (1973) theory, for example, no change in semimajor axis (a) is predicted. In another approach, Balmino (1974), there will be perturbations in a. Both are correct. At present there is an anomaly in the Lageos satellite orbit with the existence of an unexplained change in a. Rather thorough analysis has failed to provide a good physical explanation. We do not

59

E. M. Gaposchkin and B. Kołaczek (eds.), Reference Coordinate Systems for Earth Dynamics, 59–70.
Copyright © 1981 by D. Reidel Publishing Company.

suggest that the anomaly is related to the Kozai or Balmino reference frame. We do question if it may be due to reference frame related phenomena. There are also examples of phenomena in the solar system that may have the same origin: for example, the excess secular change of the obliquity of the ecliptic (0".31/century). Considerable effort has been given to seeking a geophysical explanation. However, the answer may lie in the reference frame itself.

The purpose of this article is twofold: to describe the principles and to present observational accuracies. The inexorable improvement in accuracy will certainly continue and we should be designing a reference system that will be durable, in definition, yet able to undergo successive improvements in materialization.

It is important to draw a careful distinction between a reference frame and a reference system. The former is simply a mathemtical description. For example, the use of three orthogonal unit vectors in three-space to define the coordinates of a rotating frame is a reference frame. A reference system, however, is much more. It is often based on a reference frame, and includes a prescribed procedure to materialize the system, including methods of observation and reduction of data. Therefore, once a reference frame has been defined, it is unambiguous and immutable ever after. However, a reference system can change considerably, depending upon the prescribed procedure. The Conventional International Origin (CIO), defined by the adopted latitudes of the five ILS stations, is a reference frame. It has become rather impractical to express observations with respect to this frame. On the other hand, the BIH zero meridian comes closer to being a reference system. Though it lacks the conceptual simplicity and elegance of the CIO, the BIH zero meridian is a convenient system for expressing observations of UT. We hope to maintain the distinction between frame and system in this paper.

One last point: since we are studying the physical world, we must choose the correct description, given our limited understanding of the universe. This immediately leads us to a discussion of General Relativity (GR), though not because there are relativistic effects that must now be taken into account (though there are). The development of the concept of GR for reference frames has been covered by Moritz (1979, 1981) and the relevant effects in the solar system have been discussed by Brumberg (1981). There is a growing number of books on GR, which can provide any amount of needed detail (e.g. Misner et al. 1973). We prefer to view GR as the necessary vehicle for approaching the subject because the concept is essentially involved with the fundamental nature of reference systems. In essence, GR is a denial of absolute motion, and therefore absolute reference frames, and can be used as an analysis of reference frames. In a recent review of experimental tests of GR (Will 1979), no competing theory has received any observational confirmation. The very measurements discussed here will be used as tests of GR and competing theories. Therefore, although we can presently treat the departures of GR from Newtonian mechanics as small corrections (as we do with other phenomena, such as refraction), increased measurement accuracy and new measurements will require more complete treatment. We might as well take the correct view now, and create a reference system that will remain correct in principle as the measurement accuracy increases.

MEASUREMENTS

All measurements are carried out in two steps: establishing a standard, and specifying a procedure for comparison fo the standard with the object system. Measurements are essentially four-tuple quantities, each consisting, in principle, of four numbers. Generally one accepts the standard and analyses the comparison object. However, one can reverse the process. For example, the meridian passage of a star or the return of a laser pulse from a satellite can be used to set a clock. Satellite-determined datum scale have necessitated reinter-pretation of the surveyor's scale. An algebra for discussing these four-tuple quantities has been developed (Eddington 1946).

The usual fundamental standards are the centimeter, the second, and the gram, on which all our measurements are based. In fact, the speed of light (c) has become the operational definition of our length scale. This is most easily understood by introducing some formal ideas. Having taken GR as our description of the universe, we consider the metric ($g_{\mu\nu}$) using the summation convention:

$$ds^2 = g_{\mu\nu} \, dx^\mu \, dx^\nu = (n_{\mu\nu} + h_{\mu\nu}) \, dx^\mu \, dx^\nu$$

$$\mu,\nu = 0, 1, 2, 3 \tag{1}$$

$$x^0 = ct, \quad x^1 = x, \quad x^2 = y, \quad x^3 = z \tag{2}$$

$$n_{\mu\nu} = \begin{bmatrix} 1 & 0 & 0 & 0 \\ 0 & -1 & 0 & 0 \\ 0 & 0 & -1 & 0 \\ 0 & 0 & 0 & -1 \end{bmatrix} \tag{3}$$

$g_{\mu\nu}$, $n_{\mu\nu}$, $h_{\mu\nu}$ are metric tensors.
$n_{\mu\nu}$ is the Minkowski tensor of flat space time or special relativity.
$h_{\mu\nu}$ is the contribution of the gravitational field or curvature of space time. A particularly important solution of the field equations for a spherically symmetric static (i.e., stationary) field is due to Schwarzschild. It is normally written in polar coordinates:

$$ds^2 = (1 - \frac{2GM}{r})dt^2 - \frac{dr^2}{(1 - \frac{2GM}{r})} - r^2 d\theta^2 - r^2 \sin^2\theta \, d\lambda^2 \tag{4}$$

or in its isotropic form:

$$r = r' \left(1 + \frac{2GM}{4r'}\right)^2 \tag{5}$$

$$ds^2 = \left(\frac{1 - \frac{GM}{2r}}{1 + \frac{GM}{2r}}\right)^2 dt^2 - \left(1 + \frac{GM}{2r}\right)^4 (dr^2 + r^2 d\theta^2 + r^2 \sin^2 \theta d\lambda^2) \tag{6}$$

In this case we can write approximately

$$g_{\mu\nu} = \begin{bmatrix} \left(1 - \frac{2GM}{r}\right) & 0 & 0 & 0 \\ 0 & -\left(1 + \frac{2GM}{r}\right) & 0 & 0 \\ 0 & 0 & -\left(1 + \frac{2GM}{r}\right) & 0 \\ 0 & 0 & 0 & -\left(1 + \frac{2GM}{r}\right) \end{bmatrix}$$

When the units of GM are geometrodynamic centimeters (i.e., GM/c^2), we have $GM_\odot$ = 1.476 km for the sun, and $GM_\oplus$ = 0.4438 cm for the earth. One must realize that the Schwarzschild solution is only one possible solution to the field equations. It is not correct in principle, since it represents a universe with only one body, a point mass. However, it is a very good approximation on the scale of the solar system, even though it leaves out all nonlinear and rotation terms (e.g., The Lense Thirring Effect). See the textbooks for metrics containing these terms (e.g., Ohanian, 1976).

Most modern procedures for measuring length involve transmitting a photon, or other electromagnetic radiation, and measuring a time interval. This is possible due to the fact that the path of a photon is governed by:

$$ds = 0 \tag{8}$$

We send a photon from A to B, reflect it from B back to A, and measure the transit (Δt) time with a clock at A. Since dt is the co-ordinate time of the observer, integrating 1, 4, or 6 along the path gives the relationship between the time interval and the distance in the observer's frame. In flat space time we have simply $\rho_{AB} = \Delta t/2$. In curved space time we have a more complicated line integral. The effects of space curvature on range distance measurement are illustrated in Table 1 for a simple geometry, where the photon's path is along the radius to the mass. For doppler measurements, an oscillator on a satellite or spacecraft is used to generate a radio signal. The rate of this oscillator depends upon the potential it experiences, and is also known as the "effect of time dilation". For close earth satellites,

Table 1. Space-time curvature effect on range measurement.

Mass	Range Measurement to:	
Earth	Lageos	0.62 cm
Earth	Synchronous Satellite	1.59 cm
Earth	Moon	3.63 cm
Earth	Sun	8.93 cm
Sun	Lageos	-12.5 cm
Sun	Synchronous Satellite	-62.8 cm
Sun	Moon	-741. cm
Sun	Sun ($1R_\odot$)	-16.9 km

this effect has been important for many years (Gaposchkin and Wright 1969). The time dilation effect has been used as one of the fundamental tests of GR (Vessot and Levine 1979). We know, of course, that the effect of atmospheric refraction at zenith on range measurements is 2.1 m.

By using property (8), we are essentially using a clock to measure a distance in terms of the speed of light. It is, therefore, more correct to consider the metric scale imposed by a standard clock and an adopted value of the speed of light. In doing so we have as standards the second and the gram. However, if we use an atomic clock to measure Δt, then the time will be atomic time and the length scale will be tied to the atomic second. There is no a priori guarantee that the time in 2, 4, or 5 is atomic time. Atomic clocks are based on physical processes, which occur at the atomic level. We would like to have a clock that is governed by macroscopic (dynamic) processes. Such a clock is called a geometrodynamic clock. There is a body of theory, largely unverified both theoretically and observationally, that would predict a difference between atomic time and geometrodynamic time. This theory is based on the Large Number Hypothesis (LNH) of Dirac. Wesson (1980) gives the most recent summary of the status of the idea. If valid, it predicts a difference in time scale. This is also known as a the Variable G (or $\dot{G}$) Hypothesis, but that is patently a misnomer. The LNH predicts a quadratic depature of the two time scales (a linear difference would be only a question of definition) and an increase in mass of the universe. The present upper limit, observationally, is $< 2 \times 10^{-10}/$ year. Using this limit, the systematic error in range measurement across the earth's orbit is only 0.03 cm. For the present this error can be ignored in establishing the length scale. Since the dynamics of the solar system run on geometrodynamical time, departures of the two time systems would show up in establishing the dynamic reference frame. It is in that discussion that possible differences must be reckoned with.

IDEALIZED KINEMATIC SYSTEM

Consider n points in a four-dimensional Minkowsky Space. Consider
that each point can transmit to any other point and receive a reflected
photon, or electromagnetic radiation. Each point also has a clock, and
all clocks have been synchronized. Therefore each point (i) can
measure Δt to each other point (j) and compute $\rho_{ij}(t)$. Each point can
then show on a sign or broadcast by radio, which is the same thing as
both signals going at the speed of light, the ρ_{ij} and t. Another point
can observe each of the n-1 points and similarly broadcast the result.
An external observer can make a tabulation of the ρ_{ij}. The only
problem in constructing a time history is to identify the times. If
we simply require that point i's observation of point j be the same as
point j's observation of point i, then an ordering by coordinate time
can be used (the time each clock shows for its particle). Of course
this will generally not be the same reading shown by the central
observer's clock. By identifying the coordinate time with each configuration, the observer can construct a polyhedron and its temporal
evolution in three-space, or its motion through four-space. Notice
that this polyhedron is defined, even in the presence of potential
fields, accelerations, and rotations. It rigorously defines a kinematic
reference system.

For each time there are $N = n(n - 1)/2$ measured distances. Ignoring for the moment how or why we should establish a Cartesian Coordinate
System and certain degeneracies, there are $3n$ coordinates. Five parameters would need to be imposed, (an origin and orientation). There are
many ways to select these five parameters. For example, one might
prescribe three coordinates for one point and one coordinate for two
other points. Or one might choose the origin so that the weighted
mean of the x, y, and z coordinates is zero and there is not net rotation. In either case, one can then produce a time-ordered history of
the coordinates' changes.

This type of network could, for example, be realized with a system
of satellites such as the Global Positioning System (GPS). At each
instant of time, this polyhedron would define a system quite independently of origin or orientation, motion or rotation. The relative
motion of points would be obvious, because it would be directly reflected in the changes in ρ_{ij}. This polyhedron is the ultimate pure
kinematic system and serve as the definition of a reference system. In
general, it is not too useful since the ρ_{ij} can vary with time, and
the measurements of other coordinates referred to it will also vary,
even if their scalar separations are invariant. Nevertheless, this is
the kinematic reference frame we will set up, a terrestrial polyhedron
of interstation baselines. Indeed, we expect the motion to be small
and hope they will be regular. We will set up such a system using baseline determinations drawn from the laser ranging of Lageos (Latimer,
Gaposchkin 1977), GPS (Goad 1981), or Very Long Baseline Interferometry
VLBI (Robertson 1981). Bender (1981) gives a prescription for realizing
such a terrestrial system, which Mueller (1981) would call a Conven-

tional Terrestrial System (CTS). In both cases the attempt is to model the temporal changes in ρ_{ij}, using plate tectonic models, and to refer the origin of the equivalent Cartesian Coordinates to the center of mass, using the dynamic constraints from the laser tracking of Lageos.

If we take a general polyhedron and make it very large with the observer within it, then the analysis is reduced to measuring angles τ_{ij} either optically or with VLBI. However, the time now becomes the proper time of the observer. Of course, the τ_{ij} must be corrected for the "aberration" that is due to the motion of the observer. This aberration is classically known and easily derivable from the special relativity Lorentz transformation in the neighborhood of the telescope. There is also the deflection of the earth's gravity field ($0\overset{\prime\prime}{.}0003$), and the sun's gravity field (from $0\overset{\prime\prime}{.}0041$ to $1\overset{\prime\prime}{.}75$), and even for photons passing near Jupiter ($0\overset{\prime\prime}{.}0163$).

There is an interesting and important difference between the polyhedron and the unit sphere kinematic system. Establishment of the polyhedron does not depend upon the relative motion of the observer in any way. If the system were rigid and non-rotating, the observer could reconstruct it even if he was moving at a high velocity. With the unit sphere, even if the emitters are "fixed," the observer will see variations due to his motion and to the presence of a potential field. Of course, with our knowledge of the earth's rotation, orbital motion, and the potential fields in the solar system--which will cause the geodesic precession of $1\overset{\prime\prime}{.}9$/century (Eddington 1924), one can reduce the observations to a non-rotating heliocentric system. This system would then produce a reference frame with no motion, if the sources were apparently fixed.

A kinematic celestial system can be realized optically with telescopes, whether ground-based or satellite-based (Fricke 1981, Kovalevsky 1981, or by means of VLBI observations of quasars. Quasars have considerable attraction; it is believed that among the expected 40,000 observable extra-galactic objects, we can find a sufficient number to establish a kinematic unit sphere polyhedron (i.e., a celestial reference system). Such a system, related to the distant objects in the universe, would constitute an inertial reference system that could be used as a Conventional Inertial System (CIS); Mueller (1981). If our present ideas of the universe are correct, the quasars should also be geometrically fixed, i.e. the τ_{ij} are constants. Of course this needs verification (Robertson (1981).

Finally, if the baselines monitored as the terrestrial polyhedron are also the baselines of a VLBI observatory, we will be able to obtain a direct observation of the relationship between these two purely kinematic reference frames. This then defines the link between the two systems (the CTS and the CIS), avoiding the complication of the complex theory of the earth's motion in space: precession, nutation, polar motion, and earth rotation. Of course, one can choose to present this link in terms of a conventional theory, with small observed corrections. This is in fact exactly what happens with an "adopted" precession and nutation series and published polar motion. Such a procedure has the added feature of being independent of the ecliptic.

IDEALIZED DYNAMIC SYSTEMS

Until now we have discussed motion, but only in terms of the change of relative position in time, i.e., the path of a particle in Minkowski event space. We have not needed to relate the path of a point in this space. We have needed only a clock, a transmitter, and a receiver of photons or electromagnetic radiation to construct and maintain a well-defined reference system. One consequence of a metric theory of gravity is equations of motion, which take many forms. The motion of a free particle along a geodesic is derivable from the variational principle that

$$\delta \int_{s_1}^{s_2} ds$$

is stationary. One form is the Equation of Geodesic Deviation:

$$\frac{D^2\xi}{Ds^2} + R^\alpha_{\beta\gamma\delta} \frac{dx^\beta}{ds} \xi^\gamma \frac{dx^\delta}{ds} = 0 \tag{9}$$

This gives the separation vector $\bar{\xi}$ between two nearby geodesics in terms of the path of a test particle and the Riemiann curvatures tensor $R^\alpha_{\beta\gamma\gamma}(g)$. Such an equation can be used to study, for example, the motion in an orbiting spaceship by writing g in a rotating system. In this case

$$\frac{dt}{ds} = 1 \quad , \quad \frac{dx^i}{ds} = 0 \quad , \quad i = 1, 2, 3$$

and

$$\frac{D^2\xi^\alpha}{Ds^2} + R^\alpha_{0\gamma0} \xi^\gamma = 0 \tag{10}$$

This leads to the tidal force tensor

$$R^k_{0\ell0} = \frac{1}{c^2} \frac{\partial^2\phi}{\partial x^k \partial x^\ell}$$

which has the simple form for a point mass:

$$
R_{0\ell 0} = \begin{bmatrix} \dfrac{GM}{c^2 r^3} & 0 & 0 \\[2mm] 0 & \dfrac{GM}{c^2 r^3} & 0 \\[2mm] 0 & 0 & -\dfrac{2GM}{c^2 r^3} \end{bmatrix}
\tag{11}
$$

The equation (9) can also be used to define the motion of an earth satellite. One determines the geodesic motion ($\bar{\xi}$) of the satellite with respect to the earth's center of mass ($\bar{x}$), which moves on a geodesic.

The importance of equation (9) is that it displays the fact that no inertial system has been defined. Any particle can serve as a reference so long as the metric or the Riemann curvature tensor is known. However, this just begs the question and substitutes knowledge of the metric for knowledge of the inertial frame. For example, to write (11) we used the fact that the particles' (earth's) motion around the earth (sun) was due to the space curvature caused by the mass of the earth (sun). But we do not know the masses a priori. Not even Einstein succeeded in formulating a theory that predicted masses. Then we go to the more usual form of the equations of motion

$$
\frac{d^2 x^\mu}{ds^2} + \Gamma^\mu_{\alpha\beta} \frac{dx^\alpha}{ds} \frac{dx^\beta}{ds} = 0
\tag{12}
$$

or

$$
\frac{du^\mu}{ds} + \Gamma^\mu_{\alpha\beta} u^\alpha u^\beta = 0
\tag{13}
$$

where the Christophel Symbol is

$$
\Gamma^\mu_{\alpha\beta} = \frac{1}{2} g^{\mu\nu} (g_{\nu\alpha,\beta} + g_{\beta\nu,\alpha} - g_{\alpha\beta,\nu})
\tag{14}
$$

and where the comma (,) denotes differentiation. Equation (14) is completely equivalent, but the $\bar{x}$ now refer to an "inertial reference frame." In a sense this is tautology, in that an inertial reference frame is being defined as the frame in which these equations are true. There is a final form of the equations of motion known as the Parameterized Post Newtonian Approximation (PPN). First derived by Eddington and Clark (1938), the PPN or its derivative is used for all analyses of orbits in the solar system. The complete treatment of the restricted two-body problem is given by Hagihara (1930-31).

We have equations of motion that allow us to connect points on a geodesic, provided we know the metric. We must also know the non-gravitational forces, which add terms on the right-hand side of 12 and 13. In principle we have a data analysis problem to recover the necessary parameters: the metric, forces, and initial conditions. If some forces are stochastic, then the modeling problem becomes correspondingly more difficult. Aside from random forces, one could hope to define, in an inertial reference frame, the motion of a particle (a satellite or planet). As more data become available, the parameters and initial conditions can be improved. It will be axiomatic, however, that the dynamic system will be constantly "improved." Its accuracy will depend on the accuracy of the data for the interval of time observed, and will deteriorate with extrapolation.

One can inquire what use will be made of this dynamic frame. It will be excellent for relating points in its own system. However, that will not depend upon its being inertial, only that it provides a consistent theory for tracing a point in four-space. The satellite's position in this frame can easily be realized (by definition). For example, using laser ranging to Lageos one will be able to calculate station coordinate differences that are nearly as accurate as the data (Bender 1981). Such baselines will be excellent, since the major uncertainty in this system is orientation (i.e., rotation).

If one seeks a dynamic frame as a reference for dynamic studies, the situation is less favorable. It is true that the satellite trajectory will define an inertial reference system with some accuracy (Kozai 1981). The exact frame depends, of course, upon the details of the force models, data reduction, and so forth. Such frames are getting better; the main uncertainty concerns rotation, as there is no clear way to separate the gravitational effects from the general rotation without observing celestial frame or Fermi-Walker transport with dyroscopes. Even so, a Fermi-Walker transported gyroscope in earth's orbit around the sun will show the 1".9/century geodesic precession with respect to the star background (Eddington 1924). Another possibility would be to observe synchronous satellites with respect to star background with astrometric cameras or VLBI systems.

These uncertainties are long-term effects. A number of dynamic phenomena appear to be short periodic from the point of view of a satellite. The ability to treat short periodic perturbation is therefore the critical issue. In this case, the situation is reasonably good. The short-period effects of GR are generally quite small here: 0.012 cm for Lageos, 0.146 cm for the moon, and 1.8 km for Mercury (Gaposchkin 1981). The residuals due to a shift in the center of mass of the coordinates of the observing stations will appear to the satellite as a once-per-revolution effect. The accuracy of recovering the center of mass is therefore the same as the accuracy of the data and the perturbations, depending on the true anomaly. From the point of view of a satellite, the earth's polar motion with respect to an angular momentum axis appears to be strictly diurnal. Satellite doppler and laser ranging are presently giving excellent measurements of polar motion. One must be careful not to mask this with the once-per-day perturbations that are due to errors in the tesseral harmonics of order one. The determination of UTl is not so promising, for the reasons given above.

CONCLUSIONS

In summary, one can draw the following calculations from these arguments.

1. A reference systems should be simple. That is, invariant phenomena should remain invariant when expressed in the system. Apparent motion should not reflect motion of the reference system.

2. A terrestrial reference system, one fixed to the earth, can be defined and established in a completely kinematic fashion as a polyhedron. Satellite dynamics can be used to obtain the polyhedron baselines, and the relationship of the polyhedron to the center of mass.

3. A celestial reference system can also be established kinematically. The most promising approach is to use VLBI observations of extra-galactic radio sources. These VLBI stations can also contribute to the definition of the terrestrial reference frame polyhedron baselines.

4. The motion of the terrestrial reference system, with respect to the celestial reference frame, will be monitored with VLBI.

5. Satellite or dynamic reference frames can be realized. They can permit monitoring of the terrestrial reference system with respect to the earth's angular momentum axis: nearly the rotation axis.

6. The scale of the terrestrial reference frame should be defined from a standard clock and the speed of light.

7. We should seek to define a reference system that is, in principle, a factor of 10 better than we expect to need: i.e., 0.1 cm and $0''.0001$. We may not immediately realize system of such accuracy.

REFERENCES

Balmino, G. 1974. Celestial Mechanics 10, 4: 423-436.
Bender, P. 1981. Establishment of Terrestrial References frames by New Observation Techniques. This Volume.
Brumberg, V. A. 1981. Relativistic Reduction of Astronomical Measurements and Reference Frames. This Volume.
Eddington, A. S. 1924. The Mathematical Theory of Relativity. London: Cambridge University Press.

Eddington, A. S. 1946. Fundamental Theory. London: Cambridge University Press.

Eddington, A. S., and Clark, G. 1938. Proc. Roy. Soc. A 166, 465-475.

Fricke, W. 1981. Definition and Practical Realization of the Reference Frame FK5 — The Role of Planetary Dynamics and Stellar Kinematics in the Definition. This Volume.

Gaposchkin, E. M., and Wright, J. P. 1969. Nature 221: 650.

Gaposchkin, E. M. 1981. Celestial Mechanics 10, 4.

Goad, C. 1981. Positioning with the Global Positioning System. This Volume.

Hagihara, Y. 1930-31. Japanese. Journ. Astron. Geophys. Trans. 8, 3: 67-176.

Kovalevsky, J. 1981. Celestial Reference Frame. This Volume.

Kozai, Y., and Kinoshita, H. 1973. Celestial Mechanics 7: 356-366.

Kozai, Y. 1981. Motions of Artificial Satellites and Coordinate Systems. This Volume.

Latimer, J. H., and Gaposchkin, E. M. 1977. Center for Astrophysics Preprint Series 750.

Misner, C. W.; Thorne, K.; and Wheeler, J. A. 1973. Gravitation. San Francisco: W. H. Freeman Co.

Moritz, H. 1979. Ohio State University, Department of Geodetic Science Report 294. Columbus: Ohio State University Press.

Moritz, H. 1981. Relativistic Effects in Reference Frames. This Volume.

Mueller, I. I. 1981. Reference Coordinate Systems for Earth Dynamics: A Preview. This Volume.

Ohanian, H. C. 1976. Gravitation and Spacetime. New York: W. W. Norton & Co.

Robertson, D. 1981. The Use of VLBI in Establishing Reference Coordinate Systems for Geodynamics. This Volume.

Vessot, R., and Levine, M. 1979. General Relativity and Gravitation 10, 3: 181-204.

Wesson, P. S. 1980. Physics Today 33, 7: 32-37.

Will, C. M. 1979. "The Confrontation Between Gravitation Theory and Experiment." General Relativity. S. W. Hawking and W. Israel editors. London: Cambridge Univ. Press.

REFERENCE COORDINATE SYSTEM REQUIREMENTS FOR SPACE PHYSICS AND ASTRONOMY

J. D. Mulholland
University of Texas at Austin (U.S.A.)
and Centre d'Etudes et de Recherches
Géodynamiques et Astronomiques,
Grasse (France)

ABSTRACT

   Changes in reference coordinate systems have major implications well beyond the realm of Earth dynamics. Definitions that serve geodynamic convenience may cause considerable effects for other disciplines. After presenting some typical areas in which coordinate frame definitions are important, recommendations are given for criteria to be considered as boundary conditions in discussing changes. These cover such qualities as observability, complexity, stability, internal coherence and uniqueness.

   The very existence of this 2nd International Colloquium on Reference Coordinate Systems for Earth Dynamics — your very presence here — is an evidence that high-precision observing techniques no longer permit the various aspects of dynamical astronomy and solar system physics to be treated as isolated phenomena. Viewed in the context of the continual forcing action between theory and observation, we are currently in a phase where measurement capability has far outdistanced the capacity for theoretical interpretation. Simply to provide descriptive models, we are driven to computational complexity undreamed of two decades ago. The primary explanation of this state of affairs is that the physical interdependences between effects previously treated separately produce observable motions at a level that cannot be ignored if the data are to be correctly interpreted. Even the identification of appropriate coordinate systems now plays a critical role. It is my purpose here to remind you that, despite its title, this colloquium has responsibilities and influences well beyond the restricted realm of Earth dynamics.

   In fact, if one considers the program and the list of participants objectively, it is evident that the word "Earth" in the colloquium title is more for administrative convenience than for scientific description. Our number includes many who are far less interested in the Earth as a subject of study than in the Moon, planets, asteroids, stars, and even extra-galactic objects. They (we!) are not here just

E. M. Gaposchkin and B. Kołaczek (eds.), Reference Coordinate Systems for Earth Dynamics, 71–75.
Copyright © 1981 by D. Reidel Publishing Company.

to give a neighborly helping hand to the geoscientists. Surely, both
observations and theoretical descriptions of extraterrestrial objects
are required to establish an adequate set of coordinate systems for
Earth dynamics. But as a practical matter, there will not be — there
must not be — different "fundamental" reference systems for different
applications. Even if the primary motivation for defining new fundamental systems comes from terrestrial concerns, these systems should
be designed for universal applicability. Directly or indirectly, most
observations of extraterrestrial objects will be related to terrestrial
frames for the indefinite future.

COORDINATE SYSTEMS IN ASTRONOMY AND SPACE PHYSICS

It is both impractical and unnecessary to compile an exhaustive
survey of the aspects of astronomy and space physics in which coordinate
system definitions can play a significant role. It may be useful here,
however, to give a few examples, just to emphasize the point.

— In the study of pulsars, the physical mechanism for pulsation
depends on the time derivatives of the pulsation period evaluated in an
inertial coordinate frame. Thus, the observations are normally reduced
to the solar system barycenter. The topocentric position and motion
of the barycenter are affected by the assumed planetary masses and
orbits, by the station motion, and thus by coordinate system definitions,
including the transformation between proper time and coordinate time.

— The dynamical and statistical properties of our galaxy, as well
as its dimensions, are based on observed values of both systematic and
random components of the proper motions of stars. The precession of
the Earth's equatorial plane and the rate of change of obliquity are
perfectly correlated with systematic proper motions.

— Dynamical and geometric determinations of solar oblateness
depend on coordinate system definitions in different ways. Thus, reference frame inconsistencies may introduce noise into even otherwise
perfect observational comparisons.

— Inadequate coordinate systems can introduce inconsistencies in
planetary orbits through the interaction of mass, heliocentric distance
and mean angular speed (Kepler's third law).

— Unmodelled coordinate system motions can introduce errors into
estimates of the anomalous accelerations of the Moon and artificial
satellites, thus biasing discussions of lunar evolution and terrestrial
dissipation processes.

In terminating this list, I remind you that it is far from complete, only a small sample to illustrate the scope of subjects that
may be influenced by what we do here.

REQUIREMENTS AND RECOMMENDATIONS

I will not dwell long nor in detail on questions of desired or necessary precision and accuracy. The reason is simple. From an astronomical point of view, Earth is the most closely, intensively and accurately observed of all celestial bodies. In addition to purely terrestrial measures, every Earth-based observation of an exterior body is also an observation of Earth. It is indeed from this point that geophysics was born nearly a century ago: I remind you that Chandler was an editor of the Astronomical Journal and that Love's historic work "On Some Problems in Geodynamics" was a John Couch Adams prize essay of Cambridge University. This pre-eminence of Earth as a planet means that a set of coordinate systems that provides the necessary precision and accuracy for attaching terrestrial dynamics to an internal frame will also satisfy the accuracy requirement of non-terrestrial applications. But accuracy is not the only problem. There are significant qualitative aspects which must also be addressed in any redefinition of fundamental systems, as well as the realization and use of multi-application intermediate references.

It is important to stress that we are concerned here with both fundamental and secondary reference frames. It is frequently impossible to use fundamental frames directly. Good examples of this are the use of lunar and planetary ephemerides or the analysis of range and doppler observations of artificial satellites. Thus, it is reasonable to discuss qualitative desiderata for fundamental systems while ignoring the comparable aspects of secondary systems. In my opinion, the following considerations are to be taken into account:

I. Observability — Standard coordinate systems, both fundamental and derivative, should be as close as possible to the observations. Secondary systems should be directly observable. Definitions of fundamental frames should avoid conceptual bases that are inherently inaccessible to observation. As an example, despite its advantages to theorists, a terrestrial reference frame based on the total angular momentum vector must be rejected.

II. Complexity — Fundamental coordinate systems should be conceptually as simple as the demands upon them permit. In Earth dynamics, it is evident that observing stations must be permitted to move relative to any reference frame. For many non-terrestrial applications, however, the station motions will remain trivial for the foreseeable future. For these uses, a fundamental terrestrial system with time variant "mean positions" of surface points will be an unnecessary and expensive complication.

III. Stability — In general, astronomy and astrophysics are concerned as much with phenomena over periods of eons as well as nanoseconds. Changes of fundamental systems, even when obviously required, represent a serious material nuisance and a potential source of

calculational error. Such changes must be held to a minimum. Proposed
changes must be subjected to the most minute inspection and criticism,
so that formal adoptions may have the longest useful life span possible.
A painful example at present is the astronomical nutation series. Two
aspects of this controversy should be distinguished: a) the numerical
adequacy of the adopted IAU model, and b) the manner of its adoption.

There is some opinion that the numerical coefficients adopted at
Montreal in 1979 were then already inadequate to represent the observations. If this be true, then it is best to change the nutation series
now, before it is used. In that case, we simply admit to a stupidity
which we quickly erase. This should not be done carelessly, however;
reversal of an official adoption should not be permitted to become a
light matter. If the objection is not based on a currently observable
astronomical error, then the IAU decision should stand.

IV. Departures from conventional models — Two decades ago, the
reality of space exploration introduced astronomy to the world of
"crisis science". Since then, a modus operandi has evolved that one
must recognize not only as realistic but as valid. Except for periods
of scientific stagnation, or immediately after new conventions are
adopted, conventional systems designed for multi-disciplinary use
cannot serve satisfactorily for all applications at a rapidly-evolving
frontier of physical knowledge. A conventional model should represent
as well as possible the needs and capacities of its epoch of adoption,
without being expected to anticipate the future in any detail. As
the scientific frontier is pushed outward, certain high precision applications must eventually abandon the adopted system to realize maximum
value from the observations. In such cases, the departures from
conventionality must be as explicit and as well-defined as possible.

V. Internal Coherence — Adopted sets of reference systems, whether
fundamental or derivative, should be internally coherent. A near-
trivial example is the use of planetary ephemerides as a connecting
link between terrestrial and celestial reference frames. For proper
use, the ephemeris must be used together with a set of constants (e.g.
astronomical constants, station coordinates) appropriate to that
ephemeris. Station coordinates obtained by comparing observations with
an orbital ephemeris are ephemeris-dependent, not absolute.

VI. Form of Presentation — Definition of reference frames should
include not only the conception, but also the method of realization
and application. Definitions should be realizable avoiding the sort
of impossible situation that surrounded the use of Ephemeris Time.

VII. Uniqueness — Elements of a chain of coordinate systems should
be uniquely identified. Non-uniqueness offers the opportunity for
ambiguity and miscomprehension. A classical pre-space-age example is
the difference in numerical results obtained by use of Newcomb's theory
of the Sun and Newcomb's Tables of the Sun, which were constructed from

that theory. There is in fact a current analogue to that situation. Machine-readable planetary and lunar ephemerides for space research are now distributed as polynomial series fitted to numerical integrations. A single integration can be reduced to multiple versions by choosing different parameters for constructing the series. Such multiple versions can give different numerical results. We have already experienced one case of two different ephemerides with the same identification. Great care should be exercised to avoid such ambiguities.

CONCLUSION

Happily, the Program Committee did not ask me to provide <u>solutions</u> to the problems that the needs of astronomy and space physics pose for the definition of new reference coordinate systems. I have tried simply to pose boundary conditions that I think should be taken into consideration during our discussions and deliberations, to try to minimize the RMS chaos in future influences of our actions here.

CELESTIAL REFERENCE FRAMES

J. Kovalevsky
C.E.R.G.A. Grasse, France

ABSTRACT. A celestial reference frame is based on some dynamical or
kinematic approximation of an absolute coordinate system and is materialized by a fundamental catalogue of stars. It is characterized by
the accuracy of the system and the precision of its stellar realization.

The present situation, marked by the construction of the FK5 system
and catalogue, is briefly described. It will be an hybrid system based
on the discussion of all existing observations. The prospects for improvement in accuracy are good using lunar laser observations or VLBI
determination of the positions of extra-galactic radio-sources. Three
procedures that may be used to link such a system to a star catalogue
are given. A major improvement in precision is expected from the Hipparcos programme and further extensions include essentially the densification of catalogues with stars of magnitudes 10 to 14 using photoelectric automatic meridian instruments.

1. CONCEPTUAL BACKGROUND

The purpose of a celestial reference frame is to provide a materialization of a coordinate system in order to describe various motions
observed in the Earth-Moon system, the solar system and, more generally,
anywhere in the Universe. No motion can validly be studied unless it is
referred by a known process to some coordinate system in which one can
write the dynamical equations of motion. This means that the observations must be made in such a way that they can be expressed in terms of
coordinates in this system. For instance, in the case of celestial reference frames, the observations are made with respect to a stellar background, the stars providing the materialization of the coordinate system.

Conceptually, the simplest reference system is inertial: There,
the differential equations of motion may be written without including
any rotational term. Such a system is dynamical in character and can be
constructed uniquely from the analysis of the motion of one or several
celestial bodies such as the Moon or the planets, provided that all

E. M. Gaposchkin and B. Kołaczek (eds.), Reference Coordinate Systems for Earth Dynamics, 77–86.
Copyright © 1981 by D. Reidel Publishing Company.

interacting forces and initial conditions are correctly evaluated. This
means that a correct model of the dynamical system must exist (Kovalevsky, 1975 b). This is by itself a major difficulty that has been described by Mulholland (1977) for the lunar motion or by Duncombe et al.
(1975) for planetary motions. But another difficulty of a purely dynamical reference frame is that, strictly speaking, its availability for
the observation of a given body implies a simultaneous observation of
at least one of the bodies used in the definition of the system. Except
in some specific cases, this is not convenient. Thus, it is necessary
to have the positions and motions of many more bodies in this frame and
use them as fiducial points that materialize it. If, as is usually the
case, these points are stars, we have a celestial reference frame defined dynamically. The list of stars, with their positions at epoch and
their proper motions referred to the frame constitutes the fundamental
catalogue materializing the celestial reference frame.

Another, actually simpler, definition of a celestial reference
frame is found if the defining bodies are sufficiently remote so that
they can be considered as fixed or slowly moving in the sky in a model-
lable fashion. These stars or extragalactic objects also serve as the
materialization of the celestial frame. The latter is, in this case,
defined kinematically. But again, their number may be insufficient for
many purposes and it is again necessary to increase it by adding coordinates and proper motions of secondary stars in quite a similar manner
as in the case of a dynamically defined celestial system.

The existence of two theoretically - if not always practically -
separated concepts in a celestial reference system has led us to separate also the concepts of the precision with which their realization
represents reality: Let us, therefore, propose two definitions:

(a) The accuracy of a celestial reference frame is the evaluation
    of the error of the dynamical or kinematic definition of the
    frame, due to errors in the modelling or in the determination
    of kinematic conditions. If the construction of the catalogue
    introduces new systematic errors, they are included in the
    accuracy.

(b) The precision of a celestial frame describes the errors
    that are surperimposed on the systematic errors and are due to
    the inaccuracies of the positions and proper motions of stars
    of the fundamental catalogue (random, regional and, eventually,
    systematic).

2. THE PRESENT SITUATION

Since the appearence of the first modern fundamental catalogue
(Auwers Fundamental catalogue), the idea of a purely kinematic definition of the reference frame was never seriously considered. One reason is that stars have sizeable proper motions that have non-random

components in the velocity space as for instance the solar motion with respect to the nearest stars. But the main reason was that astronomers insisted in having a fundamental system expressed in true equatorial coordinates. This implied that coordinates of stars should contain elements describing the rotation of Earth in Space (precession and nutation) and the motion of the equinox. All the dynamics that are necessary to define a reference plane and an origin using the motion of the Earth are included in this requirement. This is done through the definition of the orbit of the Earth around the Sun and a dynamical representation of the motion of the equator: The formulae giving the position of the true equinox and the true equator as functions of time in a fixed frame give the dynamical definition of the frame as well as the motions of the true equatorial coordinate system.

The methods that are used to refer star positions to the Sun or planets have been widely described in the literature (e.g.,Woolard and Clemence, 1966). They are essential too for the determination of the equinox in the catalogue. The Sun, various planets and the Moon may be used to determine the equinox correction. The discussion of the results by Fricke (1979) shows a dispersion of $\pm$ 0^s.05. This leads to an accuracy of about 0".1 per century for this part of the system. In practice, there are further difficulties in trying to impose a dynamical equinox to the origin of right ascensions of the catalogue and an inconsistency is introduced between the definition and the realization using stars, especially that this offset is correlated with the realization of Ephemeris Time (Duncombe et al., 1975).

Theoretically, the precession could also be defined dynamically. However, the practical determination of this constant (see Fricke, 1977) is biased by a systematic trend of the stellar proper motions due to the galactic rotation. Kinematic determinations of the constant of precession using a model of galatic rotation are superior to purely dynamical solutions.

The methods used to construct a fundamental star catalogue, namely the FK5, are described elsewhere (see Fricke, 1977 and herein) in detail. Let us simply iterate that in practice, dynamics and kinematics are used jointly or independently to yield the best possible results. The choice between the two approaches is based only on the expected quality of the result and not on conceptual considerations. The hybrid character of the present celestial system of reference is based on a deep comparative discussion of all the existing observations.

This pragmatic approach has permitted the ultimate accuracy possible with the presently available observations: something of the order of 0".15 per century for the FK5 system.

3. PROSPECTS FOR IMPROVEMENT IN ACCURACY

The two dominant limitations in accuracy of the FK5 are the present

difficulty in having an accurate independent dynamical determination
of the constant of precession and the limited precision of the observations of the Sun and planets as referred to stars. So it is difficult
to imagine major improvements of the FK5 system without using completely new types of data, some of which actually will become available in
the nearest future.

If we consider a system based on a dynamical definition, the most
drastic improvement in the knowledge of the motion of a celestial body
concerns the Moon, thanks to the ten year series of lunar laser observations. The motion of a fixed point near the center of mass of the
Moon is now described to better than 40 cm in radius vector, corresponding to $0".001$ or $0".002$ in longitude in an adhoc dynamical reference
frame for more than 10 years. This is already better than the accuracy
of the present celestial frames. However, until now, there has been no
explicit attempt to use the lunar laser data to improve the system
(Mulholland, 1980). But even if we assume that this is done, the main
difficulty is not to lose this accuracy when constructing the catalogue.
The most critical point lies in the knowledge of the lunar limb (see
Froeschlé and Meyer, herein) where the present errors are of the order of
$0".3$ and also in the ill determined systematic offset of the center of
the apparent figure of the Moon and its center of mass. Another difficulty lies in the fact that only zodiacal stars are occulted and that
there are not many such stars in the current fundamental catalogues.
So, if the future discussions of lunar laser data may improve significantly parameters used in a classical determination of a celestial reference system, the prospects are much worse if one tries to imagine
a completely new type of method to build a celestial reference system
based uniquely on the motion of the Moon. A very large number of very
precise occultation observations, accurately reduced with improved limb
corrections, are necessary before any such attempt is made.

If, now, we turn to a kinematic definition of the system, a drastic improvement comes from the observation of selected extra-galactic
radio-sources by VLBI. These sources are used as fiducial points of a
fixed celestial reference frame. This has been advocated for many years
(Counselman, 1976 or Elsmore, 1979) and several theoretical analyses of
the possible improvements were made (Gubanov and Kumkova, 1979 or Walter, 1978). These discussions know that it is necessary to show very
precisely the position of the equator in the system, since the observations of declination are referred to the instantaneous axis of rotation
of the Earth. Furthermore, only relative right ascensions are obtained
and this raises again the problem of an independent determination of
the equinox. This can be done using artificial radio-emittors on the
Moon or on space-probes. A more appealing method was proposed by Counselman and Shapiro (1968): In addition to the determination of the
true instantaneous declination of a pulsar by VLBI, the ecliptic longitude and latitude may be derived from the analysis of the delays of the
pulses while the Earth proceeds on its orbit around the Sun. The precision of this method is now better than $0".1$ (Cole, 1976) but does not
yet reach the accuracy that is necessary to improve the existing system.

In an alternate approach, the procedure is reversed and VLBI is
used to provide an independent, purely kinematic - actually geometric -
reference frame without any reference to the Earth motions. Conceptually, the only physical assumption is the absence of a transverse component in the cosmological expansion. Krasinsky (1975) has shown that it
is possible to construct a fixed coordinate system to an accuracy of
the order of the precision of individual VLBI observations. This is done
using the mutual angular distances between the sources that are deduced
from the observed declinations and relative right ascensions. The frame
of reference deduced from the coordinates of fixed objects is also fixed.
The axes may have any direction and be totally independent of any direction linked with the Earth or any other body. The accuracy is independent of time: There is no degradation due to the proper motions, and
accumulation of observations can only improve it.

The problem with such a system is that the defining objects cannot
serve as a materialization of the system for any observational technique
except radio. The number of extragalactic sources that are sufficiently point-like and stable to be used for astrometry is rather small:
Around 150 (Elsmore, 1979) and their optical counterparts are not always well defined. In any case, they are all very faint and it is necessary to determine their positions with respect to stars that can be
candidates for a fundamental catalogue by independent optical measurements. Several such programs have been performed or are in progress
(e.g., Walter and West, 1980 or Argue et al., 1979). But the main part
of the task, namely the construction of the system, is not yet undertaken.

Let us assume that such a VLBI reference system exists and let us
call $B_V(\vec{e}_1, \vec{e}_2, \vec{e}_3)$ the base of the corresponding coordinate system.
The problem is how to extend it to a conventional celestial system containing many stars. Three possibilities may be described.

1. A first procedure consists in determining the positions of the
basic sources and their apparent variations with time in a preliminary
system (e.g. FK4) defined by a base $B_p(t)$ $\{e'_1(t), e'_2(t), e'_3(t)\}$ which
may move with time. The programs that are under way will provide such
positions. Once these are determined, one imposes to the system B_p a
fixed rotation R and a rate R' such that one has

$$B_V = B_p(t) \cdot (R + R't). \tag{1}$$

Applying this transformation to the positions of radio-sources in both
systems will permit the determination of R and R' and therefore, transform the catalogue in the base B_p into a catalogue in the base B_V.

Of course, the practical application of this procedure will not
prove to be so simple because of the errors introduced in the determination of optical sources and errors may appear that would increase
the existing ones in the materialization of the preliminary system. It
may be necessary to introduce a procedure analogous to that proposed

by Gubanov and Kumikova (1979), but this will degrade the accuracy of
the procedure. As a matter of fact, if a precision of only 0".1 is a-
chieved in connecting stars to extragalactic sources photographically,
it is to be expected that the final accuracy will be degraded to the
same order of magnitude. This could only be overcome using much more
precise determinations of the relative positions of stars and radio-
sources. Such a possibility will exist with the astrometric use of the
space telescope (Van Altena et al., 1974 and Jefferys, 1979). However,
strong difficulties arise from the limited field of view (70 square
arc-minutes) and from the fact that the lower limiting magnitude is too
high (m = 10 or 11). As pointed out by C. A. Murray during a discussion,
there is at least one AGK3 star within 15' of most sources, so that it
is possible to tie a stellar system to VLBI or visual galactic system.

2. Another method that has been proposed (Kovalevsky, 1975, and 1979)
uses as an intermediary a terrestrial reference frame. If we assume that
the Earth rotation is continuously determined in two different celestial systems, for instance using classical astronomical instruments
(base B_p) and VLBI (base B_v), if both systems were perfect, the observed rotation of the Earth in both references should be the same. If
this is not the case, the difference represents the offset and the rotation of B_p with respect to B_v.

3. A third method implies the use of radio-stars. There exist a
certain number of stars that are known to be radio-emitters and that
are brighter than magnitude 10 so that they can easily be incorporated
in a fundamental catalogue (Walter, 1977; Débarbat, herein; or Robertson
herein). One hundred such sources well distributed throughout the sky
will ideally serve the purpose. Unfortunately, despite the improvement
of the radio techniques, we are far from reaching this number (Walter's
list has 25 radio-stars with m < 11 and a few more candidates). So
it is necessary to start a specific search for radio-stars. This has not
been done because of the mediocre return in astrophysics, but such a
programme would be very important for astrometry.

Assuming that these radio-stars are observed by VLBI at two epochs,
their positions and proper motions will be obtained in the B_v system,
possibly to better than 0".001 per year. Comparing them with the proper
motions of these stars in the B_p system, we can express the differences
in terms of an offset and a rotation of B_p with respect to B_v.

4. IMPROVEMENT IN PRECISION

If we now turn towards precision, two factors play a dominant role
in degrading the stellar materialization of celestial reference systems.

1. Random errors in positions and proper motions of stars evident-
ly reflect in the errors of realization of the system by these stars.
Presently, in the FK4, the precision is of the order of 0".08 and will
probably be reduced to 0".04 in the FK5.

2. The density of stars belonging to the fundamental catalogue also reflects itself in the practical realization of the system in a given portion of the sky. If, for some uses, the 1515 stars of the FK4 may suffice (polar motion and Earth rotation with astrolabes), in most cases, the fundamental catalogue has to be densified and this usually done at the cost of a severe degradation of the precision. If this may not be the case in some very specific instances (PZT catalogues), usually, one cannot expect a precision better than 0".25 and sometimes 0".4 (AGK 3 or SAO) probably also in conjunction with some degradation of the accuracy. In other terms, whenever the direct use of FK4 stars is not possible, an important blur is introduced by the secondary catalogue random and regional errors, preventing to reach also the full accuracy of the system.

The introduction of the FK5 system and catalogue will improve the situation, since the number of stars is tripled. However, this will not solve the difficulty for many purposes, like providing stars to photographic astrometry or a list of candidates for an occultation programme.

A very good remedy is expected to be provided in the future by the astrometric satellite Hipparcos (ESA, 1979). The programme is due to start next year, the satellite being launched in 1986. Three years of observations should provide the positions and proper motions of 100,000 stars with a precision of respectively 0".002 and 0".002 per year. In principle, all bright stars up to photographic magnitude 8.5 will be included together with many fainter stars, especially in the high galactic and ecliptic latitudes. The system of the Hipparcos catalogue will not be inertial. The internal consistency of observations should avoid any detectable regional error, but the origin of the axis will have to be defined a priori and a residual rotation of the system is unavoidable. However, the system may be linked to a VLBI system by one of the procedures described in the last section, preferably by the third one. So we may expect that the Hipparcos catalogue will indeed represent a geometrically fixed celestial sphere with an accuracy of 0".002 and a slowly degrading precision, because of the accumulation with time of random errors in the proper motions.

The problem of the origin and the direction of the axes remains open. For historical reasons, it may be near-equatorial. But by no means, should it be linked to any Earth motion parameter. The fixed Hipparcos system must be used as a reference for all motions and the formulae for the computation of what are called "true positions" should not be part of the system.

5. FURTHER EXTENSIONS

The Hipparcos system will be a drastic improvement to the present situation and to the expected situation after the FK5 will become available. Later, it is still possible to imagine further improvement in accuracy by repeated observations by VLBI of already known and also new radio-sources that may be discovered among bright stars. Furthermore, a great improvement in accuracy and precision would be provided by an eventual second launch of an astrometric satellite similar to Hipparcos.

The star distribution in the Hipparcos catalogue should be sufficient for many applications, all astronomical techniques for the determination of polar motion and Earth rotation (astrolabes, PZT, zenith telescopes), lunar occultations or reference stars for the observations of major planets. However, the number of reference stars for photographic astrometry must exceed two stars per square degree as provided by Hipparcos. Often also, in order to avoid important magnitude effects, it is necessary to have very faint reference stars with magnitudes comparable to the magnitude of the object to be measured. It will, therefore, be necessary to extend the Hipparcos catalogue to many more stars of higher magnitudes.

This situation will not be new and similar needs already exist now. The extended use of Schmidt telescopes for the measurement of accurate positions of very faint objects (optical counterparts of radio-sources, minor planets, satellites) makes it necessary to use reference stars with magnitudes in the range 10-15. The astrographic catalogue is much too inaccurate to provide good positions and the stars of present reference catalogues (AGK 3, SAO, Yale zones) are too bright. So, their extension to fainter magnitudes is already necessary. The only but very important difference with the post-Hipparcos situation is that the extension will now be based on very imprecise catalogues that are already mediocre extensions of the fundamental catalogue, while later, we shall be able to start directly from the fundamental catalogue to build the extension. There will be one imprecise intermediary less.

Automatic photoelectric meridian circles seem to be now the best instruments for a precise extension and densification of catalogues. These instruments are now entering an operational stage. They may be quite precise, with observational error of the order of 0".10 up to magnitude 13 (Requième, 1979). They also may be very efficient, since it is expected that the Carlsberg automatic meridian circle will observe 100 000 stars per year with a precision of 0".20 (Fogh Olsen and Helmer, 1979). It is possible to expect that future instruments will combine both precision and rapidity. The construction of an extension to the Hipparcos catalogue including half a million stars to a precision of 0".10 - 0".15 and 0".015 per year is feasable and would suffice for practically all needs of photographic astrometry using classical astrograph or Schmidt telescopes. Because of the need of two epochs separated by at least ten years, it is desirable that observations start soon and that the Carlsberg meridian in the Canary Island not be the only instrument to perform this type of observations.

6. CONCLUSION

While the classical constructions of celestial reference
frames are expected to provide sizeable improvement over the
presently existing FK4 system, the advent of new observing techniques like VLBI and the astrometric satellite makes us expect,
for the early nineties, a drastic improvement of a celestial reference frame in accuracy, precision and density of materialization by stars. In parallel, an effort of further densification
to fainter magnitudes is also starting using automatic photoelectric meridians. All these new developments will contribute to
make the forthcoming decade very important in the construction of
better celestial reference frames.

7. ACKNOWLEDGMENT

I thank Drs. Meyer and Mulholland for having very carefully
read the manuscript and for their useful remarks.

REFERENCES

Argue, A. N., Clements, E. D., Harvey, G. M., Murray, C. A.:
    1979, in "Modern Astrometry", IAU Coll. 48, F. V. Prochazka
    and R. H. Tucker ed., p. 155.
Cole, T. W.: 1976, The Observatory, 96, p. 244.
Counselman, C. C.: 1976, Annual Rev. Astron. and Astroph., 14, p. 197.
Counselman, C. C. and Shapiro, I. I.: 1968, Science, 162, p. 352.
Duncombe, R. L., Seidelman, P. K., Van Flandern, T. C.: 1975, in
    "Reference Coordinate Systems for Earth Dynamics", IAU Coll.
    26, Torùn, p. 223.
Elsmore, B.: 1979, in "Modern Astrometry", IAU Coll. 48, F. V.
    Prochazka and R. H. Tucker ed., p. 93.
ESA: 1979, "HIPPARCOS, a Phase A Study", European Space Agency
    document SCI(79) 10, Paris.
Fogh Olsen, H. J. and Helmer, L.: 1979, in "Modern Astrometry",
    IAU Coll. 48, F. V. Prochazka and R. H. Tucker ed., p. 219.
Fricke, W.: 1977, Veröff. Astron. Rechen Institüt, Heidelberg,
    No. 28.
Fricke, W.: 1979, in "Dynamics of the Solar System", IAU Sympo-
    sium No. 81, R. L. Duncombe ed., p. 133.
Gubanov, V. S. and Kumkova, I. I.: 1979, in "Modern Astrometry",
    IAU Coll. 48, F. V. Prochazka and R. H. Tucker ed., p. 135.
Jefferys, W. H.: 1979, in "European Satellite Astrometry", C.
    Barbieri and P. L. Bernacca ed., Padova, p. 111.
Kovalevsky, J.: 1975a, in "Space Astrometry", ESRO Symposium,
    Frascati, October, 1974, Ref. ESRO SP-108, March, 1975.
Kovalevsky, J.: 1975b, in "Reference Coordinate Systems for Earth
    Dynamics", IAU Coll. 26, Torùn, p. 123.
Kovalevsky, J.: 1979, in "Time and the Earth's Rotation", IAU
    Symposium No. 82, D. D. McCarthy and J. D. M. Pilkington ed.,
    p. 151.

Krasinsky, C. A.: 1975, in "Reference Coordinate Systems for
    Earth Dynamics", IAU Coll. 26, Torùn, p. 381.
Mulholland, J. D.: 1977, in "Scientific Applications of Lunar
    Laser Ranging", Reidel Publ. Co., J. D. Mulholland ed., p. 9.
Mulholland, J. D.: 1980, Reviews of Geophysics and Space Physics,
    in print.
Requième, Y.: 1979, in "Modern Astrometry", IAU Coll. 48, F. V.
    Prochazka and R. H. Tucker ed., p. 227.
Van Altena, W. F., Franz, O. G., Fredrick, L. W.: 1974, in "New
    Problems in Astrometry", IAU Symposium No. 61, W. Gliese,
    C. A. Murray and R. H. Tucker ed., p. 283.
Walter, H. G.: 1977, Astron. and Astroph., Suppl. 30, p. 381.
Walter, H. G.: 1978, Mitt. Astron. Geselshaft, 43, p. 202.
Walter, H. G. and West, R. M.: 1980, Astron. and Astroph., 86,
    p. 1.
Woolard, E. W. and Clemence, G. M.: 1966, "Spherical Astronomy",
    Academic Press, New York.

SOME GEODETIC ASPECTS OF THE PLATE TECTONICS HYPOTHESIS

Kurt Lambeck
Research School of Earth Sciences, Australian National
University, Canberra 2600 Australia.

Geodetic observations of gravity, body tides, the Earth's rotation
and crustal motion and deformation potentially provide important constraints in the general inversion of geophysical data for determining the
structure and evolution of the Earth. More specifically, the geodetic
data provide constraints on the rheology of the planet in the frequency
range intermediate between geological and seismic frequencies, on the
geologically instantaneous kinematics of the Earth and on the mechanisms
responsible for the motions within the Earth, results that are intimately
related to the plate tectonics hypothesis. The discussion is limited
here to only a few aspects of these "geodetic" aspects of this hypothesis,
including deformation along plate boundaries, intraplate tectonics and
vertical motions.

Introduction

A general objective of the solid Earth sciences is to understand the
structure – both physical and chemical – of the Earth as it is at present
and to deduce from this the planet's evolution since its formation. The
geophysical contribution to this broad problem is the measurement on or
near the Earth's surface, of a number of quantities: Travel times,
amplitudes and frequencies of seismic waves, the magnitude and direction
of gravity, the flux of heat through the Earth's surface, magnetic and
electrical field properties or strain and deformation. It is the inversion of these data, together with physical and chemical arguments,
that provides the basis for the earth models and for the speculations on
how these models may have evolved with time.

If the term geodesy is rather broadly defined its contributions to
geophysics fall into the rather traditional categories of the Earth's
gravity field, its solid tides and rotation, and its crustal motions and
deformations. Such a subdivision is more a consequence of the manner in
which the geodetic measurement techniques have evolved than of the underlying geophysical phenomena which themselves are closely related. In the
above context, of these observations contributing one more piece of

E. M. Gaposchkin and B. Kołaczek (eds.), Reference Coordinate Systems for Earth Dynamics, 87–101.
Copyright © 1981 by D. Reidel Publishing Company.

information to a complex geophysical inverse problem, it is more
appropriate to consider the geodetic quantities as providing two main
inputs: (i) a measure of low frequency motions and deformations of
points fixed to the Earth's crust, and (ii) the gravitational potential –
and its low frequency temporal variations – on and outside the Earth's
surface.

Following Kaula[1] it is convenient to describe the low frequency
motions and deformations in terms of their spatial and temporal spectra
(Figure 1). In the frequency domain the geodetic measurements lie
roughly between the seismic observations at high frequencies and the
palaeomagnetic and geological observations at the low frequency end of
the spectrum. Spatially, the geodetic observations span the entire
spectrum, from global observations based mainly on satellite and other
extra-terrestrial techniques, to local observations based mainly on conventional geodetic measurement techniques. At low frequencies and long
wavelengths the dominant geophysical phenomenon is mantle convection and
its surface reflection, plate tectonics. The geological record provides
evidence for extensive horizontal offsets of geological features along

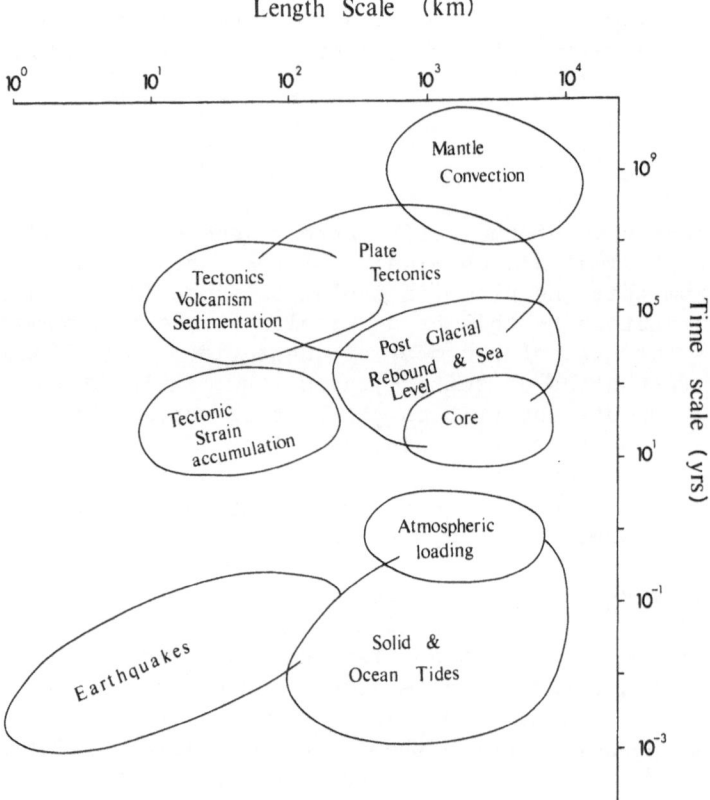

Figure 1. Space-time spectrum of geodynamic processes (adapted from Kaula

fault systems and for vertical uplifts as seen in raised beaches and terraces. While the palaeomagnetic record provides the principal evidence for a large scale orderly and global displacement of the continents and ocean floor on a typical time scale of 10^6 years and at rates of a few centimeters per year. At the other end of the time-space spectrum, the dominant phenomena are the local displacement fields associated with earthquakes, the "instantaneous" expression of plate tectonics and mantle convection. Obviously the two extremes of the bi-spectra are intimately related.

Figure 2 illustrates some specific areas to which the geodetic observations may contribute. Broadly, these contributions fall into two categories. The first is where the measurements provide a measure of the response of the Earth to either known or unknown forces. In some

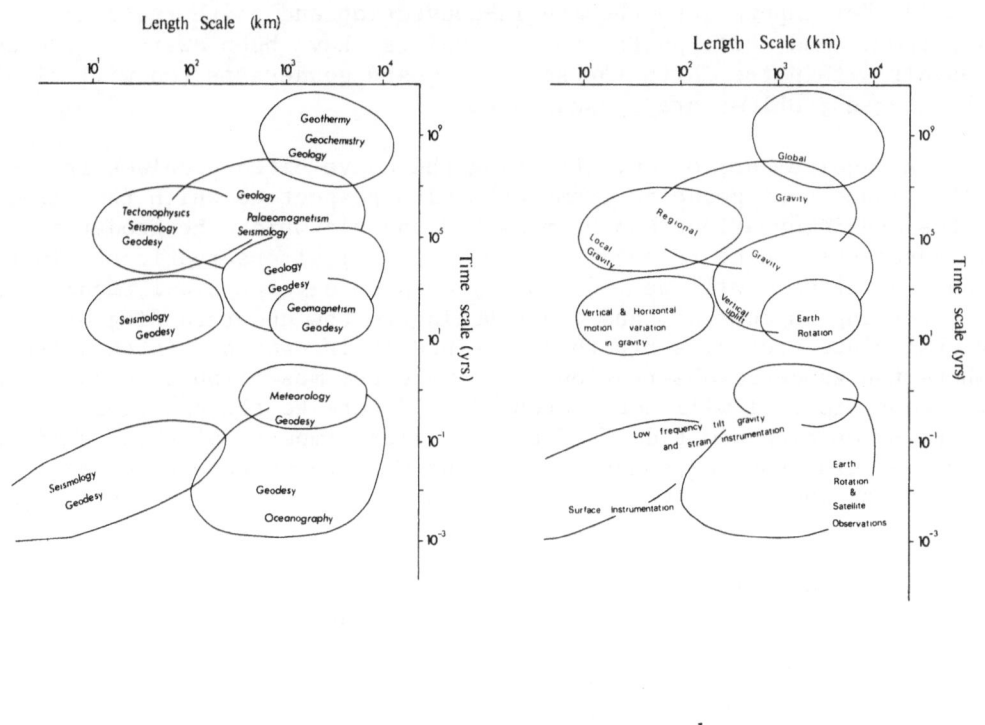

a.   b.

Figure 2. a. Disciplines of the Earth Sciences that contribute to the dynamic processes sketched in Figure 1.

b. Geodetic techniques or observations that contain potential information on the dynamic processes of Figure 1.

instances, such as surface loading problems where the load history is partially known or in tide problems where the applied force is known, the deformation observations provide constraints on the rheological properties of the crust and mantle. Vertical uplift observations over areas previously subjected to extensive glaciation, provide estimates of mantle viscosity, while the tide and rotation observations provide estimates of the global elastic and inelastic response at frequencies that are high in geological terms but low by seismologists' standards.

In the second class of problems the observations provide constraints on the forces themselves. Examples of this include motions and deformations associated with a variety of tectonic problems. Another example is given by the global gravity field where the implied lateral density anomalies provide evidence for the non-hydrostatic state of the mantle and for some form of mantle convection. Shorter wavelength gravity anomalies will reflect mainly crustal structure and will provide constraints on local and regional tectonic processes and on crustal rheology. Both classes of problems, often of considerable intrinsic interest, provide key inputs into the mantle-convection and plate tectonics problems. And while poets[2] and geologists[3] have been aware of the upheavals within the Earth for some 150 years geodesists too will find that these problems are inescapable.

A requirement for investigating the above-cited problems is the establishment of geodetic frameworks with respect to which the strain rates and deformations can be measured and via which the geodetic measurements can be related to other geophysical observations. In this paper an attempt will be made to lay down some geophysical requirements for testing aspects of, and contributing to the understanding of, the global plate tectonics hypothesis, since it is probably in this area where the geodetic observations will make the most important contributions to geophysics. I will not address myself here to the problems of the Earth's rotation and tides. The geophysical imports of these subjects has been discussed elsewhere[4,5] and the discerning geodesist can draw from these books his own conclusions about appropriate geodetic reference system requirements for these matters.

A not entirely unrelated problem, both in terms of the geodetic measurements and in terms of physical consequences, is the input that these measurements have in physical oceanography. The problem here is to deduce the circulation in the oceans from a variety of observations (e.g., acoustic travel times, currents, temperatures, density) and the geodetic measurements provide a boundary condition at the free surface[6]. Intellectually the problem, is not very different from the geophysical one except that the ocean can be sampled down to all depths, and that the time scale involved is relatively short, leaving much less room for speculation than is permitted of the geophysicist.

"Geodetic" aspects of the plate tectonics model

The plate tectonics hypothesis has been clearly defined by LePichon

et al.[7] and needs little elaboration here. The proposed model is
essentially a kinematic one in that it does not specify the dynamic
mechanism responsible for the motions and neither does it define the
nature of the motions below the crust. Built mainly on geological and
palaeomagnetic observations, the hypothesis only represents average
motions at the Earth's surface, averages over time spans of the order
of 10^6 years. Studies of seismicity along plate boundaries confirm the
overall global tectonic motions provided that the results are averaged
over a time interval of 50-100 years or longer[8] but only geodetic
observations will give an "instantaneous" picture of the motions as
well as provide further and possibly more dramatic observational support
for the hypothesis, should such still be required to convince its
recalcitrant objectors. The National Aeronautical and Space Administration's geodynamics program considers as one of its principal objective
the "testing" of this plate tectonics hypothesis by direct measurements[9].
Convinced "drifters" may not share this concern and may instead see
these measurements as fulfilling a scientifically more important role,
namely in contributing to understanding several closely related problems;
the rheology of the crust and upper mantle, the mechanism responsible for
the motions and deformations of the plate, and ultimately the prediction
of earthquake activity.

Figure 3[10] illustrates schematically the average relative motions
of the major plates for the last 5-10   million years, with the motions
being determined mainly from the marine magnetic anomalies, seismic slip
vectors and transform fault directions. Typically the plates move at

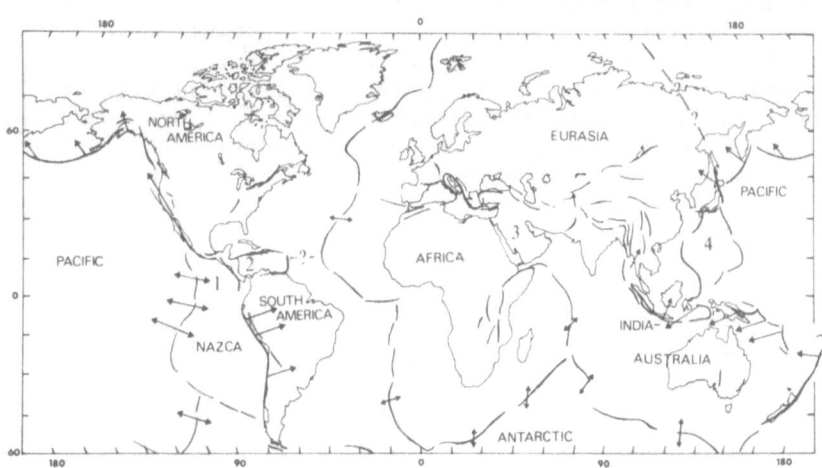

Figure 3. Motions at the boundaries of the tectonic plates[10]. Arrows in-
        dicate direction and rate of motion (see scale in bottom right
        hand corner). Spreading margins are indicated by the double
        arrows and motion at converging boundaries by single arrows.

rates of a few cm per year with the oceanic plates such as the Pacific
and Nazca moving faster than the predominantly continental plates of
Eurasia and North America. This presumably implies that viscous drag
forces are operating on the base of the plate and that they play a
regionally variable role[11]. Maximum relative velocities of nearly 20 cm
yr^{-1} occur along the Pacific-Nazca boundary while the lowest velocities
of 1-2 cm yr^{-1}, occur between the mainly continental plates of Africa
and Europe. By providing an instantaneous picture of the global tectonics
regime and by comparing these results with the seismic and magnetic
evidence for motion, insight may be gained into certain aspects of the
rheology of the lithosphere and anthenosphere and into the mechanism
driving the plates. Matters upon which these measurements may throw
some light include the following.
(i)   Discrepancies are often seen at plate boundaries between slip-rates
based on seismic data and on the magnetic-anomaly data. Are these
evidence for present motions being different than those over the past few
million years? Is it due to aseismic deformation? Or is the distortion
occuring over a broad zone along the plate boundary?
(ii) What is the response of the interiors of the plates during their
relative motions? Do the plates behave as rigid units during their
motion or are they subject to deformation?
(iii) What is the nature of the deformations along the plate boundaries?
Over what distances do these deformations occur? Is the motion uniform
or jerky? Is there evidence for slowly propagating stress waves?
(iv) What is the relation between horizontal and vertical motions,
particularly along plate boundaries?

     An area where the geodetic measurements will not contribute is in
resolving the geophysically vexing problem of absolute motion, a concept
that has a different meaning here than in geodetic usage. One day it
may be possible for geodesists to detect motions of their tracking
stations with centimeter accuracy within an absolute celestial reference
frame but in geophysics it is the motions of these stations relative to
the inaccessible interior of the Earth that is of importance[7]. An overall westward motion of the lithospheric plates over the underlying mantle,
for example, would not be detected by the geodetic measurements. One
would erroneously interpret such a coherent drift as a change in the
speed of rotation of the Earth although it is only the crust that is
involved. It is the same as the older problem of the separation of
continental drift from polar wander[4,12]. Geophysicists have resolved the
problem in a seemingly ad-hoc way with the introduction of Wilson's hot
spot model and with the associated axiom that the hotspots are fixed
relative to each other and to the deep mantle[13]. The result is apparently
self-consistent[10,14] and the model is useful but it cannot be tested by
external, geodetic measurements. Other attempts at establishing a fixed
reference frame have been based on calculations in which the plate
motions have been constrained to plausible physical properties[15].

     The above comments are, in many respects, mere platitudes, drawing
attention once again to aspects of the plate tectonics models that are
more talked about than quantitatively analysed and geodetically observed.

One reason for this state of affairs is the lack of observational evidence in between the frequency bands provided by seismology and geology, a sufficiently large and reliable geodetic data base not yet being available to permit an unambiguous evaluation of some of the above aspects.

Deformation along plate boundaries

Very large earthquakes account for most of the energy release and fault slip within a seismic zone. Events smaller than "very large" are not significant contributors and are usually considered to be local phenomena, either aftershocks or local reactions to the stress redistributions accompanying the larger quakes[16]. One may expect that the large earthquakes should follow a systematic pattern if the plates are moving regularly but seismicity records suggest instead that motion at the plate boundaries is not continuous but occurs mainly in jerks separated in time by a few decades to a few centuries[18]. The recurrence interval between large earthquakes along a particular section of a plate boundary is variable, about 60 years along the Alaska-Aleutian region, about 100 years for Chile[17] and longer at the India-Eurasian boundary. Seismicity studies have also revealed that large earthquakes occur in sequence around a sector of a plate and several migration patterns have been observed[19]. For example, Mogi[19] noted a progressive migration of seismicity from Japan to Alaska in about 35 years and also from Central America to Southern Chile in about the same time interval. The migration rates are of the order 150-300 km year^{-1}. These migrations of seismic activity have been attributed to the presence of long-period stress waves or stress diffusion[18].

The stresses and strain history at a zone of continental-oceanic convergence can be qualitatively described in the following terms (Figure 4). During intervals between great earthquakes along a particular segment of a subduction zone the lithospheric plate, overlying a viscous asthenosphere, is relatively stationary at the trench boundary. The plate is under stress due to the more-or-less steady plate tectonics driving force - either gravitational forces or viscous drag forces at the base of the plate[21] - and the state of compression may extend well into the interior of the plate. At the trench, the lithosphere is held fixed by frictional forces along the interface of the continental and oceanic plates but to the sea-ward side of the trench there is an upward bulging of the oceanic lithosphere[22]. With time the stresses and deformation increase until a critical stage is reached and one or both of two things can occur. Underthrusting of the oceanic plate may occur if the frictional force is exceeded at the trench and the boundary between the underthrusting plate and the adjacent restraining plate is broken and a temporary decoupling of the converging plates occurs. Secondly, the bending moment of the oceanic plate may become excessive, resulting in a tensile fracture in the upper boundary of the lithosphere where the bulge and bending moment are a maximum. This gives rise to two types of major earthquakes at the zones of convergence[23]; decoupling earthquakes resulting in the underthrusting, and tensile lithospheric earthquakes that break the lithosphere seaward of the trench. Simple plate models

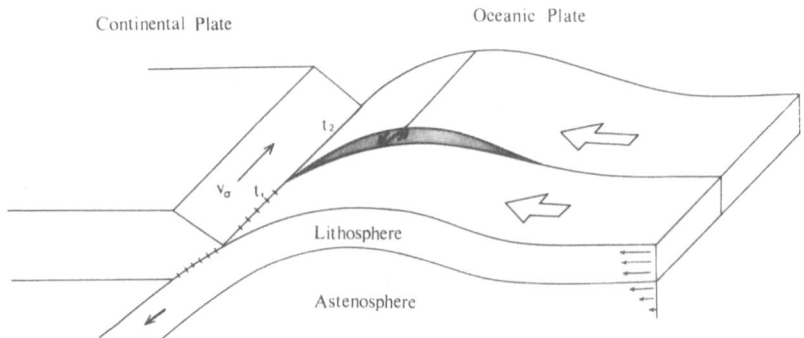

Figure 4. A simple model of stress accumulation at a subduction zone[18].
          The oceanic plate is under compression but its motion is
          locked at the trench. When the stress concentration is
          sufficiently high a decoupling earthquake occurs at time t,
          decreasing the stresses in the segment of the subduction
          zone closest to the viewer, but increasing it elsewhere (in
          the section away from the viewer). This leads to a sub-
          sequent earthquake at time t_2. v_σ denotes the velocity of
          the propagation of the stress wave along the subduction zone.

suggest that the latter may occur at distances of about 100 km from the
trench[18].

     The deformation phenomena associated with these earthquakes, other
than the obvious co-seismic deformations, are several and include an
accelerated plate motion in the vicinity of the boundary before the
decoupled lithospheric boundary heals. During this time interval the
rate of underthrusting of the down-going slab and the rate of approach
of the oceanic plate to the trench will increase while the stress
previously built up in the plate as a whole is relieved. On the
continental side of the trench a sinking of the lithosphere may occur
while on the oceanic side there will be a reduction in the elevation
of the lithospheric bulge. This cycle of crustal movements has long
been recognized although variations from this general model do occur[24].

     At the time of the earthquake the stress will diffuse partly at
elastic wave velocities and adjacent segments of the arc will feel
immediately the consequences, triggering further seismic activity along
other parts of the arc. But, because of the viscous coupling of the

lithosphere to the underlying asthenosphere, kinematic stress waves are
also generated[18],[20] which serve to decrease the stress in the interior
of the plates but increase it elsewhere along the plate boundary. The
post seismic deformations may be of the order of 1 - 5 m/year within
about 100 km of the boundary - the rate decreasing with time - and for
these waves to travel the length of the plate may take several decades.

The comparison of seismic slip-rates with the rates inferred from
the instantaneous plate motion models indicates that agreement along
many boundaries is satisfactory only when the seismic data is averaged
over a long time period[8]. Elsewhere substantial discrepancies have been
found between the computed seismic slips and the inferred rates and
either much of the motion is aseismic or the current plate motion is
significantly different from that of the past[24]. Along the Marianas
trench, for example, the motions are predicted by the plate models to be
of the order of 10 cm year^{-1} but no large-moment earthquakes appear to
have occurred there during the last few centuries. Elsewhere, near Japan
the disparity between the two estimates is a factor of about 5 [17] and
between Eurasia and India it is a factor of about 3 [25] while along the
San Andreas fault system it is perhaps a factor of 2 [26]. The discrepancies
along the subduction zones may be a consequence of insignificant coupling
between the continental and oceanic lithospheres so that the subducted
plate is not locked by frictional forces and is permitted to subduct
more easily resulting perhaps in "slow" earthquakes[27] or perhaps in no
seismic activity at all. At the continent-continent interaction between
India and Eurasia the discrepancy may well be a consequence of the
deformation being taken up over a major part of the two plates since the
large scale tectonics of Asia as a whole appears to be a result of this
collision[28] while in California, where a major part of the deformation
occurs along well defined faults, substantial deformation may occur well
into the Basin and Range province of the western U.S.[10],[29].

This brief summary of seismic events along plate boundaries clearly
indicates that the deformation associated with an earthquake is more
complex than suggested by the elastic dislocation theory which describes
only the instantaneous deformation at a time of faulting. What the
above comments indicate is that pre-seismic deformation occurs and it is
important to understand this, not only because it is a premonitory
phenomena but also because it is informative on the rheology of the crust
and on the mechanism driving the plate. Geodetic evidence of these pre-
seismic movements abound and go back at least as far as the 1920s[30]. But
the interpetation of these data has not always been clear partly because
adequate systematic surveys are generally not available[31]. Following the
actual earthquake, accelerated crustal deformation may persist for
several years and total post-seismic deformations may reach several
meters[32]. These deformations are a maximum near the plate boundaries
but may still be significant at the 10 cm level hundreds of km away from
the boundaries. The geophysical interest of these observations is that
they provide insight into the nature of coupling between the crustal
layer and the underlying material. Some simple models have been proposed
to explain these post-seismic deformations and stress diffusion[18],[20],[33],

models whose central aspects may be tested by a few well-planned measurements taken over a 5 - 10 year period immediately preceding and following a large earthquake.

Intraplate tectonics

An important axiom of the plate tectonics hypothesis is that the plates behave as essentially rigid entities. It would be truly remarkable if the large, irregularly shaped tectonic plates can be moved without deformation over large distances along the surface of the ellipsoidally-shaped earth[34]. But what permits geologists and geophysicists to make this assumption and make the hypothesis reasonable is that the time scale of their model is quite different from that of the average geodesist's lifetime. It works because the motions are averages over long time periods but in the "snap-shot" picture provided by the seismic and geodetic evidence some deformation of the plates undoubtedly occurs, as is indicated by the stress-waves travelling away from the plate boundaries and by intraplate seismicity. The magnitudes and time constants of internal deformations of the plates are clearly important in understanding intraplate tectonics and intraplate seismicity as well as in interpreting the geodetic measurements; for how can one be sure that plate motion is occurring when any internal deformations of the plate are not monitored?

In recent years there has been a substantial increase in the information available on the state of stress within the continental areas of the plate, information provided by in-situ stress measurements and by earthquake fault-plane solutions[35]. The resulting stress patterns are considerably more complicated than those predicted by the simple plate tectonics models due to a super positioning of local and regional stress fields on the global field where the former may be associated with loading of the crust by sediments, by igneous activity or by non-hydrostatic mass distributions associated with past geologic events.

Evidence that continents as a whole are subject to differential stresses is readily seen in the distribution of seismic activity within the continents of North America,[36] or of Australia[37]. Sykes[38] summarized the distribution of the intraplate earthquakes and of igneous activity and one of the principal conclusions he reached was that in continental areas seismic activity tends to be concentrated along pre-existing zones of weakness - along unhealed faults - within areas affected by the youngest orogenesis that predated the opening of the present oceans and led to the present cycle of plate tectonic activity. This seismicity is presumably activated in response to the present-day stress regime in the plate but which is not necessarily the same as that which created the zone of weakness in the first place. Examples abound; the Rhine graben in Europe[38], the New England - Ottawa zone of weakness (postulated to be a continental continuation of the Kelvin seamount chain[36,38]) or the Adelaide Geosyncline in South Australia[39]. Where the lithosphere is thick, cold and strong - as for the cratons and the older ocean basins -

seismic activity is generally much reduced and there is usually little
evidence for geologically recent break-up or rifting.

Vertical motions

    The plate tectonics hypothesis as it is commonly accepted is
concerned predominantly with horizontal motions and largely ignores
vertical deformations; an omission that has sometimes been seized upon
as a demonstration of failure of the model in part or as a whole[40]. But
vertical motions obviously accompany the plate motions, particularly
along the boundaries and height and gravity data provide an important
quantity for constraining stress and deformation models. For example,
at the subduction zones the accumulation of stress will result in an
elevation of the lithosphere on the continental side of the trench and
in a growing bulge ocean-ward of the trench. When decoupling of the
lithosphere occurs such that the stress at the boundary is released,
there may be a sinking of both features. A well documented case history
of the vertical motions during the pre-seismic, coseismic and post-
seismic deformation cycle is provided by the Nankaido earthquake of 1946
in southern Japan where extensive first-order leveling and triangulation
networks have been resurveyed on several occasions over the last 100
years. Differential uplifts of 1 m over distances of about 100 km
occurred during the co-seismic phase. The pre-seismic vertical motions
over some 30 years prior to the earthquake were of the order of 10 - 20
cm over the same distance, while the post-seismic deformations over two
decades following the event were of similar magnitude. Equally well
documented case histories are provided for other regions in Japan[41].
Apart from these measurements evidence for secular uplift is also seen
in uplifted marine terraces, both in southern Japan and along most of
the compressive margins of the Pacific plate.

    Elevation changes have also been noted along transform faults and a
widely discussed example is provided by the still enigmatic Palmdale
bulge[42] where repeated levelling from 1961 to 1971 revealed height
changes of 25 cm over distances of about 150 km. These vertical motions,
by pointing to anomalous areas of the crust, may be important premonitory
indicators of seismicity.

    Within the plate interiors vertical motions in response to variable
surface loading conditions are common, although these are not usually
associated with Earthquake activity. Well known examples include the
post-Pleistocene rebound of Fennoscandia and the Laurentide region[43]
where the presently observed uplift rates are of the order of a few mm
per year. These rates, taken together with geological observations of
vertical motion, have provided useful information on the viscosity of
the mantle below the lithosphere[45]. Viscosity estimates of the mantle
below the oceans are more difficult to come by. One possible example
is for Iceland where extensive glaciation and subsequent rebound has also
occurred[46] but here the problem is complicated by the presence of an
active spreading center which may also be associated with vertical motions

Elsewhere in the oceans subsidence of some islands has also been noted, a few mm/year in the case of the Hawaiian islands[47] for example, and this has been interpeted in terms of a viscous relaxation of the stress in the oceanic lithosphere due to the volcanic load[48]. Island uplift and subsidence may also occur in association with the stress cycle near the subduction zones as discussed previously. I am unaware of observations that support this aspect of the model but simple order-of-magnitude calculations demonstrate that it may not be insignificant - of the order of centimeters.

Geodetic Requirements

In one sense the geodetic requirements for monitoring the motions and deformations associated with the plate tectonics hypothesis are readily stated. A highly accurate dense network of stations, along plate boundaries and within the plate interiors, whose positions in three dimensions are determined at regular and frequent time intervals. If the realities of resources and geography are considered the specifications become considerably more difficult to detail and the requirements for each problem will probably need to be investigated separately. At this stage I am not prepared to lay down specific requirements as it is the function of this meeting to do so. All I can do is offer some general points that may be considered in subsequent discussions.

As indicated earlier, I do not consider that the measurement of present-day plate tectonics is the most important goal. The reason for this is related to the complex behaviour of the plates at and near their boundaries and the possible presence of long period strain waves that travel deep into the interiors of the plates. In order to glean a coherent picture from the observations of motion and deformation it will be necessary to integrate motions over at least several decades if some of the finer aspects of the plate response to stress is not fully under-stood. Instead, I consider that the important contributions are to be made in areas relatively close to and across plate boundaries, particularl at the continent-ocean lithosphere collision zones, at transform faults and at spreading centers. Such contributions would lead to a better understanding of earthquake mechanisms and stress propagation as well as of the lithosphere and mantle rheology.

At the subduction zones the geodetic points should extend well into the plate interiors (up to 1000 km?) to monitor the pre- and post-seismic deformations associated with the large decoupling earthquake and the net-work should extend along large segments of the plate boundary to provide the basis for studying the nature of stress propagation along the boundary. Geographical considerations will probably limit the nature of the network that can be established since the physically more tractable solutions involve ocean plates and few stable island sites will be available. The required density of stations will be high - with a separation of less than 100 km - since deformations may vary rapidly over short distances and a lesser density could readily lead to misleading interpretations of the results. Already established geodetic networks should be fully integrated into the new networks for while these measurements may not meet the accuracy standards quoted for the new measuring techniques, they are all we have at the moment and it will be a long time before we have anything that is better. A complete geological record and complementary geophysical surveys and instrumentation should also be available.

For the question of understanding the plate boundary deformations "absolute" scale and orientation of the network is not essential and the essential criterion is one of repeatability so that strains can be determined. To fit the regional observations into the framework of the global plate tectonics hypothesis some link to the centers of plates, to stations on the stable regions, must also be considered so that while a single global reference frame is not entirely necessary it is a desirable feature. What is a stable region remains problematical for while old cratons are generally not subject to active seismicity they are often subject to vertical motions.

Several areas are appropriate for geodetic studies of the kind outlined. Examples of accessible transform faults include the New Zealand Alpine fault and the Californian fault system, two areas where the geological and geophysical records are relatively complete. Of a spreading center we have Iceland or the Gulf of Aden where considerable work has been done in recent years[49]. For a subduction zone, southern Japan provides a well documented record of both past geodetic and seismic results although here it will be difficult to monitor the deformations on the seaward side. Possibly the New Hebrides - Tonga - Fiji region merits a close investigation.

References

[1] Kaula, W.M. 1978. Rept. No.280, Ohio State Univ., Dept.Geod.Sc. 345-351.

[2] Tennyson, A. Lord 1850. In Memoriam, sections 123-128.

[3] Lyell, C. 1830. Principles of Geology 1. John Murray, London.

[4] Lambeck, K. 1980. The Earth's Variable Rotation. Cambridge University Press.

[5] Melchior, P. 1978. The Tides of the Planet Earth. Pergamon Press, Oxford.

[6] Mather, R.S.,R. Coleman and B. Hirsch 1980. J.Phys.Ocean. 10, 171-185.
Coleman, R. 1980. Marine Geodesy (in press).
Wunsch, C. and E.M. Gaposchkin 1980: Rev.Geophy. & Space Phys. 18, 4, 725-745.

[7] LePichon, X. J. Francheteau and J. Bonnin 1973. Plate Tectonics. Elsevier, Amsterdam.

[8] Lomnitz, C. 1970. Geol.Rundschau 59, 938-960.
Plafker, G. 1976. J.Geophys. Res. 81, 309-18.
Imamura, A. 1928. Jap.J.Astron.Geophys. 6, 119-37.
Ando, M. 1975. Tectonophysics 27, 119-40.

[9] Applications of Space Technology to Crustal Dynamics and Earthquake
     Research (or In Search of a Justification of a Space Program?)
     NASA Technical Paper 1464, 1979.

[10] Based on J.B. Minster and T.H. Jordan's reconstruction (Geophys.Res.
      83, 5331-54 1978).

[11] Richardson, R.M, S.C. Solomon and N.H. Sleep 1979. Rev.Geophys.
      Space Phys. 17, 981-1019.
     Davies, G.F. 1978. Geophys.Res.Lett. 5, 161-4.
     Richardson, R.M. 1976. J.Geophys.Res. 81, 1847-56.

[12] Goldreich, P. and A. Toomre 1969. J.Geophys.Res., 74, 2555-67.

[13] Wilson, J.T. 1963. Nature 197, 536-7.
     Morgan, W.J. 1971. Nature 230, 42-3.
     Minster, J.B., T.H. Jordan, P. Molnar and E. Haines 1974. Geophys.J.
      36, 541-76.

[14] McDougal, I. and R.A. Duncan 1980. Tectonophysics 63, 275-95.
     Burke, K., W.S.F. Kidd and J.T. Wilson 1973. Nature 245, 133-7.
     Molnar, P. and J. Francheteau 1975. Geophys.J. 43, 763-74.

[15] Minster et al. 1974 (loc.cit., note 10).
     Kaula, W.M. 1975. J.Geophys.Res. 80, 244-8.
     Solomon, S.C. and N.H. Sleep 1974. J.Geophys.Res. 79, 2557-67.

[16] Gutenberg, B. and C.F. Richter 1954. Seismicity of the Earth,
      Princeton University Press.
     Brune, J.N. 1968. J.Geophys.Res. 73, 777-84.

[17] Kanamori, H. 1977. Island Arcs, Deep Sea Trenches and Back-Arc Basins,
      Am.Geophys.Union, Maurice Ewing Series Vol.1, 163-173.

[18] Anderson, D.L. 1975. Science 187, 1077-1079.

[19] Mogi, K. 1968. Bull.Earthquake Inst. 46, 53-74.
     Imamura, A. 1946. Proc.Acad.Japan 22, 284-288.
     Terada, T. and N. Miyabe 1928. Bull.Earthquake Inst. 6, 333-343.
     Chinnery, M.A. and T.E. Landers 1975. Nature 258, 490-493.

[20] Bott, M.H.P. and D.S. Dean 1971. Nature, 243, 339-341.

[21] Hales, A.L. Earth Planet Sc.Lett. 6, 31-4.
     Richardson et al. (loc.cit, note 11)

[22] Hanks, T.C. 1971. Geophys.J. 23, 173-89.
     Forsyth, D.W. 1979. Rev.Geophys.Space Phys. 17, 1109-14.

[23] Kanamori, H. 1971. Tectonophysics 12, 187-198.

[24] Imamura, A. 1930. Publ.Imp.Earthquake Inv.Comm. 25.
     Tsuboi, C. 1933. Jap.J.Astron.Geophys. 10, 93-248.
     Scholz, C.H. 1972. Tectonophysics 14, 201-217.
     Fitch, T.J. and C.H. Scholz 1971. J.Geophys.Res. 76, 7260-92.

[25] Chen, W.P. and P. Molnar 1977. J.Geophys.Res. 82, 2945-69.

[26] Thatcher, W. 1979. J.Geophys.Res. 84, 2283-95.

[27] H. Kanamori and J.J. Cipar (Phys.Earth Planet.Int., 9, 128-36, 1974)
      observed a slow deformation over a period of about 15 minutes
      in the epicentral area prior to the major shock of the 1960
      Chili event. Other evidence for the "silent" events is very
      limited A.M. Dziewonski and F. Gilbert (Nature 247, 185-8, 1974)
      have reported stress release just before two deep events and
      W. Thatcher (Science 184, 1283-5) finds aseismic slip associated
      with the 1906 San Francisco earthquake.

[28] Molnar, P. and P. Tapponnier 1975. Science 189, 419-26.

[29] Atwater, T. 1970. Bull.Geol.Soc.Am. 81, 3513-36.

[30] See, for example, papers in some of the early volumes of the Bulletin of the Earthquake Institute of Japan.

[31] See, for example, some of the conflicting conclusions reached about strain accumulation in Southern California by W. Thatcher (Science 194, 691-695, 1976 and loc.cit. note 26) and J.C. Savage and W.H. Prescott (J.Geophys.Res. 84, 171-177, 1979).

[32] Nur, A. and G. Mavko 1974. Science 183, 204-5.
Fitch, T.J. and C.H. Scholz 1971. J.Geophys.Res. 76, 7260-92.
Thatcher, W. (loc. cit.)

[33] Savage, J.C. and W.H. Prescott 1978. J.Geophys.Res. 83, 3369-76.
Rundle, J.B. and D.D. Jackson 1977. Pure Applied Geophys. 115, 401-411.

[34] Turcotte, D.L. and E.R. Oxburgh, 1973. Nature, 244, 337-9.

[35] See, the recent journal volumes devoted to the state of stress in the crust such as Pure and Applied Geophysics Vol.115, No.1-2, 1977 and the Journal of Geophysical Research 85, (in press) 1980.

[36] Sbar, M.L. and L.R. Sykes 1973. Geol.Soc.Am.Bull. 84, 1861-82.

[37] Fitch, T.J., M.H. Worthington and I.B. Everingham 1973. Earth Planet. Sci.Lett. 18, 345-356.
Denham, D., L.G. Alexander and G. Worotnicki 1979. J.Aust.Geol. Geophys. 4, 289-95.

[38] Sykes, L.R. 1978. Rev.Geophys.Space Phys. 16, 621-688.
Illies, J.H. and G. Greiner 1978. Tectonophysics 52, 349-359.

[39] Brown, D.A., K.S.W. Campbell and K.A.W. Crook 1968. The Geological Evolution of the Australian Continent, Pergamon Press, Sydney.

[40] E.V. Artyushkov, 1978. Geodinamita. V.V. Beloussov, Trans.Am.Geophys. Un. 60, 207-211.

[41] See, for example, C.H. Scholz and T. Kato 1978. J.Geophys.Res. 83, 783-797.

[42] R.O. Castle, J.P. Church and M.R. Elliott 1976. Science 192, 251-3.

[43] J.T. Andrews (ed). Glacial Isostasy, Dowden, Hutchinson and Ross (1974).

[44] Cathless, L.M. 1975. The Viscosity of the Earth's Mantle. Princeton University Press.
Peltier, W.R. and J.T. Andrews 1976. Geophys.J. 46, 605-46.

[45] Crittenden, M.D. 1967. J.Geophys.Res. 68, 5517-30. Geophys.J. 14, 261-79.

[46] Einarsson, T. 1966. Jökoll 3, 157-66.

[47] Moore, J.G. 1970. Bull.Volcan. 34, 562-76.

[48] Lambeck, K. and S.M. Nakiboglu 1980. Submitted to J.Geophys.Res.

[49] Courtillot, V.E. 1980. Phys.Earth Planet.Int. 21, 343-50.
Daignieres, M., V.E. Courtillot, R. Bayer and P. Tapponnier 1975. Earth Planet.Sc.Lett. 26, 222-32.
Ruegg, J.C. 1975. Ann.Géophys. 31, 329-60.

THE THEORETICAL DESCRIPTION OF THE NUTATION OF THE EARTH

Martin L. Smith
CIRES
University of Colorado/NOAA
Boulder, Colorado 80309, USA

1.   INTRODUCTION

This talk was intended, I think, to be an opportunity to recite the conventional wisdom of the class of geophysicists who are interested in investigating theoretically the Earth's wobble and nutation. It has become clear since the Kiev meeting in 1977, however, that we have not yet agreed upon the contents of our conventional wisdom and it would be premature, and presumptuous, of me to pretend to recite it. What I can do is to tell you what I think the conventional wisdom ought to be. In doing so I shall give free rein to my prejudices and little consideration to opposing points of view (they must speak for themselves). Nothing in this talk is noticeably original; virtually all of it is extracted from the work of others or from the folklore of this topic.

I am mostly going to discuss how we view the results of theoretical calculations of the Earth's nutation (all of which currently come from computers) and how we try to tie those to observation. I originally claimed that I was going to cover both polar motion and nutation (and most of what follows could be readily extended to polar motion). However since nutation is by far the most predictable disturbance of the Earth's rotation it is presumably the element of principal interest in designing coordinate systems. In deference to time and space limitations, then, we will give short shrift to polar motion (and also to changes in the length-of-day). I should offer a word of warning: my astronomy stops at the Big Dipper and I do not understand geodesy at all. Some but I hope not too much, of what I say will surely be in error.

2.   THE ELEMENTS OF THEORETICAL NUTATION AND POLAR MOTION

We shall work exclusively in a reference frame which rotates rigidly with the unchanging angular velocity

$$\vec{\Omega} = \Omega_0 \, \hat{z} \qquad\qquad (1)$$

103

E. M. Gaposchkin and B. Kołaczek (eds.), Reference Coordinate Systems for Earth Dynamics, 103–110.
Copyright © 1981 by D. Reidel Publishing Company.

where Ω_0 is a constant scalar and $\hat{z}$ is a constant vector. This frame continues its invariant rotation no matter what happens to the Earth. Call this frame M_I. We suppose that we have chosen Ω_0 and $\hat{z}$ sufficiently cleverly that the Earth, as seen in our frame, deviates only slightly from equilibrium. There is, of course, no such frame and we shall, from time to time, have to adjust both the length and orientation of $\vec{\Omega}$ to account for secular changes in the length-of-day and orientation of the mean Earth (as from the precession). On the time scales of interest to us, however, the pleasant fantasy that such a frame exists is quite useful and very nearly true.

One of the useful features of M_I is that it bears an unchanging relation to inertial space. Consequently the motion of an observatory in M_I is, apart from the easily-handled effects of the unvarying rotation $\vec{\Omega}$, the same as its motion with respect to inertial space. This makes the connection between theory and observation relatively straightforward.

Let $\mathbf{x}$ denote position in our frame. Since we suppose that the Earth has a unique equilibrium position, we can regard $\mathbf{x}$ as also specifying a particle in the Earth. If the Earth is disturbed by some agency, internal or external, then in general it will depart from equilibrium. Each particle $\mathbf{x}$ will move to a new point $\mathbf{r}$ given by

$$\mathbf{r}(\mathbf{x},t) = \mathbf{x} + \mathbf{s}(\mathbf{x},t), \tag{2}$$

where $\mathbf{s}$ is what we call the <u>Lagrangian particle displacement</u>. So $\mathbf{r}(\mathbf{x},t)$ is the current location of the particle which is normally (in equilibrium) at $\mathbf{x}$. Expressing the Earth's state as

$$\{ \text{current state} \} = \{ \text{equilibrium state} \} + \{ \text{perturbation} \} \tag{3}$$

is useful as long as the <u>perturbation</u> is small compared to the <u>equilibrium state</u>. Fortunately that is true for nutation, as well as a whole lot of other things of geophysical interest.

Suppose that given an external tidal potential of frequency ω, say, we can compute $\mathbf{s}$ and to adequate accuracy for geophysically reasonable models of the Earth. The question then arises how we connect our theoretically convenient description to some observationally useful quantity. This seems to be a question of arriving at a suitable decomposition of $\mathbf{s}$.

Suppose that we express $\mathbf{s}(\mathbf{x},t)$ as a sum of two terms.

$$\mathbf{s}(\mathbf{x},t) = \vec{\Theta}(t) \times \mathbf{x} + \mathbf{s}'(\mathbf{x},t) \tag{4}$$

where $\vec{\Theta}$ is a spatially constant vector representing a time-dependent rigid rotation and $\mathbf{s}'$ is simply everything left over. We can think of $\vec{\Theta}$ as being the "nutation part" of $\mathbf{s}$, and $\mathbf{s}'$ as being the "body tide part" (assuming $\vec{\Theta}$ represents something sensible). Let v be some por-

tion of the Earth (such as all of it, or maybe just the mantle, etc.).
Let us choose $\vec{\theta}$ so that

$$\int_V \rho |s(x,t) - \vec{\theta}(t) \times x|^2 \, dv = \text{minimum} \quad . \tag{5}$$

This is simply a specialized flavor of Tisserand's mean axes of body.
Then, in the particular sense defined by this integral, $\vec{\theta}$ is the
instantaneous mean rigid rotation of v. Clearly, $\vec{\theta}$ will be different
for different choices of v. What choice shall we make?

A number of possibilities immediately present themselves:

(i) v = the whole Earth

This is the most obvious choice but not a very useful one. In many
cases of interest, and in particular for the nutations, the core and the
mantle undergo greatly different, even opposing, mean rigid rotations.
This choice for v leads to a value of $\vec{\theta}$ which is not the actual mean
rigid rotation of anything.

(ii) v = the crust and mantle

This is a pretty good choice. Molodensky and others have used it. The
reason it is a good choice for the Earth is because the crust and mantle
of our planet very nearly rotate together. I think, however, that this
is something of a cosmological coincidence and one which we do not
really have to rely on. Suppose that the crust and mantle did not move
together. This wouldn't bother us much; we would simply choose that
definition of v which avoided the uncooperative (and unnecessary) mantle, to wit:

(iii) v = the crust

By "crust" we in fact mean the Earth's solid outer surface. This is
where our instruments reside and this is the platform whose orientation
we wish to know. Here I reveal my partisan colors and for the remainder
of this talk I shall take $\vec{\theta}$ to be the instantaneous mean rigid rotation of the Earth's surface:

$$\int_{\text{Crust}} \rho |s(x,t) - \vec{\theta}(t) \times x|^2 \, d\sigma = \text{minimum} \quad . \tag{6}$$

The more traditional position is to use the mantle or the mantle
plus crust as the reference body. I think that this view is based on
the notion that the mantle is a more stable and somehow more fundamental
piece of matter than this torn and heterogeneous crust of ours. I claim
that this is not so; the most we can say for the mantle is that

> it is bigger than the crust,
> it is removed from direct observation,
> and its tectonics are poorly understood.

It is true as we have already noted, that its rigid rotation is very nearly the same as that of the crust, but that does not constitute a reason to use the mantle as a reference. For any observational purpose known to me the salient quantity is the motion of the crust.

Theory, of course, delivers the analytically exact mean rotation of the surface while observation delivers the mean rotation of a set of observatories. Between these two quantities lie the effects of station distribution in space and time and, most important, processes in the solid Earth and oceans beyond the reach of our models. The observation and understanding of just those processes is, of course, one of the goals of this province of science.

3.   OTHER QUANTITIES

$\vec{\theta}$ is now the instantaneous rigid rotation vector of the crust (assuming you agree with our choice of v). This quantity can be converted to more familiar measures of the Earth's rotation, such as motion of the instantaneous angular velocity vector or of the figure axis. In this section we will make some of those connections.

To be a little more specific, we shall assume that $\vec{\theta}$ has the form

$$\vec{\theta}(t) = \theta_0 \, (\hat{x} + i\hat{y})e^{i\omega t} \tag{7}$$

which corresponds to a rigid rotation of θ_0 radians about an axis which is rotating in the Earth's equatorial plane with angular frequency ω. With these conventions, nutation corresponds to $\omega \simeq \Omega_0$, and polar motion (such as the Chandler wobble) corresponds to $\omega \simeq -\varepsilon\Omega_0$ where ε is some measure of the Earth's ellipticity of figure and is small ($\simeq 1/300$).

The geographic axis. Suppose that when the Earth is at rest we define a reference system by measuring the position of a very large number of globally distributed stations and that we use this reference system to define a particular direction which we call the geographic axis and which passes through a large painted X near the North Pole. We call the point where the axis pierces the surface the geographic pole. When the Earth nutates, or whatever it happens to do, all of the reference stations get pushed around. If at some instant we try to determine the geographic axis by measuring the locations of all of the stations we will somehow have to accommodate the fact that in addition to rigidly rotating, our network has gotten all "squished up" and deformed. If we use least-squares techniques to fit our old reference system to the new station positions we will find that the geographic pole has moved by an amount

$$\mathbf{P} = \vec{\Theta} \times \hat{\mathbf{z}} = \theta_0 (i\hat{\mathbf{x}} - \hat{\mathbf{y}}) e^{i\omega t} \quad . \tag{8}$$

Note that this pole does not in general intersect the X any longer (by an amount given by $\mathbf{s}'$). $\vec{\Theta}$, then, corresponds directly to the mean rigid rotation of a dense crustal station network in M_I.

The rotation axis. The instantaneous rotation axis of the crust is offset from $\hat{\mathbf{z}}$ by

$$\mathbf{R} = \frac{i\omega}{\Omega_0} \vec{\Theta} = \frac{\omega}{\Omega} \theta_0 (i\hat{\mathbf{x}} - \hat{\mathbf{y}}) e^{i\omega t} \quad . \tag{9}$$

Note that this quantity is scaled by ω/Ω_0 which is about unity for nutation and very small for polar motion.

The forced nutations are sometimes described in terms of the motion of the instantaneous rotation axis. We actually observe $\mathbf{P}$, however, and not $\mathbf{R}$. The difference is sometimes called "diurnal polar motion" and is

$$\text{polar motion} = \mathbf{R}-\mathbf{P} = \left| \frac{\omega - \Omega_0}{\Omega_0} \right| \vec{\Theta} \tag{10}$$

which is small for the nutations ($\omega \approx \Omega_0$) but unfortunately does not exactly vanish. ($\mathbf{P}$ shows no such motion, of course.) The rotation axis is not directly observable and, so far as I can tell, has usually confused matters when used as an intermediary. It would seem to be more direct to specify nutation in terms of the mean crustal rigid rotation (or equivalently the geographic axis, above). See Fedorov, 1963, and Jeffreys' foreword thereto for further discussion.

The figure axis. I take this to be the instantaneous axis of greatest inertia of the Earth. This quantity is not usefully defined in terms of a specific region v; we have to compute it for the whole Earth. For the record, the instantaneous figure pole is offset from $\hat{\mathbf{z}}$ by the amount

$$\mathbf{F} = \frac{1}{C-A} [\delta I_{xz} \hat{\mathbf{x}} + \delta I_{yz} \hat{\mathbf{y}}] e^{i\omega t} \quad . \tag{11}$$

where δI_{xz} etc. are the instantaneous perturbations in the Earth's inertia tensor due to $\mathbf{s}$. (Because of our definition of M_I the δI are partially due to deformation and partially due to rigid rotation.) For a rigid body the figure and geographic axes coincide. For the Earth they do not. As the above equation implies, $\mathbf{F}$ is quite sensitive to the values of δI_{ij}. A perfectly spherical Earth does not have a unique figure axis. Since the Earth is very nearly spherical, its figure axis is quite sensitive to the deformational portion of $\mathbf{s}$, and consequently, $\mathbf{F}$ is not a very stable quantity. So far as I can tell, it is not a very useful one either.

The angular momentum axis. The instantaneous angular momentum vector is well-determined without resort to complex calculations from a knowledge of the Earth's shape and of the external torques exerted upon the Earth. It is wholly impervious to internal influence. This quantity is theoretically important, and in fact is computationally useful as a check on our calculations (see Wahr, 1980). It is not observable and its role in discussing nutation is the subject of debate. (See, again, Fedorov (1963) and Jeffreys for two sides of this question.)

4.   HOW TO COMPUTE **s**:

In order to find **s** for, say, a particular tidal component we have to solve the elastic gravitational equations of motion for a rotating, slightly elliptical, self-gravitating Earth with a stratified, compressible fluid outer core and a stratified, elastic mantle. In general we don't know how to do this.

There have, however, been a fruitful series of steadily improving approximate assaults described by Jeffreys and Vicente (1957a, 1957b), Molodensky (1961), Shen and Mansinha (1976), Sasao et al. (1980), and Wahr (1980). These studies vary in several respects but, so far as I am aware, all seek to model the same physics and all are essentially correct at their various levels of approximation. Apart from an occasional numerical or algebraic error, there have been no great surprises over the two-plus decades covered by these authors. There has been, I think, a general improvement in the accuracy of our theoretical results due to substantial refinements in geophysical Earth models and the availability of more powerful computing machinery, and there has also been an improvement in the clarity and completeness of the theoretical underpinnings of this effort (as I think we might expect from twenty-three years of experience). I would not characterize this process as the development of new and improved nutation theories but rather as the extension and refinement of the theory of the Earth's nutation.

These calculations are, in detail, extremely complex. I shall take the liberty of summarizing the central features which, in my view, are necessarily common to all of them. Those features arise from two obstacles faced by every nutation calcuation:

(1) How can we compute the response of a rotating Earth which is initially in perfect hydrostatic equilibrium to an external gravitational potential?

(2) How can we correct for the fact that the real Earth is not in perfect hydrostatic equilibrium?

The first question arises because the rotating Earth is not spherically symmetric and vector surface spherical harmonics no longer provide separable basis functions for representing **s**. Our only escape from this to date has been to approximate **s**; usually this means representing **s** by a spherical harmonic series which (we hope) will converge fairly rapidly. In fact, as we can show from both a priori argument and a posteriori example, that this seems to be the case. The results of such a calculation give us the response of an Earth model whose equilibrium state is purely hydrostatic.

The Earth is not quite in hydrostatic equilibrium. For a modern geophysical Earth model with the correct mass and moment of inertia, Clairaut's equation yields (Smith and Dahlen, in press)

$$\frac{C - A}{C} = \frac{1}{308.8}$$

while the correct value for the Earth (Kinoshita, 1977) is

$$\frac{C - A}{C} = \frac{1}{305.4} \ .$$

The difference is of order 1 percent and is non-negligible. We have not dealt with this by extending the theory principally because we would have to know (but do not) the internal deviatoric stresses which keep it out of equilibrium. The source, magnitude, and distribution of these stresses is currently a mystery. The precise treatment of this dilemma varies from author to author but they are all logically (and practically) equivalent to a single scheme.

This scheme goes as follows: Let $\vec{\Theta}$ be the quantity of interest given by some theoretical calculation; to be specific we might suppose that it is the surface rigid rotation associated with some circular nutation term. Let $\tilde{\Theta}$ be the corresponding quantity for a model which differs only by being perfectly rigid. Because the rigid but hydrostatic Earth does not have the same value for (C-A)/C as the real Earth, $\tilde{\Theta}$ will differ from the value predicted for the quantity $\tilde{\Theta}$ by a modern rigid-body nutation calculation. Let $\vec{\Theta}_R$ be the value predicted by such a calculation (such as Kinoshita, 1977). Our corrected estimate for the Earth's predicted nutational motion is given by

$$\vec{\Theta}_E = \vec{\Theta} + (\vec{\Theta}_R - \tilde{\Theta}) \tag{12}$$

This achieves a simple correction for the difference between the rigid-body response of the model Earth and that of the real Earth. Some studies have used an explicit form of the Liouville equation but that is simply a different flavor of this same correction process.

Ironically, of all the improvements in the theory which have occurred since Molodensky published his results in 1961 the most important seems to have been the refinement of geophysical Earth models. Wahr (1980) discusses how some results due to Shen and Mansinha (1976), who repeated Molodensky's calculations with a more modern Earth model, may be interpreted to show that the principal difference between Molodensky's original results and those of Sasao et al. (1980) and Wahr (1980) is due to changes in the Earth models used. (This does not, of course, alter the fact that the new calculations are better, but it does reflect well on 1960-style intuition.)

There is, fortunately, some reason to expect that the next twenty years will not rearrange our results as much as the last twenty have. That reason lies in the nature of the great improvement in geophysical Earth models beginning in the late 1960's. Models constructed since that time have been constrained to fit the observed long-period elastic-gravitational normal modes or free oscillations. These observations first became available following the 1960 Chilean earthquake but were not systematically used in constructing Earth models until much later in the decade (see for example Gilbert and Dziewonski, 1975). The gravest of these normal modes has a period of about one hour and describes the global elastic-gravitional response of the Earth associated with spherical harmonic terms of degree l=2. Since it is precisely these terms which dominate the effects of elasticity on the forced nutations and since the observations are not likely to change, I think we have reason to expect that future Earth models will nutate (and wobble) about the same as our present ones do. We would not expect this to be the case for models available in 1960.

REFERENCES

Fedorov, E.A., 1963. Nutation and Forced Motion of the Earth's Pole, MacMillan, New York, 152 pages.

Gilbert, F. and A.M. Dziewonski, 1975. An application of normal mode theory to the retrieval of structural parameters and source mechanisms from seismic spectra, Phil. Trans. R. Soc. Lond., Ser. A, 278, 187-269.

Jeffreys, H. and R.O. Vicente, 1957a. The theory of nutation and the variation of latitude, Mon. Not. R. astr. Soc., 117, 142-161.

Jeffreys, H. and R.O. Vicente, 1957b. The theory of nutation and the variation of latitude: The Roche model, Mon. Not. R. astr. Soc., 117, 162-173.

Kinoshita, H., 1977. Theory of the rotation of the rigid Earth, Celest. Mech., 15, 277-326.

Molodensky, M.S., 1961. The theory of nutation and diurnal earth tides, Comm. Obs. R. Belgique, 288, 25-56.

Sasao, T., S. Okubo, and M. Saito, 1980. A simple theory of the dynamical effects of a stratified fluid core upon nutational motion of the Earth, in Proceedings of IAU Symposium No. 78: Nutation and the Earth's Rotation, ed. Fedorov, E., Smith, M., and Bender, P., D. Reidel, Dordrecht, Holland.

Shen, P-Y and L. Mansinha, 1976. Oscillation, nutation, and wobble of an elliptical rotating earth with liquid outer core, Geophys. J. R. astr. Soc., 46, 467-496.

Smith, M.L. and F.A. Dahlen, 1981, in press. The period and Q of the Chandler wobble, Geophys. J. Roy. Astron. Soc.

Wahr, J.M., 1980. The forced nutations of an elliptical, rotating, elastic, and oceanless Earth, Geophys. J. R. astr. Soc., in press.

THE DEFINITION OF THE TERRESTRIAL COORDINATE FRAME BY LONG BASELINE INTERFEROMETRY

W. H. Cannon
Centre for Research in Experimental Space Science and Physics
Department, York University, Toronto, Ontario, Canada.

M. G. Rochester
Department of Physics, Memorial University of Newfoundland,
St. Johns, Newfoundland, Canada.

ABSTRACT

This paper examines the question of the definition of the celestial and terrestrial coordinate frames by the technique of long baseline interferometry. It demonstrates how the celestial coordinate frame may be usefully defined in terms of basis 1-forms associated with the advancing phase fronts of the radiation fields from compact radio sources using only interferometer observables. The paper then proceeds to show how the terrestrial coordinate frame could be usefully defined, incorporating fully the effects of plate tectonics and secular motion of the observatories, by an application of the theory of continuum mechanics to interferometer observations.

If we consider a long baseline interferometer with baseline $\vec{B}$ observing radiation at angular frequency ω_o emanating from a source which lies in the direction of the unit vector $\hat{s}$, then the delay observable τ is a scalar quantity which can be expressed as $\tau = \frac{1}{c} \hat{s} \cdot \vec{B}$ where c is the velocity of light in m.sec^{-1} (Thomas 1972, Cannon 1978). The phase Φ of the interferometer is related to the delay τ by $\Phi = \omega_o \tau$. We may also express the interferometer phase as $\Phi = \nabla\phi \cdot \vec{B}$ where $\nabla\phi$ is the gradient of the phase ϕ of the radiation field from the source which lies in the direction $\hat{s}$. Associated with the gradient vector $\nabla\phi$ there is a 1-form $\tilde{\omega}$ such that the surfaces of the 1-form $\tilde{\omega}$ correspond physically to the constant phase fronts of the radiation from the source $\hat{s}$. This permits us to express the interferometer phase Φ and delay τ in terms of the 1-form $\tilde{\omega}$ as $\Phi = \langle \tilde{\omega}, \vec{B} \rangle$ and $\tau = \frac{1}{\omega^o} \langle \tilde{\omega}, \vec{B} \rangle$.

We may now introduce a space-fixed coordinate frame by choosing three arbitrary radio sources lying in the directions of the unit vectors $\hat{s}_1$, $\hat{s}_2$ and $\hat{s}_3$ and defining these unit vectors as the basis vectors of the space fixed frame. In general this space fixed coordinate frame will not be orthogonal and we shall have

$$\hat{s}_i \cdot \hat{s}_j = G_{ij}$$

E. M. Gaposchkin and B. Kołaczek (eds.), Reference Coordinate Systems for Earth Dynamics, 111–118.
Copyright © 1981 by D. Reidel Publishing Company.

where $G_{ij} = \cos \widehat{S_i S_j}$ are the covariant components of the metric tensor of the space fixed frame and where $\widehat{S_i S_j}$ indicates the angular distance between the radio sources lying in the directions $\hat{S}_i$ and $\hat{S}_j$. The basis vectors $\hat{S}_1$, $\hat{S}_2$, and $\hat{S}_3$ have associated with them a covariant set of basis 1-forms $\tilde{\Omega}_1$, $\tilde{\Omega}_2$, and $\tilde{\Omega}_3$ defined such that

$$\langle \tilde{\Omega}_i, \hat{S}_j \rangle = G_{ij} .$$

The phase $\Phi(i)$ of the interferometer which is measured when observing the radio sources in the directions $\hat{S}_i$ is a direct measure of the space fixed covariant components B_i of the baseline vector $\vec{B}$. This is seen as follows:

$$\Phi(i) = \langle \tilde{\Omega}_i, \vec{B} \rangle = \langle \tilde{\Omega}_i, B^1 \hat{S}_1 + B^2 \hat{S}_2 + B^3 \hat{S}_3 \rangle$$

$$= \langle \tilde{\Omega}_i \cdot \hat{S}_1 \rangle B^1 + \langle \tilde{\Omega}_i, \hat{S}_2 \rangle B^2 + \langle \tilde{\Omega}_i, \hat{S}_3 \rangle B^3 = G_{ij} B^j = B_i .$$

The familiar contravariant components B^i follow from $B^i = G^{ij} B_j$ where G^{ij} is defined by $G^{ik} G_{kj} = \delta^i_j$.

This approach is ideally suited to geodetic applications of long baseline interferometry for it makes no appeal to traditional celestial or terrestrial coordinate frames and depends entirely on interferometer observables including the angles $S_i S_j$ which define the elements of G_{ij} and its matrix inverse G^{ij}.

In the case where the interferometer is not observing one of the sources $\hat{S}_1$, $\hat{S}_2$, or $\hat{S}_3$ which define the basis vectors of the space fixed frame but is instead observing some arbitrary radio source which is lying in the direction $\hat{s}$, we introduce a 1-form $\tilde{\omega}$ defined such that $\langle \tilde{\omega}, \hat{s} \rangle = 1$. The unit vector $\hat{s}$ can be expressed as a linear combination of the basis vectors, $\hat{s} = c^k \hat{S}_k$, where $c^k = \cos s \, S_k$ and the 1-form $\tilde{\omega}$ can be expressed as a linear combination of the basis 1-forms $\tilde{\omega} = A^k \tilde{\Omega}_k$.

The phase of the interferometer while observing this source is $\Phi(\hat{s}) = \langle \tilde{\omega}, \vec{B} \rangle$. This reduces as follows:

$$\Phi(\hat{s}) = \langle \tilde{\omega}, \vec{B} \rangle = \langle A^k \tilde{\Omega}_k, B^j \hat{S}_j \rangle = A^k B^j \langle \tilde{\Omega}_k, \hat{S}_j \rangle$$

$$= A^k B^j G_{kj} = A^k B_k .$$

To determine the space fixed covariant components B_k of the baseline vector requires observations on three sources $\hat{s}_i$ and the solution of the equations $\Phi(\hat{s}_i) = A^k_i B_k$, $i = 1,2,3$. The values of the elements A^k_j follow from the property that $\langle \tilde{\omega}_j, \hat{s}_j \rangle = 1$. This reduces as follows:

$$\langle \tilde{\omega}_j, \hat{s}_j \rangle = \langle A^k_j \tilde{\Omega}_k, c^\ell_j \hat{S}_\ell \rangle = A^k_j c^\ell_j \langle \tilde{\Omega}_k, \hat{S}_\ell \rangle$$

$$= A^k_j c^\ell_j G_{k\ell} = 1 .$$

From this it follows that

$$A_j^k = [\cos(\widehat{s_j S_1}) G_{k1} + \cos(\widehat{s_j S_2}) G_{k2} + \cos(\widehat{s_j S_3}) G_{k3}]^{-1} .$$

Once again the procedure is independent of any traditional celestial or terrestrial coordinate frame and depends entirely on interferometer observables namely the angles $\widehat{S_k S_\ell}$ which constitute the elements of $G_{k\ell}$ as well as the angles $\widehat{s_j S_\ell}$ appearing in the above formula.

The rotating body fixed frame of the earth has traditionally been spanned by an orthogonal set of body fixed basis vectors $\hat{e}_i^o$ which are defined by the mean axis of figure of the earth, the mean equator of figure of the earth, and the mean prime meridian of longitude or the "mean observatory". A space fixed coordinate frame has traditionally been spanned by an orthogonal set of space fixed basis vectors $\hat{E}_i$ defined dynamically (inertially) by the mean pole of the ecliptic, the mean plane of the ecliptic and the mean equator of a given epoch. The time dependent rotation of the set $\hat{e}_i$ relative to the set $\hat{E}_j$ can be expressed as $\hat{e}_i = P_{ij}(t) N_{jk}(t) S_{k\ell}(t) W_{\ell m}(t) \hat{E}_m$ when the orthogonal matrices $P_{ij}, N_{ij}, S_{ij}, W_{ij}$ represent the effects of earth precession, nutation, rotation (spin) and polar motion (wobble) respectively. (Woolard & Clemence 1966, Mueller 1969).

In so far as the earth is rigid and of a known figure the elements of all matrices can be predicted in advance on the basis of known astronomical forcing functions. The finite strength of the earth and the interaction between the earth's solid and fluid portions has complicated this transformation somewhat (Rochester 1973) but its continued use in the present day implicitly carries the assumption that an observer who remains locally fixed relative to the material of the solid earth will, on average, also remain globally fixed relative to the material of the solid earth.

The discovery that an observer who is fixed locally relative to the material of the solid earth will not, on average, remain fixed globally relative to the material of the solid earth but will, according to the theory of plate tectonics, exhibit secular motion relative to the global distribution of earth material at rates varying from 1 to 10 cm yr^{-1} has rendered this assumption invalid. This fact together with the development of modern space techniques of geodesy such as long baseline interferometry, which are expected to yield intercontinental baseline measurements with precisions of a few centimeters, have made it necessary to re-examine the question of the definition of the body fixed frame and to develop rigorous operational definitions and computational procedures which are capable of incorporating secular deformability and continuing fracture of the earth.

In our treatment we shall be guided by the well known procedures of continuum mechanics (cf. Prager 1973) in which the deformation field imposed on a continuous medium is described by a differentiable vector function of position $\vec{u}(\vec{r})$. The components of the relative displacement $\vec{\Delta}$ of two mass elements at positions $\vec{r}_1$, $\vec{r}_2$ separated, before deformation, by the infinitesimal separation vector $\vec{dx}$ is given by

$$\Delta_i = e_{ij}\,dx_j + \Omega_{ij}\,dx_j$$

where

$$e_{ij} = \frac{1}{2}\left[\frac{\partial u_i}{\partial x_j} + \frac{\partial u_j}{\partial x_i}\right] \quad \text{and} \quad \Omega_{ij} = \frac{1}{2}\left[\frac{\partial u_i}{\partial x_j} - \frac{\partial u_j}{\partial x_i}\right]$$

are the strain tensor and rotation tensor respectively. The strain tensor has the property that the variation in the squared distance $|\vec{dx}|^2$ between the mass elements accompanying deformation is given by $\delta|\vec{dx}|^2 = 2e_{ij}dx_i dx_j$. The rotation tensor has the property that the components Ω_i of rigid body rotation $\vec{\Omega}$ imparted to the separation vector $\vec{dx}$ by the deformation field are given by $\Omega_i = -\frac{1}{2}\varepsilon_{ijk}\Omega_{jk}$ where ε_{ijk} is the alternating tensor.

We proceed by considering a long baseline interferometer baseline $\vec{B}(t) = b_1(t)\hat{e}_1^o + b_2(t)\hat{e}_2^o + b_3(t)\hat{e}_3^o$. The procedure for determining the traditional body fixed components $b_i(t)$ involves: (i) a determination of the contravariant space fixed components $\vec{B}(t) = B^1(t)\hat{S}_1 + B^2(t)\hat{S}_2 + B^3(t)\hat{S}_3$ given with respect to the space fixed basis vectors $\hat{S}_i(i=1,2,3)$, (ii) a transformation using the celestial coordinates (right ascension and declination) of the radio sources $\hat{s}_i(i=1,2,3)$ to express the base-line components in terms of the space fixed basis vectors $\hat{E}_i(i=1,2,3)$, or $\vec{B}(t) = \beta_1(t)\hat{E}_1 + \beta_2(t)\hat{E}_2 + \beta_3(t)\hat{E}_3$, (iii) a transformation $b_i(t) = W_{ij}^T(t)S_{jk}^T(t)N_{k\ell}^T(t)P_{\ell m}^T(t)\beta_m(t)$ where the superscipt "T" denotes a matrix transpose.

Repeated measurements of the baseline $\vec{B}(t)$ at times $\ldots t_{m-1}, t_m, t_{m+1}\ldots$ yield observatory relative displacement vectors $\vec{D}(t_m) = \vec{B}(t_m) - \vec{B}(t_{m-1})$ with body fixed components $d_i(t_m) = b_i(t_m) - b_i(t_{m-1})$. We introduce the quantities

$$C_{ij}(t_m) = \frac{d_i(t_m)b_j(t_m)}{b_\ell(t_m)b_\ell(t_m)}$$

to serve as analogs to the tensor $\frac{\partial u_i}{\partial x_j}$ of continuum mechanics. We may decompse C_{ij} into symmetric and antisymmetric parts as $C_{ij} = E_{ij} + \Theta_{ij}$ where $E_{ij} = \frac{1}{2}[C_{ij}+C_{ji}]$ and $\Theta_{ij} = \frac{1}{2}[C_{ij}-C_{ji}]$. It can be readily shown that the body fixed components of the relative displacement vector can be written as $d_i = E_{ij}b_j + \Theta_{ij}b_j$ which is the analog of the expression $\Delta_i = e_{ij}dx_j + \Omega_{ij}dx_j$ from the theory of continuum mechanics.

Further similarities to the theory of continuum mechanics follow. It may be shown that the variation $\delta|s^2|$ in the squared distance s^2 between the observatories which occurs as a result of the relative displacement $\vec{D}$ is given by $\delta|s^2| = 2E_{ij}b_i b_j$ which is the analog of the expression $\delta|\vec{dx}|^2 = 2e_{ij}dx_i dx_j$ from the theory of continuum mechanics. It may also be shown that the body fixed components of the rigid body rotation $\vec{\theta} = \theta_1\hat{e}_1^o + \theta_2\hat{e}_2^o + \theta_3\hat{e}_3^o$ imparted to the interferometer baseline by the relative displacement of the observatories D are given by $\theta_i = -\frac{1}{2} \epsilon_{ijk} \theta_{jk}$ which is the analog of the expression $\Omega_i = -\frac{1}{2} \epsilon_{ijk} \Omega_{jk}$ from the theory of continuum mechanics.

These procedures may be applied to the problem of the definition of the terrestrial frame in the presence of arbitrary earth deformation fields producing arbitrary relative motions of the observatories. We presuppose a global network of N nonredundant interferometer baselines $\vec{B}^\alpha$ (α = 1, 2, 3 ... N) and their measured strain tensor analogs E_{ij}^α and rotation tensor analogs θ_{ij}^α α = 1, 2, 3 ... N. We may define a unique global strain tensor analog Γ_{ij} as being that which minimizes in a weighted least squares sense the total departure between the observed baseline length variations δs^{α^2} given by

$$\delta s^{\alpha^2} = 2E_{ij}^\alpha \, b_i^\alpha \, b_j^\alpha$$

and the predicted baseline length variations $\overline{\delta s}^{\alpha^2}$ given by

$$\overline{\delta s}^{\alpha^2} = 2\Gamma_{ij} \, b_i^\alpha \, b_j^\alpha \; .$$

We then choose to minimize P^2 where

$$P^2 = \sum_{\alpha=1}^{N} w^\alpha [\delta s^{\alpha^2} - \overline{\delta s}^{\alpha^2}]^2$$

or

$$P^2 = 4 \sum_{\alpha=1}^{N} w^\alpha [(E_{ij}^\alpha b_i^\alpha b_j^\alpha)^2 - 2E_{ij} \Gamma_{\ell m} b_i^\alpha b_j^\alpha b_\ell^\alpha b_m^\alpha + (\Gamma_{\ell m} b_\ell^\alpha b_m^\alpha)^2]$$

where w^α are appropriate weights. A choice of $\Gamma_{\ell m}$ which minimizes P^2 is given by

$$\frac{\partial P^2}{\partial \Gamma_{pq}} = 0$$

which leads to

$$4 \sum_{\alpha=1}^{N} w^\alpha \left[-2E_{ij}^\alpha + 2\Gamma_{ij} \right] b_i^\alpha b_j^\alpha b_p^\alpha b_q^\alpha = 0.$$

Since this condition should be fulfilled independently of the particular choice of baseline network we conclude that we shall in general require

$$\Gamma_{ij} = \frac{1}{\displaystyle\sum_{\alpha=1}^{N} w^\alpha} \sum_{\alpha=1}^{N} w^\alpha E_{ij}^\alpha \; .$$

We may also define a unique global rotation tensor analog Λ_{ij} as being that which minimizes in a weighted least squares sense the vector norm of the net departures between the observed baseline rotation $\vec{\theta}^\alpha$ whose components are given by

$$\Theta_i^\alpha = -\frac{1}{2}\,\epsilon_{ijk}\,\Theta_{jk}^\alpha$$

and the predicted baseline rotation $\vec{\Lambda}$ whose components are given by

$$\Lambda_i = -\frac{1}{2}\,\epsilon_{ijk}\,\Lambda_{jk}\,.$$

We then choose to minimize

$$Q^2 = \sum_{\alpha=1}^{N} w^\alpha |\vec{\Theta}^\alpha - \vec{\Lambda}|^2$$

or

$$Q^2 = \frac{1}{2}\sum_{\alpha=1}^{N} w^\alpha [\Theta_{\ell m}^{\alpha 2} - 2\Lambda_{\ell m}\,\Theta_{\ell m}^\alpha + \Lambda_{\ell m}^2]$$

where w are appropriate weights. A choice of Λ_{pq} which minimizes Q^2 is given by

$$\frac{\partial Q^2}{\partial \Lambda_{pq}} = 0$$

which leads to

$$\sum_{\alpha=1}^{N} w^\alpha [-2\Theta_{pq}^\alpha + 2\Lambda_{pq}] = 0$$

which has the solution

$$\Lambda_{ij} = \frac{1}{\sum_{\alpha=1}^{N} w^\alpha}\sum_{\alpha=1}^{N} w^\alpha\,\Theta_{ij}^\alpha.$$

The tensor analogs E_{ij}^α and Θ_{ij}^α for each interferometer baseline can be expressed as the sum of the global quantities Γ_{ij} and Λ_{ij} plus a "local" residual ϵ_{ij}^α and ω_{ij}^α respectively.

$$E_{ij}^\alpha = \Gamma_{ij} + \epsilon_{ij}^\alpha$$

$$\Theta_{ij}^\alpha = \Lambda_{ij} + \omega_{ij}^\alpha$$

where, by definition, the weighted global means of the local residuals vanish

$$\frac{1}{\sum_{\alpha=1}^{N} w^\alpha}\sum_{\alpha=1}^{N} w^\alpha\,\epsilon_{ij}^\alpha = 0$$

$$\frac{1}{\sum_{\alpha=1}^{N} w^\alpha}\sum_{\alpha=1}^{N} w^\alpha\,\omega_{ij}^\alpha = 0.$$

The "local" residuals $\varepsilon_{ij}^{\alpha}$, ω_{ij}^{α} $\alpha = 1, 2, 3 \ldots N$, given relative to the traditional body fixed basis vectors $\hat{e}_1^0$ $\hat{e}_2^0$ $\hat{e}_3^0$, represent strains and rotations of individual baselines relative to the network as a whole and will presumably contain information about regional tectonics including inter- and intraplate geologic processes.

The global quantities Γ_{ij} and Λ_{ij} represent global strain and global rigid rotation of the network as a whole relative to the traditional body fixed basis vectors $\hat{e}_1^0$ $\hat{e}_2^0$ $\hat{e}_3^0$. The global network of interferometers and the global quantities Γ_{ij}, Λ_{ij} can be used to redefine the body fixed coordinate frame. In general we may express the new body fixed basis vectors $\hat{e}_i$ in terms of the traditional body fixed basis vectors $\hat{e}_i^0$ by a transformation of the form

$$\hat{e}_i = [\delta_{ij} + \sigma_{ij} + \alpha_{ij}]\hat{e}_j^0$$

where σ_{ij} and α_{ij} are small quantities and where:

(i) $\sigma_{ij} = \sigma_{ji}$ is the symmetric part of the transformation,

(ii) $\alpha_{ij} = -\alpha_{ji}$ is the antisymmetric part of the transformation.

The symmetric and antisymmetric parts of the transformation are independent of each other and must be determined separately. In general the antisymmetric part of the transformation, given by α_{ij}, represents a rigid rotation of the basis vectors $\hat{e}_i$ relative to the basis vectors $\hat{e}_i^0$ while the symmetric part of the transformation, given by σ_{ij}, represents a deformation of the basis vectors $\hat{e}_i$ relative to the basis vectors $\hat{e}_i^0$.

We shall define the body fixed frame to be spanned by new body fixed basis vectors $\hat{e}_i$ such that if the material of the earth were subjected to a uniform strain Γ_{ij} and a uniform rotation Λ_{ij} relative to the basis vectors $\hat{e}_i^0$ then the body fixed coordinates of the mass elements of the earth would remain constant when referred to the basis vectors $\hat{e}_i$. Accomplishing this in the presence of the global rotation Λ_{ij} fixes the antisymmetric part of the transformation and accomplishing this in the presence of the global strain Γ_{ij} fixes the symmetric part of the transformation.

It is clear that the antisymmetric part of the transformation requires

$$\alpha_{ij} = \Lambda_{ij}$$

while it can be shown (cf. Brillouin, 1964, pp. 287 ff.) that the symmetric part of the transformation can be deduced from the requirement that to preserve the coordinates of the mass elements in the presence of the deformation Γ_{ij} we require a nonorthogonal set of basis vectors $\hat{e}_i$ which have the property that

$$\hat{e}_i \cdot \hat{e}_j = g_{ij}$$

where

$$g_{ij} = \delta_{ij} + 2\Gamma_{ij}$$

is the metric tensor of the body fixed coordinate frame. This leads to the result that

$$\sigma_{ij} = \Gamma_{ij} \, .$$

This gives

$$\hat{e}_i = [\delta_{ij} + \Gamma_{ij} + \Lambda_{ij}] e_j^o$$

and

$$\hat{e}_i = [\delta_{ij} + \Gamma_{ij} + \Lambda_{ij}] \, P_{jk} \, N_{k\ell} \, S_{\ell m} \, W_{mn} \, \hat{E}_n$$

as the transformation from the dynamical space fixed frame to the body fixed frame of the interferometer network.

REFERENCES

Brillouin, L.: 1964, Tensors in Mechanics and Elasticity, Academic Press.

Cannon, W. H.: 1978, Geophys. J. R. astr. Soc., 53, 503.

Mueller, I. I.: 1969, Spherical and Practical Astronomy as Applied to Geodesy, F. Ungar Publ. Co.

Prager, W.: 1973, Introduction to Mechanics of Continua, Dover Publications.

Rochester, M. G.: 1973, E⊕S, Transactions of the American Geophysical Union, 54, 769.

Thomas, J. B.: 1972, An Analysis of Long Baseline Radio Interferometry, Deep Space Network Progress Report, Jet Propulsion Laboratory, Pasadena, California Technical Report, 32-1526, Vol. VII.

Woolard, E. W., and Clemence, G. M.: 1966, Spherical Astronomy, Academic Press.

A NEW PARAMETERIZATION OF POLAR MOTION

Haim B. Papo
Department of Civil Engineering
Technion, Haifa, Israel

ABSTRACT

The rotational motion of the earth is decomposed into spin,polar motion and local motions. The rotation vector components are associated to phenomena such as precession,nutation,diurnal spin,polar motion and local motions. The above decomposition is accomplished without refering to an earth-fixed CIO pole or BIH zero meridian. The time-like variations of the coordinates of a surface point in a geocentric equatorial reference frame are presented as a function of the rotation vector components. In the rigid earth approximation three scalar parameters are necessary for evaluating point coordinate variations,namely spin rate of the earth, polar motion magnitude and spin rate of the polar motion vector. Two numerical examples are given as an illustration.

INTRODUCTION

The polar motion phenomena has been continuously studied,measured and analysed for over a century by an ever-increasing number of devotees. The past two decades have been marked by extensive refinements in the dynamical analysis of the rotation of the earth and by the advent of new observational systems of superior accuracy (VLBI,LASER ranging). Grand designs are under way with a promise of polar motion determination with centimeter level accuracy. Somehow,the objectives of all those efforts have not changed. The ultimate results sought are still the age-old x,y angles. The purpose of this paper is to challenge the uniqueness of the classic x,y polar motion parameters. It is suggested that rather than the x,y angles,their time rates be determined,presented as a vector and related to the diurnal spin vector of the earth.

REFERENCE FRAMES AND THE ROTATION VECTOR

The basic reference frames needed in the present analysis are briefly defined,where a more complete discussion may be seen in Grafarend et.al. (1979). The _inertial_ frame e is defined by a star catalog system,where the e^1,e^3 axes are identified respectively with the mean equinox and

E. M. Gaposchkin and B. Kołaczek (eds.), Reference Coordinate Systems for Earth Dynamics, 119–123.

pole of the catalog at the standard epoch. The equatorial frame e_3 is
defined by the observable spin axis of the earth (See Atkinson,1975 and
Leick,1978) and the mean ecliptic at epoch T. The time-like variations
of e_3 vs. e are modeled by the parameters of precession and forced nutation (without diurnal frequencies). The conventional frame e_c defined
by a reference pole and zero meridian is explicitly related to a net of
stations at a particular epoch. An arbitrary point p on the earth surface is defined by its geocentric position vector $\bar{p}$ which is refered to
the e_3 frame by geocentric colatitude σ and hour angle h.

The rotation vector $\bar{\omega}$ of a material point p on a deformable earth is
defined as the vorticity of the local velocity field (See Grafarend et.
al.,1979). This definition is instrumental for partitioning the time-
like variations in position of p into rotations,translations and deformations. The vector $\bar{\omega}$ representing the complete spectrum of rotational
motions of p vs. a geocentric inertial frame is partitioned into the
following components:

$$\bar{\omega}_{pn}$$ - precession + nutation   derived from theory

$$\bar{\omega}_s$$ - diurnal spin

$$\bar{\omega}_f$$ - polar motion   determined from observations

$$\bar{\omega}_1$$ - local motions

The first three components of $\bar{\omega}$ are global, i.e., space-invariant, same
any p point. At a given epoch they represent the rotation of the earth
crust in an average sense. Regarding σ,h of a station as the directly
observable quantities we are interested in the spin,polar motion and
local motions components of the rotation vector.

REPARAMETERIZATION OF POLAR MOTION

Universal time UT1 or equivalently Θ. The apparent sidereal angle and
the polar motion parameters x,y are essentially transformation angles
through which the conventional reference frame e_c is related to e_3 the
equatorial frame at a given epoch. The time-like variations of those
angles,denoted respectively by $\dot{\theta},\dot{x},\dot{y}$,represent the rotational motion
of the conventional vs. the equatorial frames. But these same motions
were defined above by the diurnal spin $\bar{\omega}_s$ and polar motion $\bar{\omega}_f$ rotation
vector components. From Figure 1 the functional relationship between
the two sets of parameters is derived as follows:

$$|\bar{\omega}_s| = \dot{\theta} = \omega_s$$ - spin rate of the earth (vs. e_3)

$$|\bar{\omega}_f| = \sqrt{\dot{x}^2 + \dot{y}^2} = \omega_f$$ - magnitude of polar motion vector

$$h_f = \theta - \arctan(\dot{y}/\dot{x}) - \Pi/2$$ - hour angle of polar motion vector

$$\dot{h}_f = \dot{\theta} + (\ddot{x}\dot{y} - \ddot{y}\dot{x})/\omega_f^2$$ - spin rate of polar motion vector

where $\ddot{x},\ddot{y}$ are the respective second derivatives of x,y vs. time.
As the proposed parameters $\omega_s,\omega_f,\dot{h}_f$ are functions of the time derivatives of UT1,x,y they also have irregular time-like variations and have
to be determined empirically from observations on a day by day basis.

Given a discrete series of $\omega_s,\omega_f,\dot{h}_f$ values, UT1,x,y can be evaluated for
any desired epoch by quadrature. The set of initial values $(UT1,x,y)_o$
at the zero epoch is equivalent to the selection of a particular conventional reference frame (CIO/BIH being only one out of many possibilities)
Thus we see that the proposed set of parameters is fundamentally different from the one currently in use by being completely independent of
the particular choice of conventional frame.

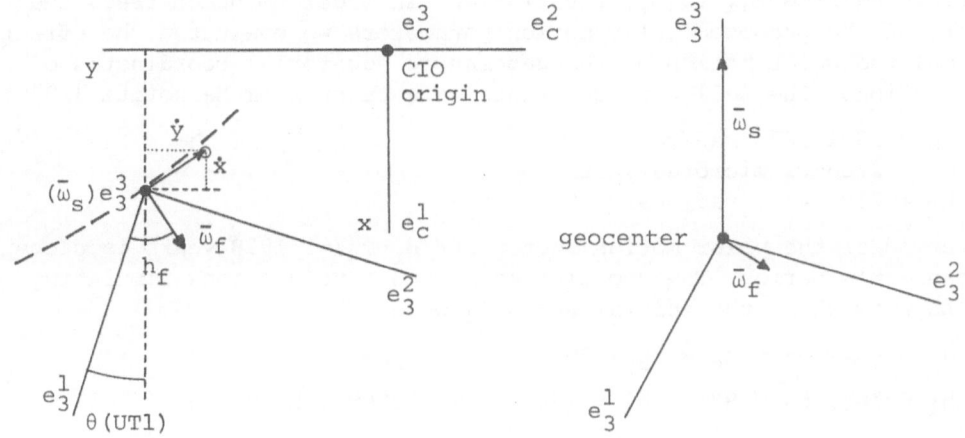

Figure 1. The polar motion vector

A simple numerical exercise was performed to obtain some feeling for the
magnitudes of the proposed polar motion parameters. Table 7 in the 1977
BIH Annual Report (x,y, UT1-AT1 smoothed values at 5 day intervals) was
differentiated numerically and transformed into $\omega_s,\omega_f,\dot{h}_f$. Table 1 lists
the results at 40 day intervals.

Table 1.   Spin and polar motion parameters for 1977.

Epoch day	$\omega_s-2301.$ rad/year	$\dot{h}_f-\omega_s$ rad/year	ω_f micrad/year
0	0.167305	10.7319	6.3278
40	308	4.1028	8.0486
80	295	7.6559	6.7288
120	293	5.5007	6.6373
160	315	3.3389	7.9860
200	330	5.3783	8.1570
240	314	7.1653	7.3203
280	295	8.9587	7.1117
320	296	3.7296	6.4164
360	292	7.7184	7.8234

VARIATION OF COORDINATES DUE TO POLAR MOTION

The time-like variation of $\bar{p}$ in e_3 due to $\bar{\omega}_s$ and $\bar{\omega}_f$ is given by:

$$\dot{\bar{p}} = \bar{\omega}_s \times \bar{p} + \bar{\omega}_f \times \bar{p} \quad .$$

From equivalent matrix expressions for the vector product or using spherical trigonometry one can easily derive the following equations:

$$\dot{\sigma} = 0 \quad - \omega_f \sin(h-h_f)$$

$$\dot{h} = \omega_s - \omega_f \cos(h-h_f) \cot\sigma$$

The terms in the above expressions due to $\bar{\omega}_f$ can be obtained also from
Mueller (1969,pp.87) by differentiating equations 4.39 and 4.40 and
transforming from $\dot{x},\dot{y}$ to ω_f,h_f variables. In order to demonstrate the
utility of the proposed polar motion parameters we evaluated the effect
of simulated polar motion on the geocentric equatorial coordinates of
five stations. The following constants were taken from Markovitz (1976):

$$\omega_s = 2301.1676 \text{ rad/year}$$
$$\dot{\omega}_f = 3.86309 \text{ microrad/year}$$
$$\dot{h}_f = 2306.4797 \text{ rad/year}$$

The period of the polar motion vector $2\Pi/(\dot{h}_f-\omega_s)=1.1828$ years is close
to Chandler's period. The coordinates of the five stations simulating
the ILS network at the initial epoch T_O were:

$$\sigma_1 = \sigma_2 = \sigma_3 = \sigma_4 = \sigma_5 = 50^O$$

$$h_1 = 10^O; \ h_2 = 82^O \ ; \ h_3 = 154^O \ ; \ h_4 = 226^O \ ; \ h_5 = 298^O$$

The value of h_f at T_O was set to be Π. Using an ordinary numerical integration procedure σ_i,h_i were evaluated over a period of 440 sidereal
days. Table 2 shows the variations in σ_i and h_i at 40 day intervals.
In Figure 2 we plotted the varying coordinates of station 1 at the beginning of each sidereal day,using polar stereographic projection with
differend scales for σ_1 on one hand and for $\Delta\sigma_1,\Delta h_1$ on the other. We
plotted also the varying position of a reference pole which is at equal
distances from the five stations. The resulting circle with a radius of
$0\overset{..}{.}15$ is the inverse of polar motion as it shows the path of the referencpole with respect to the equatorial e_3 frame.

Table 2. Variations in equatorial coordinates of five stations($0\overset{..}{.}01$

Epoch	Δh_1	$\Delta\sigma_1$	Δh_2	$\Delta\sigma_2$	Δh_3	$\Delta\sigma_3$	Δh_4	$\Delta\sigma_4$	Δh_5	$\Delta\sigma_5$
0	0	0	0	0	0	0	0	0	0	0
40	7	-1	3	8	-5	6	-6	-4	1	-8
80	13	-6	9	12	-7	14	-13	-4	-1	-16
120	15	-15	16	12	-5	22	-19	2	-7	-21
160	13	-23	22	7	1	27	-22	10	-14	-22
200	7	-28	25	-1	8	28	-20	18	-20	-17
240	0	-30	24	-9	15	24	-15	24	-23	-9
280	-6	-26	19	-15	18	16	-7	25	-22	-1
320	-10	-18	12	-17	17	8	-1	22	-18	6
360	-10	-10	5	-14	13	1	3	15	-11	8
400	-5	-3	1	-7	6	-2	3	6	-4	5
440	1	0	0	1	-1	1	-1	-1	1	-1

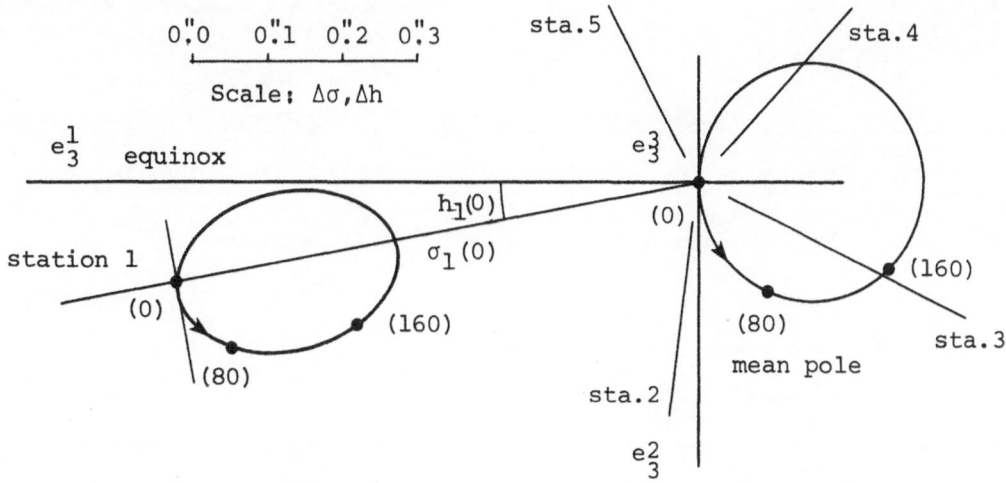

Figure 2. Variation of coordinates due to polar motion.

SUMMARY

The proposed reparameterization of polar motion is not just another
mathematical gag. It should be seriously considered for analysis and
representation of earth rotation. Its primary merit is in avoiding any
dependence on a fixed conventional origin (pole) and zero meridian. At
any time and by an explicit formulation a conventional reference frame
may be derived from the varying positions of a network of monitoring
stations. This seems the right way of analysing modern, high accuracy
observations without blurring their accuracy by artifacts which are in
open contradiction to the behaviour of the real earth.

ACKNOWLEDGMENT

Most of the study reported in the above paper was carried out while the
author was on sabbatical leave with Prof. I.I. Mueller at OSU. Working
with Prof. Mueller was exciting,challenging and highly rewarding.Partial
support for the study was provided by NASA/GSFC Grant No. NSG 5265.

REFERENCES

Atkinson R.d'E. :1975,"On theEarth's Axes of Rotation and Figure"
     Mon.Not.R.Astr.Soc.,71,pp.381-386.
Grafarend E.W.,Mueller I.I.,Papo H.B.,Richter B. :1979,"Concepts for
     Reference Frames in Geodesy and Geodynamics",Bull.Geod.,53,No.3.
Leick A. :1978,"The Observability of the Celestial Pole and its Nutations"
     OSU Dept. of Gedetic Science,Report No.262,Columbus.
Markovitz W. :1976,"Comparizon of ILS,IPMS,BIH and Doppler Polar Motion
     with Theoretical",Report to IAU Comm. 19 and 31,XVI General
     Assembly,Grenoble,France.
Mueller I.I. :1969,"Spherical and Practical Astronomy as Applied to
     Geodesy",Frederick Ungar Publ.Co.,New York.

COMMENTS ON THE TERRESTRIAL POLE OF REFERENCE, THE ORIGIN OF THE LONGITUDES, AND ON THE DEFINITION OF UT1

Bernard Guinot
Bureau International de l'Heure
Paris, France

ABSTRACT. The definition of a terrestrial reference pole (TRP) and of a Terrestrial Longitude Origin (TLO), by fixing the stations coordinates of an adopted network seems unrealistic. It is proposed that TRP and TLO be the realized origins of some specified service working in accordance with some stated principles. It is also proposed to clarify the definition of UT1 and to base it on some physical properties.

INTRODUCTION

A terrestrial reference system (T_O) is defined by the center of masses T of the Earth, an arbitrary axis Tz_O, and another axis Tx_O in the plane perpendicular to Tz_O. For practical reasons, Tz_O is chosen close to the terrestrial rotation axis. When it is needed to relate a terrestrial direction to space fixed directions, the common practice is to use the coordinates of the pole in both the Earth and space reference systems, and the value of an angle around the rotation axis, which is conventionally linked to UT1. Therefore the coordinates x and y of the pole and UT1 (designated as the Earth rotation parameters, ERP, in the following) should refer to (T_O).

We are presently far from this ideal situation, the main reason being that the ERP are determined by instruments which use, as a terrestrial reference, not a direction linked to the body of the Earth, but the vertical. Thus the coordinates of the stations which determine the ERP, which should be the primary reference for geodetic networks, cannot be used for accurate purposes.

A number of other defects also affect the present system for referring the ERP. Although, under the instigation of Markowitz (1968), an attempt was made to define a realizable polar origin, known as the Conventional International Origin, CIO, by specifying station coordinates, observational and computational procedures, strictly speaking, this origin will be lost, because some of the stations of the Interna-

125

E. M. Gaposchkin and B. Kołaczek (eds.), Reference Coordinate Systems for Earth Dynamics, 125–134.
Copyright © 1981 by D. Reidel Publishing Company.

tional Latitude Service (ILS) which realize it, will close. On the other hand, no precise definition of the origin of the longitudes was given. This origin should be a point on the equator of CIO, so that the prime meridian should pass through a given point of the old Greenwich Observatory ; in practice, this condition is not realizable and the adopted origin is implicitly linked to the longitude system of the Bureau International de l'Heure (BIH).

Thus, we are faced to a number of problems.

(a) Since precise terrestrial coordinates are obtained by methods which refer to the body of the Earth, and not to the vertical, the method for determining the ERP should also refer to the body of the Earth. Even if more precise, the classical astronomical methods could remain a good tool for geophysical studies concerning the Earth rotation irregularities and for interpolation, but could not satisfy alone the geodetic needs.

(b) The choice of the coordinates of the stations which determine the ERP is a difficult task, in particular because fixed coordinates cannot be acceptable, in the long term.

(c) As the ILS network will be dismantled before the official organization of new networks for measuring the ERP, an interim reference pole is needed to replace CIO.

(d) A decision is also desirable for an interim definition of the longitude origin.

(e) A somewhat independent problem is the definition of UT1, which, to my opinion, should be re-considered.

These problems are most difficult. I do not pretend to have made their exhaustive study. I only offer some comments, as a modest contribution to their solution.

FRAMES OF REFERENCE FOR THE EXISTING EARTH ROTATION SERVICES

Concerning problems (b), (c) and (d), the consideration of what has been accomplished by existing services might provide some guidance, and we will briefly summarize it.

The ILS solution

The ILS network consists in 5 stations observing latitude. The observations are solved for 3 unknowns : the two coordinates of the pole and an auxiliary quantity, the z-term. The 5 initial latitudes define the CIO.

Three stations would have been geometrically required. The redundancy is nevertheless acceptable because of the simple dependence of the pole

errors on the measurement errors at the stations: the proportionality.
Therefore, if there are linear latitude drifts at the stations (non
common), there is a linear drift of the CIO, which is not too serious.
However, this property is obtained at the cost of some limitations.

The linear dependence on station errors requires a specific
method of computation. It is made possible by the linear relationship
between the pole coordinates and the local latitudes, by the properties
of the least squares method, and by the fact that fixed weights for the
observed data are adopted, in spite of their variable quality and amount.
It therefore requires averaging the observed data over long intervals
(one month) to smooth out the effects of the weather,and forbids fine
time resolution and rapid service.

The BIH solution

The BIH method of computation of the ERP, adopted in 1967 (BIH,
Rapport Annuel pour 1967 ; Feissel, 1980) was devised with the aim of
using all existing measurements, to have short averaging time (5 days),
and to allow a rapid publication of the ERP. This requires that any
subset of observational data should lead to the same solution, except
for the effects of the random errors. It was thus necessary to maintain
at any time a coherent system of initial longitudes and latitudes. This
was accomplished by an algorithm giving predicted values of the systematic corrections to the initial coordinates.

In this algorithm, the coordinates are revised every year by
applying a new set of corrections. It is done in such a way that, in
the yearly average, there is no systematic change of the ERP when using
the old or the new corrections. Therefore, the BIH algorithm realizes
the condition that the variable initial coordinates bring no rotation
in the BIH reference system for the ERP, although this has never been
explicitly said.However, when realizing this no-rotation condition, the
stations receive weights, according to their observation uncertainties,
which is questionable.

The IPMS solution

The International Polar Motion Service uses the same observational
data as the BIH, but processes them independently. First,fixed initial
coordinates were used, with some exceptions (Yumi, 1975). But, in a
recent revision (Yumi, 1980), variable coordinates have been adopted.
The general principle according to which the variations are obtained is
not given.

The DMA Doppler solution

The Doppler solution for the pole coordinates by the Defense Mapping
Agency (see, for instance, Oesterwinter, 1979) uses fixed reference
positions of the Doppler stations. However, it was found by Anderle
(1978) and by McLuskey (1979) that station residuals in latitude,

longitude, and height have significant variations. Although the results
are derived from an all-weather observing network, and that it can be
expected that each 2-day pole position is derived from the data of the
quasi-complete network, one can wonder whether a constantly coherent
system of stations positions would improve the results, especially when
poor conditions of reception occur. This raises again a question about
the principles which should govern the choice of the zero of the
adopted rate corrections.

Comparisons between the origins of the existing services

        Different methods have been used for maintaining the pole and
the longitude origin. Nevertheless it is interesting to compare these
origins and their relative rates : it will give an idea of the size of
the drifts which can occur, due to the different effect of the plate
motions and/or to other causes. In these comparisons, the following
series are used

ILS : revised results (Yumi and Yokoyama, 1980).

IPMS : revised results (Yumi, 1980),
       1967-1978, solution from time and latitude, with τ,
       1962-1978, solution from latitude only.

BIH : BIH Global solution, BIH Ann. Rep. for 1978, D-68 to D-76
                          and BIH Ann. Rep. for 1979, D-79.
       The intervals 1967-1978 and 1962-1978 are considered, the
       results for 1962-1966 being less accurate than the following ones.

DMA : the latest results available in the BIH files and used in the BIH
       Ann. Rep. for 1979. Results prior to 1972, which were obtained
       by a different method are not used.

The results of the comparisons are given by Table I.

Interim definition of the origins for the pole and the longitudes

        The BIH pole is not as close to CIO (the ILS pole) as the IPMS
pole, probably because it was adjusted to CIO over a shorter interval :
1964 - 1966. However, the BIH pole, which was realized several years
before the IPMS pole and which is still available with slightly shorter
delays, is generally adopted as the current reference for geodetic
purposes.

        There is no observable relative drift of the IPMS and BIH poles
since 1967 -at the level of $0\overset{..}{.}0007$ per year (2cm y^{-1}), one sigma. The
drift of these two poles relative to CIO might be of the order of $0\overset{..}{.}001$
per year (to be compared to the drift of the ILS mean pole towards
America by $0\overset{..}{.}003$/year).

        The IPMS and BIH systems for UT1 show a small relative drift,
which might be due to the effect of different weighting, combined with
the stars proper motion errors.

Table I. Difference between the Earth rotation parameters by
   various services :
      Service - BIH = A + B (T - 1975.0)
   Units are : 0''001 for x and y, 0.0001s for t = UT1. The variations
   B are yearly rates.

Services	Interv.	A_x	B_x	A_y	B_y	A_t	B_t
ILS - BIH	62-78	36 ± 4	-0.2 ± 0.9	-2 ± 3	1.0 ± 0.5		
	67-78	39 ± 4	2.3 ± 1.3	-4 ± 3	-0.2 ± 0.8		
IPMS(L)	62-78	29 ± 2	-0.3 ± 0.3	2 ± 2	1.1 ± 0.5		
- BIH	67-78	29 ± 1	0.0 ± 0.4	1 ± 3	-0.4 ± 0.7		
IPMS(L+T,τ)	67-78*	30 ± 1	0.3 ± 0.3	3 ± 2	0.0 ± 0.6	-8 ± 3	-1.9 ± 1.0
- BIH							
DMA - BIH	72-79	-15 ± 2	-1.6 ± 0.6	-2 ± 3	1.6 ± 1.1		

   * Interval 1967-77 for UT1.

The pole of the satellite Doppler system has a drift with respect
to IPMS, ILS and BIH, but still slow.

We can conclude that the IPMS and BIH poles are equally acceptable
as the "official" reference pole, in replacement of CIO, until modern
networks be operational, and that the dismantling of the ILS network
would have no adverse consequences. However if it is planned to instal
new optical instruments, such as PZT's, their location on the international ILS parallel would be much better than a location "at random" .

CONSIDERATIONS ON THE REFERENCES FOR THE FUTURE EARTH ROTATION NETWORKS

The evolution of the services processing the data of the optical
astronomy is, to many respects, exemplary. The initial plan of the ILS,
based on geometric properties, is being abandonned and replaced by a
statistical solution which allows taking advantage of instrumentation
progress. But, for decades, valuable observations were not used to
derive polar motion for fear of spoiling the ILS system.

In the future organization of the measurements of the ERP, one
could be tempted to rebuilt an ILS, i. e. to freeze a technique and a
network so that the origin could be defined by a set of adopted
coordinates. Such an organization would not be realistic. No technique
appears much superior to others, and competition between
techniques will continue. On the other hand, for a given technique, it
should be possible to profit from theoretical developments,

instrumental improvements, and increased participation. The organization of new networks should be flexible enough to allow this progress without loosing the reference frame . This leads to the following considerations.

(a) Ideally, the coordinates of the observing stations measuring the ERP should be "absolute". They should belong to a worldwide homogeneous system, representing the real shape of the Earth, which should define both the reference system for geodesy and for the ERP. Leaving in the station coordinates some foreign quantities which appear in the data processing (to use them as a "garbage box") should be avoided. An example is the mixing of the longitude and right ascension errors.

(b) Fixed coordinates of stations, or even variable coordinates according to some conventionally adopted model of plate motions, cannot be convenient, in the long term, for the operation of a network determining the ERP. The evolution of the network (in particular, closing of some stations, changes of weighting station data on account to instrumental modifications, and variable influence of each station in the global solution for the ERP due to meteorological effects and failures) requires that, at every instant, the station coordinates constitute a coherent system. This is also needed for a correct evaluation of the random uncertainties. I suggest that the corrections to the initial coordinates be established in accordance with the condition that they should bring no rotation in the reference system for the ERP obtained by the considered network.

(c) The above "no-rotation" condition cannot be strictly specified. In principle, if the station motions were random, following a common distribution law, non-weighted averages of the motions should be consider· ed. However, the systematic effects due to plate motion and the anomalous local effects prevent one from giving precise rules. I suggest that the conventionally adopted frame of reference for the ERP be the one realized by the service in charge of the final evaluation of the ERP, working in accordance with the general "no-rotation" condition. An example of such a definition already exists for the International Atomic Time, TAI. The 15th General Conference on Weights and Measures adopted the following text (non-official English translation from French)

> "The International Atomic Time is the time coordinate
> established by the BIH on the basis of the readings
> of atomic clocks operating in various establishments,
> in conformity with the definition of the second, time
> unit of the SI".

A possible definition of the reference frame for the ERP is given in the Conclusions.

DEFINITION OF UT1

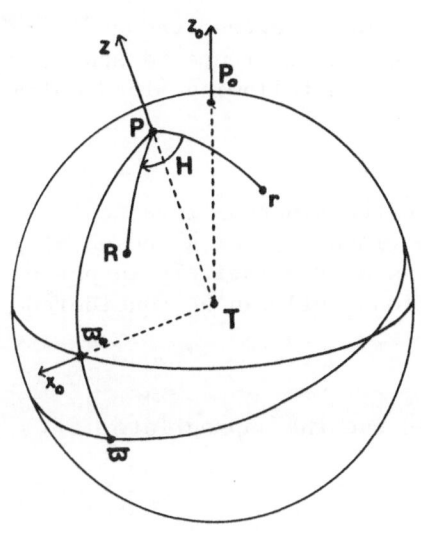

Fig. 1

Let us represent directions by points on a celestial sphere accompanying the Earth rotation (fig. 1). P_O is for Tz_O; $\bar{w}_O$, for Tx_O; P for Tz, the instantaneous rotation axis (more precisely P is the newly defined celestial Ephemeris Pole). One needs to know the angular position of (T_O) around Tz. This can be accomplished by giving the angle $H = r\hat{P}R$, as a function of TAI, where Tr is a known direction in (T_O), and TR a known direction in a space non-rotating reference system.

In practice r is ϖ_O, and R is ♈ the mean equinox of date, H being then known as the Greenwich Mean Sidereal Time, GMST. Let us investigate the reasons of these choices.

The necessity of having the Earth reference on the equator is clear. Thus, when P moves, as it remains close to P_O, the $P\varpi_O$ meridian has no appreciable rotation around the Pz-axis : GMST is independent of the polar motion in (T_O). The origin of the longitudes in the instantaneous system (T) can be defined as the intersection ϖ of $P\varpi_O$ with the instantaneous equator.

The choice of the space reference ♈ is traditional. As ♈ does not move uniformly on the equator, GMST is not proportional to the sidereal rotation of the Earth. This has no importance for geodetic applications since the observed relation GMST/TAI would satisfy the needs. However, for geophysics, one requires the departure of the Earth rotation from uniformity. Both the geodetic and geophysical needs are fulfilled by
- the UT1/TAI relation (obtained from measurements),
- the UT1/GMST relation (conventional),
where the latter relation should ensure the proportionality of UT1 to the sidereal rotation of the Earth.

Equivalently one can define a fictitious point moving on the equator, the Mean Sun MS, in such a way that UT1 be equal to the hour angle of MS from the prime meridian, + 12h, and that UT1 be proportional to the sidereal rotation of the Earth. But this definition has no operational usefulness since the position of the celestial bodies are referred to ♈, not to MS, which rapidly moves among the stars.

Guinot (1979), using the concept of a non-rotating origin on the mean equator of date, has found that the proportionality of UT1 to the sidereal rotation was not ensured neither by the current UT1/GMST relation, nor by the proposed relation in conjunction with the adoption of the IAU (1976) System of Constants.

In order to clarify the concept of UT1, and to extend the definition of UT1 to possible non-rotating systems of reference in space, in which the equinox could not be determined, the following broad definition is suggested.

"UT1 should be an angle proportional to the sidereal rotation of the Earth, the coefficient of proportionality being chosen so that 12 h UT1, in the long term, remains approximatively in phase with Greenwich Noon. In some applications, UT1 can be considered as a non-uniform time scale."

To implement this definition, one could use the appropriate UT1/GMST relation (Aoki and Kinoshita, 1980).

However, I believe that, even in the case of classical determinations of UT1 by optical observations of fundamental stars, it would be better to define explicitly a non-rotating origin (NRO) on the true equator. In case of changes of conventional equator and ecliptic, the NRO can be transferred on the newly adopted equator using the non-rotation condition. Thus, if we call θ the hour-angle of the NRO from the prime meridian, a fixed linear relation UT1/θ could be adopted, not affected by revision of constants and ensuring the continuity of UT1 in phase and in rate, when these revisions occur.

The NRO could be made available through a γ/NRO relationship. As the initial NRO should be chosen conventionally, as γ, and as the subsequent position of γ is obtained from conventionally adopted series and the position of the NRO is rigourously computable, the relation γ/NRO would be exactly known permanently. But Guinot (1979) believes that it would be preferable to use the NRO instead of γ for referring star positions. This more general use is not discussed in this paper.

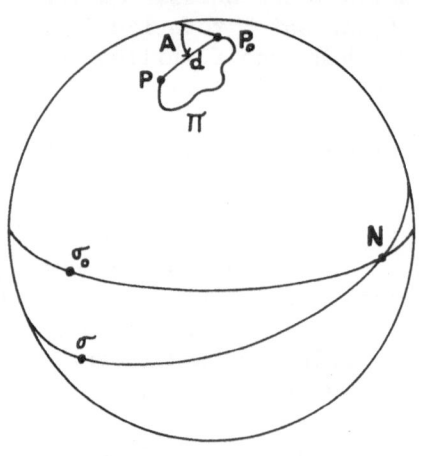

Fig. 2

The position of the NRO is computed in the following way. If P_O and P are two positions of the pole on the non-rotating celestial sphere at dates t_O and t, A and d are the polar coordina of P from P_O (fig. 2), A being reckoned from a fixed direction in space, σ_O

and σ are the NRO at t_0 and t, π is the path of P, and N is the ascending node of the equator of P on the equator of P_0, it is easily demonstrated that:

$$s = \sigma_0 N - \sigma N = \int_\pi (\cos d - 1) \, dA .$$

The effects of the nutation, which should be included in π, are not entirely negligible, since they are of the order of 0''0001 per year.

CONCLUSIONS

We summarize the main suggestions of this paper

(a) The primary methods for determining the Earth rotation parameters should use, as terrestrial reference, directions tied to the body of the Earth -not the vertical. This is needed for expressing these parameters in a worldwide geodetic network.

(b) The definition of the Terrestrial Reference Pole,TRP, and the Terrestrial Longitude Origin, TLO, could have the following form.

> The TRP is the origin of the polar motion derived by XXX. The TLO is the point of the equator of TRP used by XXX for deriving UT1. In these derivations, the assumption is made that the progressive changes of the reference coordinates of the stations contributing to the determination of the instantaneous pole, and of UT1 do not represent statistically a rotation.

XXX represents the future officially designated service for issuing the values of the Earth rotation parameters. But, provisionally, such a definition could be used for a specified existing service, even if the coordinates of the observing stations are the astronomical latitudes and longitudes.

(c) The definition of Universal Time UT1 should be based on a physical property : UT1 should be proportional to the rotation of the Earth in space. The explicit definition of a non-rotating origin on the true conventional equator would help in realizing this definition.

REFERENCES

Anderle, R.J. : 1978, NSWC/DL Techn. Rep. 3884, Dahlgren Virginia, Naval Surface Weapons Center.

Aoki, S., and Kinoshita, H. : 1980, submitted to Celestial Mechanics.

Feissel, M. : 1980, Bull. géod., 54, pp. 81-102.

Guinot, B. : 1979, in D.D. McCarthy and J.D. Pilkington (eds) "Time and the Earth's Rotation", IAU Symp. 82, pp. 7-18.

Markowitz, Wm. : 1968, in Wm. Markowitz and B. Guinot (eds),
    "Continental Drift, Secular Motion of the Pole, and Rotation
    of the Earth", IAU Symp. 32, pp. 25-32.

McLuskey, D. : 1979, Report presented at the 17th IUGG General
    Assembly, Canberra.

Oesterwinter, C. : 1979, in D.D. McCarthy and J.D. Pilkington (eds),
    "Time and the Earth's Rotation", IAU Symp. 82, pp. 263-278.

Yumi, S. : 1975, Ann. Rep. of the IPMS for the year 1973, Central
    Bureau of IPMS, Mizusawa.

Yumi, S. : 1980, Ann. Rep. of the IPMS for the year 1978, Central
    Bureau of IPMS, Mizusawa.

Yumi, S. and Yokoyama, K. : 1980, Results of the International
    Latitude Service, Central Bureau of IPMS, Mizusawa.

THE EFFECTS OF THE REFERENCE FRAMES AND OF THEIR REALIZATION ON
THE EARTH ROTATION PARAMETERS COMPUTED FROM DIFFERENT
OBSERVATIONAL TECHNIQUES

Nicole Capitaine and Martine Feissel
Bureau International de l'Heure
Paris
FRANCE

ABSTRACT

The inaccuracies in the reference frames actually realized by the
different techniques for measuring the Earth's rotation are
theoretically investigated. The intercomparison of the available
series of measurements provides numerical estimations of these
defects. Using data corrected for reference frame effects high
frequency fluctuations of UT1 are detected.

INTRODUCTION

The present methods for measuring the Earth rotation parameters
are based on the determination of the Earth's orientation in
space by terrestrial observation of the position of celestial
objects. The geometric relations appearing in the reduction of
these observations depend, indeed, on the orientation (given
through precession, nutation, polar motion and UT1) of a terrestrial frame denoted (x,y,z), in which the observing stations lie,
with respect to the non-rotating one, denoted (X,Y,Z), in which
celestial objects are fixed or have well known motions. This is
true for classical optical astrometry and satellite tracking (by
Doppler or laser measurements), lunar laser ranging and interferometry on radio sources. All these kinds of measurements are
now used to determine the Earth rotation parameters (BIH, 1980).

Each method, using one kind of measurement, has its particular
problems in realizing these conventional reference frames. The
differences in their realization can give rise to fictitious
differences between the Earth rotation parameters computed by the
different methods.

E. M. Gaposchkin and B. Kołaczek (eds.), Reference Coordinate Systems for Earth Dynamics, 135–144.

The aim of this paper is, in Part 1, to investigate all the defects in realization of the above reference frames by each method
and, in Part 2, to perform a numerical estimation of these defects by a comparison between the available series of Earth rotation parameters.

1. EXPECTED DIFFERENCES DUE TO THE REALIZATION OF THE REFERENCE
    FRAMES

The theoretical ideal frames (x,y,z) and (X,Y,Z) used for computing Earth rotation parameters are both rectangular Cartesian
frames centered at the Earth's center of mass G and such that the
z and x axes are respectively close to the mean position of the
Earth's rotation axis and towards the conventional origin of
longitude. The Z and X axes are respectively the Earth's rotation axis and an arbitrary line of the true equatorial plane.
Each type of measurement used in the computation leads to a non-
ideal realization of these frames that will be analyzed now.

1.1 Optical Astrometry

The observations give the components of the local vertical in the
celestial reference frame adopted as (X,Y,Z) system. The Earth
rotation parameters are derived from the observations by a net of
stations and the terrestrial reference frame is then realized by
a set of conventionally adopted astronomical latitudes and longitudes of the observing stations.

Because of the non-coincident directions of the vertical and the
normal to the geodetic ellipsoid, the center of this terrestrial
frame cannot be the Earth's center of mass.

Because of plate motion, local deformations and tidal effects,
the astronomical coordinates of the stations are not constant.
In order to avoid rotation of the terrestrial frame, the number
of observing stations must be sufficient and their adopted coordinates must be regularly updated for tectonic motions and
corrected for tidal periodic variations.

The non-rotating reference frame is given by the conventional
positions and proper motions of the observed stars in a stellar
catalog (which, due to observational constraints, is not always
the most precise and accurate one) and by a model for the pre-
cession-nutation rotations. The celestial reference frame so
realized inevitably reflects the errors in the star coordinates
and proper motions as well as in the representation of the pre-
cession-nutation. The minimization of these perturbations can
be obtained by some applied corrections, which are described by
Feissel (1979).

1.2 Satellite Tracking

The observations give the distance or line of sight velocity of
the satellite with respect to a set of terrestrial stations and
are used to compute an Earth-based orbit and the Earth rotation
parameters. The terrestrial reference frame is then realized by
a set of geodetic station coordinates. The variation of these
coordinates due to global or local Earth's crustal motions is
not actually taken into account in the Doppler and laser reductions, except for the tidal effect (Anderle, et al., 1975, Smith,
1980) and that can be a source of deformation of the (x,y,z)
frame.

The effect of inaccuracies of the station coordinates is minimized by a high density and an uniform distribution of the observing stations. This terrestrial frame is centered at G and
is referred neither to CIO nor to the conventional longitude.
The bias so obtained in the Earth rotation parameters derived
by this method have been minimized in the case of the Doppler
results by a fitting to the BIH data at an initial date (Anderle,
et al., 1975).

The non-rotating reference frame is the one in which the satellite orbit is computed from the observations using a theoretical
model of forces and would be the true equatorial system at the
date of the beginning of the observations. It is practically obtained after a few iterations and is very dependent on the imper-
fections of the model of forces (errors in the representation of
the Earth's potential, atmospheric drag, solar radiation forces,
no consideration of the oceanic and atmospheric tidal forces) and
to a lesser extent on the precession and nutation representation.
This is responsible for linear and periodic errors and possible
discontinuities in the Earth rotation parameters derived by this
method which are specific to a given satellite, and of the used
theoretical model of forces.

1.3 Lunar Laser Ranging

Each observation gives a time of aberration corresponding to the
distance between the reflector on the Moon and the terrestrial
observing station. A lunar ephemeris in the celestial reference
frame and the corresponding observational residuals are then
computed from such observations, using a theoretical model of
forces and of lunar rotation, and nominal values for the Earth's
orientation in space; these residuals can then be used for computing corrections to some parameters, such as the geocentric
coordinates of the station in an Earth-fixed reference frame and
UT1-UTC (Calame, 1980).

When several stations are operating, the terrestrial reference
frame is realized by a set of geodetic station coordinates of
great accuracy. Prior to 1978, UT1 was derived from the observations of a single station (Calame, 1980), thus some local effects
can perturb the results.

The non-rotating reference frame is the one in which the ephemeris of the lunar reflector is computed and would theoretically
be the true equatorial system of the date of the observation. It
is practically realized through the theoretical representation of
the lunar orbital and rotational motions and thus reflects their
errors. This is responsible for linear and periodic errors.

1.4   Radio Interferometry

The measurement consists of a phase difference between signals
from a radio source when received by two terrestrial stations.
These phase differences are temporal functions of the baseline
and source parameters varying with the Earth's orientation in
space. The Earth rotation parameters can be deduced from such
measurements of one or several baselines of observing stations by
two different observational techniques which are respectively
connected element interferometry (McCarthy, et al., 1980) and
very long base interferometry (Fanselow, et al., 1979).

The terrestrial reference frame is realized by the very accurate
coordinates of radio interferometric stations as computed from
these observations at an initial date and referred to the BIH
origins of pole and longitude using the BIH Earth parameters at
this date (BIH A.R. for 1979). The accuracy of this terrestrial
frame can be perturbed by some local effects in the case of a
single or too short baseline (as in the case of a single connected interferometer).

The non-rotating reference frame is realized by the coordinates
of the observed radio sources in a catalogue which is very accurate in the case of a global determination but can be deterio-
rated by local effects in the case of one baseline determination.

2.   OBSERVATIONAL EVIDENCE

The data analysed are those present in the BIH files, and published in the Annual Report of the BIH for 1979 (the pages are
indicated in brackets), except for the results of IPMS, taken
from Yumi (1980), p. 119-123.

Optical astrometry: Two computing centers, same observations.
       IPMS: x, y (smoothed values at 0.05 y interval), and UT1
              (monthly means)
       BIH (AST): x, y, UT1 from astrometry only, at 5-day and
                   0.05 y intervals (pp. B-17, D-3).

<u>Satellite Doppler tracking</u>: Three computing centers, one common
                                satellite, some common stations.
    DMA: x, y at one-day interval (p. D-9)
    MEDOC: x, y at 2-day interval (p. D-27)
    NSWC: UT1 at one-day interval (p. D-11)

<u>Lunar laser ranging</u>: One computing center, two lunar ephemerides,
                          same observations.
    EROLD: UT1 at irregular interval, using the JPL ephemeris,
        DE 86, or the GERGA-Texas Ephemeris, ECT 18,
        (p. D-35)

<u>Satellite laser ranging</u>: Two computing centers, same observations.
    GSFC: x, y, l.o.d. at 5-day interval (p. D-47)
    IASOM: x, y, l.o.d. at 5-day interval (Fanselow, 1980)

<u>Very long base interferometry (VLBI)</u>: One computing center, one
                                          network.
    DSN: x, y, UT1 at irregular interval (p. D-75)

<u>Connected interferometry (CERI)</u>: One computing center, one
                                      interferometer.
    GBI: UTO, ϕ at irregular interval (p. D-67)

2.1 Long Term

The relative drifts are given in Table 1.

Table 1. Drifts relative to BIH (AST). (Units: 0$''$001 for x,y;
          0^{s}001 for UT1)

Series		Years	x	y	UT1
IPMS		67–78	−0.1 ± 0.2	+0.1 ± 0.3	−1.7 ± 0.4
DMA		72–79	−2.0 ± 0.4	+1.9 ± 0.5	
EROLD	DE 86	71–78			−2.5 ± 0.7
EROLD	ECT 18	71–79			−4.0 ± 0.7
DSN		71–78			+0.6 ± 0.7

The drift of IPMS is due to the implementation of different algorithms for the long term stability by the two services. The
drift of LLR is partly due to the long term errors on the right
ascension of the moon.

2.2 Intermediate (6 c/y to 1 c/y)

The dominant features in optical astrometry and CERI are annual
(and semi-annual) terms. Table 2 gives the amplitude of these
terms for several series compared to BIH (AST). BIH (AST) is
expressed in the 1979 BIH System, which annual terms have been
calibrated by DMA for polar motion and EROLD (DE 86) for UT1.

Table 2. Annual and semi-annual terms relative to BIH (AST)
(Units: $0\overset{''}{.}001$ for x, y; $0\overset{s}{.}001$ for UT1). The differences are expressed as b sin2πt + c cos2πt + d sin4πt +
e cos4πt, t in years.

Series	Years	x b c d e	y b c d e	UT1 b c d e
IPMS	67–78*	+5 0 0 −2	+13 +2 +1 −3	+1 +5 −5 +5
DSN	71–78			+12 −7 −12 +8
GBI	78–79	(Latitude:	+38 −80 0 +6)	+14 +67 +10 +10

*67–77 for UT1

UT1 (NSWC) is subject to large periodic errors in this domain due
to the neglect of oceanic and atmospheric tides (Anderle, 1980).
EROLD (ECT 18) shows no significant difference with EROLD (DE 86)
at these frequencies. The analysis of GSFC and IASOM shows some
signal in this frequency domain. This might be due to some imperfection in the modelling of non-gravitational forces in DMA, or it
could be an effect of the changes in the effective network in SLR.
The annual term of 2 ms amplitude in UT1 (DSN) is obtained from
a small number of observations and needs further confirmation.

2.3 Short Term (periods under 60 days)

The data at 5-day and 2-day intervals have been analysed, after
removing strong smoothing by the Vondrak's method (see Figure 1).

The spectral analysis of BIH (AST), GSFC, and IASOM show only
noise in this domain. The Doppler method (DMA, MEDOC) shows
periodic terms, due to inaccuracies in the resonance terms of the
force field model, around 12d and 6d. The adjustment of the
parameters of these terms is somewhat hazardous, as they are not
high above the noise, and also they vary slightly with time.
Figure 2 shows an example of the spectra obtained. Such terms
are not present when using the Lageos satellite.

Universal time. Prior to the analyses, the series have been
corrected for the variation of UT due to the zonal Earth tides

with periods shorter than 32d. BIH (AST) has only noise in this
domain. NSWC has a perturbing term at 13,6d due to the neglect
of ocean tides. EROLD (ECT 18) and DE (86) have some perturbing
terms due to the ephemerides used and to indirect effects of the
interruption of observations at new moon. The amplitudes of
these terms are 1 to 2 ms for both ephemerides.

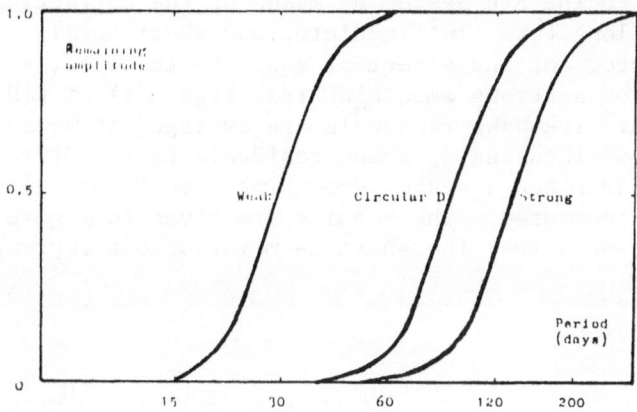

Figure 1. Filters corresponding to the smoothings
           used in the study.

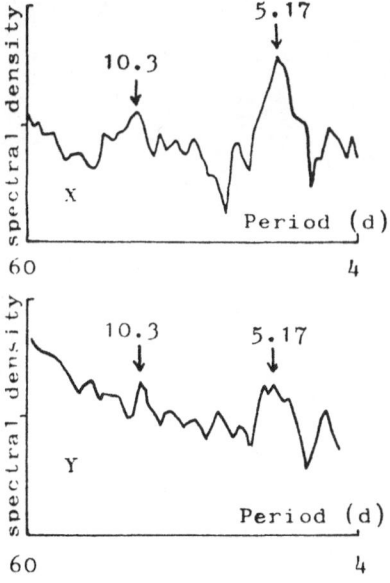

Figure 2. Doppler method for
           polar motion.
           Resonance terms.

2.4 Comparison of corrected results. An example.

In order to show the improvement that can be obtained by correcting the original data with the systematic terms listed in this
study, the short term variations of UT have been evaluated from
May to December, 1979, from the independent sets of results
available: BIH (AST), NSWC, EROLD (ECT 18), and GBI. The latter
three are brought to the BIH System by means of the corrections
determined above (long term, intermediate, and short term), all
results are corrected for the effect of zonal Earth tides, and
their residuals from a strong smoothing (see Figure 1) of BIH
(AST) are computed. The NSWC residuals are averaged at 5-day
intervals. The rms distances of these residuals to two different
smoothings (Circular D and a weaker smoothing, see Figure 1) of
BIH (AST) are then computed. The results are given in Figure 3
and Table 3. They show that the short term variations represented

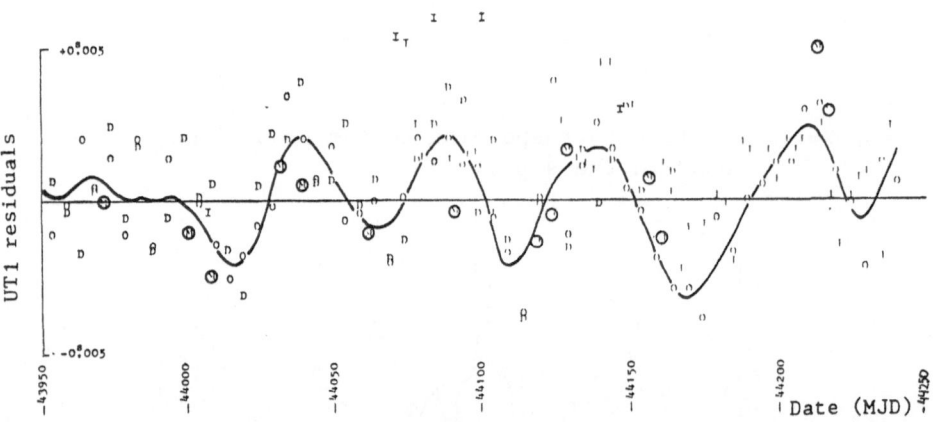

Figure 3:   UT from May through December, 1979. Independent
            determinations (O: BIH (AST), D: NSWC, (M): EROLD
            (ECT 18), I: GBI) and weak smoothing of BIH (AST)
            (solid line). The reference is a strong smoothing
            of BIH (AST).

Table 3.   Independent measurements of UT1, May–December, 1979:
rms distances to two different smoothings (units:
0^s.0001).

	rms distance to smoothings				Correlation coeficient of distances to smoothings	
	BIH (AST)	NSWC	EROLD (ECT 18)	GBI	BIH/NSWC	BIH/GBI
Circ. D	16	17	14	23	0.52	0.52
weak sm.	11	15	11	21	0.03	0.02
Nb of values	58	43	15	32	43	32
st. error	8	8	11	11		

by the weak smoothing are real. Such temporary perturbations in
the Earth's rotation were already suspected, in connection with
motions in the atmosphere, but as long as the optical astrometry
was the only set of available data, no evidence of their reality
was possible. The existence of such variations is probably responsible for the difficulties in determining the permanent
short term variations (zonal Earth tides) and in detecting the
abrupt changes in the length of day (Guinot, 1970). Another conclusion of Table 3 is that the standard error of BIH (AST), published yearly in the BIH Annual Report, Table 6B, is a good
estimate of the precision of the series.

CONCLUSION

There is not actually one method able to provide at the same time
an accurate geodetic network linked to the Earth, a perfect realization of the non-rotating celestial reference frame, and a continuous monitoring of the Earth's rotation. Each method has its
own strengths and weaknesses. A calibration of the different
methods by one another in order to express them in a common
accurate system can be obtained by evaluating and correcting the
effects of the perturbations present in each method.

REFERENCES

Anderle, R. J. and Tanenbaum, M. C.: 1975, On Reference Coor-
    dinates Systems for Earth Dynamics, p. 341–380, ed. B.
    Kolaczek, Varsaw.
Anderle, R. J.: 1980, Spring Americ. Geoph. Un. Toronto, Canada,
    22–27 May, 1980.
BIH Annual Report for 1979, 1980, Paris.
Calame, O.: 1980, BIH Annual Report for 1979, p. D35–D38.
Fanselow, J. L., et al.: 1979, in Time and the Earth's Rotation,
    p. 199–210, ed. D. D. McCarthy and J. D. H. Pilkington, Reidel
    Publishing Company.
Fanselow, J. L.: 1980, Private Communication.
Feissel, M.: 1979, in Time and the Earth's Rotation, p. 109–114.

Guinot, B.: 1970, Astron. Astrophys. 8, p. 26.
McCarthy, et al.: 1980, BIH Annual Report for 1979, p. D67-D70.
Smith, D. E.: 1980, BIH Annual Report for 1979, p. D47-D48.
Yumi, S.: 1980, Ann. Rep. of the IPMS for 1978, Mizusawa, Japan.

ON THE ADOPTION OF A TERRESTRIAL REFERENCE FRAME

Dennis D. McCarthy
U. S. Naval Observatory
Washington, D.C. 20390

ABSTRACT. The report of the IAU Working Group on Nutation endorsed by
Commissions 4, 8, 19 and 31 at the 1979 General Assembly points out
that "... the complete theory of the general nutational motion of the
Earth about its center of mass may be described by the sum of two components, astronomical nutation, commonly referred to as nutation,
which is nutation with respect to a space-fixed coordinate system, and
polar motion, which is nutation with respect to a body-fixed system
...". Unlike the situation for the space-fixed frame, there is not an
adequate, formally accepted, body-fixed system for this purpose. The
Conventional International Origin (CIO) as it is presently defined is
no longer acceptable because of recent improvements in observational
techniques. The effective lack of this type of terrestrial reference
frame limits the complete description of the general nutational motion
of the Earth. In the absence of a terrestrial reference frame suitable
for specifying the orientation of the Earth, it is suggested that a
body-fixed system could be represented formally in a manner analogous
to that used to represent the space-fixed frame. This procedure would
be quite similar to methods employed currently by the International
Polar Motion Service and the Bureau International de l'Heure, and would
allow for the use of observations from new techniques in the definition
of a terrestrial reference frame to be used to specify the complete
nutational motion of the Earth.

1. INTRODUCTION

The theories of precession and nutation along with the observed polar
motion and rotation angle of the Earth are required to describe the
orientation of users' local, body-fixed reference directions in the
space-fixed frame. These directions are reproducible reference
vectors used to make observations. Examples include local plumb
lines, baseline vectors of radio interferometers, vectors directed
from the Earth's center of mass to the observer's location, etc. The
mathematical models and observations seek to describe the orientation
of a terrestrial frame with respect to the space-fixed frame in such a

145

E. M. Gaposchkin and B. Kołaczek (eds.), Reference Coordinate Systems for Earth Dynamics, 145–153.
Copyright © 1981 by D. Reidel Publishing Company.

way that an observer may account for the changing orientation of the
local reference direction.

Commissions 4, 8, 19 and 31 of the International Astronomical Union
have recently specified the theory of nutation to be used to model the
motion of the Celestial Reference Pole with respect to a space-fixed
reference frame (IAU Commission 4 Report; Kinoshita, et al, 1979). The
space-fixed frame is realized in practice by the adopted precession
along with the positions and proper motions of a set of fundamental
stars (FK4 or FK5) or by positions and proper motions of stars which
can be considered to be "in the fundamental system" (Fricke, 1975).
Other realizations of a space-fixed system are considered to be a set
of positions of distant radio sources or the dynamic reference frame
defined by the ephemerides of solar system objects or Earth satellites
(Kovalevsky, 1979). Important considerations regarding the rigorous
definition of a space-fixed or non-rotating system remain to be discussed. However, for practical purposes we may consider the directions
to the stars of the FK4 or its successor, FK5, to be the realization of
the space-fixed system implied in the theories of precession and nutation.

To complete the description of the Earth's orientation to meet users'
needs, then, we require a terrestrial system which can be related to
the space-fixed frame through observations using the theories of precession and nutation. We assume that the individual reference vectors may be related to a global reference frame. If the orientation
of this frame is determined, the observer can account for the changing
orientation of the local reference. Traditionally, the Earth-
fixed vectors are specified by the adoption of numerical constants
at some epoch and are generally related to a terrestrial coordinate
system defined by the Conventional International Origin (CIO) and a
longitude reference (e.g., Bureau International de l'Heure zero meridian). In view of the advent of new, more precise observational techniques and the need for higher accuracy, it is important to re-examine
the concept of the terrestrial reference system, particularly in conjunction with the IAU precession and nutation theories to be adopted
in 1984.

2. OBSERVATIONAL PRACTICE

Presently we assume that the reference directions are "fixed to the
Earth" and that these directions do not change in time with respect
to some Earth-fixed reference frame. These directions are defined
at a reference epoch, T_0, in the space-fixed system. This is done by
making astronomical observations of the reference direction at an
epoch, T, and rotating the observed direction vector through assumed
rotation angles to account for the change in orientation of the Earth
in the space-fixed system from T_0 to T. The orientation at T with respect to T_0 is determined using the adopted positions and proper motions of the fundamental stars, the adopted precession and nutation
theories, and an assumed knowledge of the polar motion and the change

in rotational speed from T to T_O. Since the reference directions are
fixed to the Earth, observations of the changes of the direction of
the vectors in the space-fixed system as a function of time may be used
to determine the changing orientation of the terrestrial reference system with respect to the space-fixed system. In this process the Celestial Ephemeris Pole or the rotational pole in Woolard's (1953) nutation
serves as an intermediate reference direction. Note that the Celestial
Ephemeris Pole is not Woolard's rotational pole.

Because classical astro-geodetic instruments have been used to monitor
the orientation of the Earth, the reference directions until now have
generally been specified by the astronomical latitudes and longitudes
of observing sites. This has been done to make use of the plumb line,
or gravity vector, as the local reference direction. Astronomical coordinates, however, in general do not specify the location of a site
with respect to a particular origin, but they do provide an orientation
reference useful to measure changes in the Earth's orientation in space.
To determine the changing orientation of the Earth a set of coordinates
specifying station location with respect to an origin (e.g., Earth's
center of mass) is not required. Only the specification of local reference directions is needed (Fedorov, 1979; Grafarend, et al, 1979).

2.1. DEFINITION OF THE CIO

To define the equator of the currently adopted terrestrial system, the
astronomical latitudes of the five International Latitude Service (ILS)
visual zenith telescopes are specified. Since the astronomical latitude defines the angle between the local vertical and the equatorial
plane, the equator is defined and the CIO is the direction perpendicular to that plane (Fig. 1). Note that this does not define a unique
pole on the surface of the Earth (Fedorov, 1979), but rather a direction perpendicular to the plane of the equator. The definition of this
equator (and the direction of the CIO) contains the effects of any
errors in the adopted astronomical latitudes of the rather limited set
of the five ILS instrumental reference vectors. This definition is
extended informally through the adoption of observatory "coordinates"
for a number of observing locations. These coordinates are related
to the CIO by the IPMS and the BIH.

2.2. DEFINITION OF THE ZERO MERIDIAN

Analogously, we define a reference direction lying in the plane of the
equator by the adopted astronomical longitudes of a number of stations
which observe the times of transit of stars (Fig. 2). The stars have
an assumed angular relationship to the fiducial point defining UT1
through their adopted positions and proper motions. The procedure by
which the longitudes are defined is based on the combination of observations from a large number of instruments which is done in an effort
to eliminate systematic errors. Operationally this function is performed by the BIH. This procedure, while adequate to determine UT1-
UTC until the present time, is not clearly defined in terms of a for-

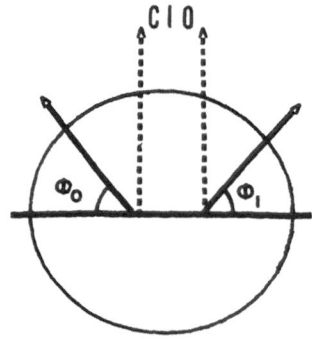

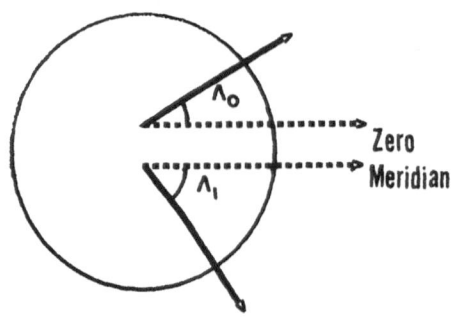

Fig. 1. Relationship between astronomical latitudes and the CIO.

Fig. 2. Relationship between astronomical longitude and the zero
meridian.

mally adopted terrestrial reference frame.

2.3. NEED FOR IMPROVEMENT

Recently, new techniques and improvements in classical techniques have
made it possible to determine Earth orientation parameters with unprecedented precision. At the same time new demands have been made on
the accuracy required by users of Earth orientation information. The
precision already achieved and the accuracy demanded exceeds the precision with which the present formal realization of a terrestrial system is defined. Because of the age of the instruments, the ILS visual
zenith telescopes may no longer be capable of providing polar motion
data. In addition, the present definition does not formally incorporate the possibility of including new techniques in the definition of
the terrestrial frame. Therefore, it is important to reconsider the
realization of the terrestrial frame in order to make use of the precision currently available and to produce a clearly defined system for
users of Earth orientation information that may continue to be useful
in the future. Any useful reference system must also be easily accessible operationally.

3. SUGGESTED IMPROVEMENT

A possible improvement is to define a Conventional Terrestial System by
a set of reference directions similar to the way the Fundamental Star
System is defined by the directions to fundamental stars. This is
quite similar to what is done now in practice by the BIH and the IPMS
through the adoption of the astronomical coordinates of contributing
optical observatories. However, as we have seen above, the adopted
astronomical coordinates are suitable only for those techniques which
utilize the gravity vector for the local reference direction. The
present situation could be improved and generalized by the adoption
of reference vectors suitable for each technique and by including a
larger number in the formal definition of the system. This would be

analogous to the Fundamental Star System where the directions to a
number of stars are specified in order to define it. The direction
perpendicular to the Conventional Earth System equator could continue
to be called the CIO. The adopted reference vectors can be chosen so
that there would be no discontinuity in the polar coordinates and
UT1-UTC. This procedure would allow the formal definition of a terrestrial system to incorporate a larger number of reference directions
with improved precision and accuracy. It would also involve continued
dedicated observations of the defining reference directions.

3.1. IMPLEMENTATION

Consider a body-fixed reference frame with unit base vectors E_1, E_2
E_3 defined by the direction of reference vectors observed in the
space-fixed frame at an epoch, T_0. These reference vectors include
plumb lines, interferometer baselines, vectors directed from the center of mass of the Earth to the observer, etc. Assume that the reference frame is rotating about E_3 with an angular speed of Ω, and that
E_3 is the direction of the pole implied in the theory of precession
and astronomical nutation. E_1 and E_2 are orthogonal to E_3 and to each
other (Fig. 3).

Assuming that we know Ω and the theory of nutation and precession, we
can predict the orientation of (E_1, E_2, E_3) as a function of time
$[E_1(T), E_2(T), E_3(T)]$ in the space-fixed system. However, if the observed effects of the Earth's variable rotation rate and polar motion
are not accounted for, this will not accurately specify the orientation of the "Earth-fixed" reference vector, R, in the space-fixed
frame. To do this we define the Terrestrial Reference Frame (unit
base vectors e_1, e_2, e_3) to be identical with (E_1, E_2, E_3) at T_0.

This system is defined in reality by the definition of the reference
vectors,

$$R(T_0) = R_1 e_1(T_0) + R_2 e_2(T_0) + R_3 e_3(T_0) = R_1 E_1(T_0) + R_2 E_2(T_0) + R_3 E_3(T_0).$$

We assume that the (e_1, e_2, e_3) system is rotating with angular speed
$\omega(T) = \Omega + \delta\omega(T)$, $\delta\omega(T)/\Omega$ being small. Let us also assume that polar
motion is represented by a counterclockwise rotation about E_2 by an
angle $x(T)$ and a counterclockwise rotation about E_1 by an angle $y(T)$.
Also let

$$\theta = \int_{T_0}^{T} \delta\omega(t) \, dt$$

represent the counterclockwise rotation of (e_1, e_2, e_3) about E_3.
Then :

$$e_1(T) = E_1(T) + \theta(T) E_2(T) + x(T) E_3(T),$$
$$e_2(T) = E_2(T) - \theta(T) E_1(T) - y(T) E_3(T),$$
$$e_3(T) = E_3(T) - x(T) E_1(T) + y(T) E_2(T).$$

At an epoch, T, we can repeat the observations of R in the space-fixed
frame and obtain

$$R(T) = [R_1 + \delta R_1(T)]\ E_1(T) + [R_2 +\ \delta R_2(T)]\ E_2(T) + [R_3$$
$$+ \delta R_3(T)]\ E_3(T).$$

We also have the expression:

$$R(T) = [R_1 + m_1(T)]\ e_1(T) + [R_2 + m_2(T)]\ e_2(T) + [R_3$$
$$+ m_3(T)]\ e_3(T)$$

where the terms m_1, m_2, m_3 represent possible effects of the relative
motion of the reference vectors. For a simple rigid-Earth model such
as is used currently, these terms are zero. Combination of the above
expressions leads to:

$$\begin{pmatrix} \delta R_1(T) \\ \delta R_2(T) \\ \delta R_3(T) \end{pmatrix} = \begin{pmatrix} m_1(T) \\ m_2(T) \\ m_3(T) \end{pmatrix} + \begin{pmatrix} -R_3 & 0 & -R_2 \\ 0 & R_3 & R_1 \\ R_1 & -R_2 & 0 \end{pmatrix} \begin{pmatrix} x(T) \\ y(T) \\ \theta(T) \end{pmatrix}$$

The solution for all three components of the Earth orientation parameters using observations from one reference vector is indeterminate.
If more than one suitably conditioned reference vector is available
the solution is possible. It can be shown that the use of the above
matrix expression leads to the expression for the variation in interferometer baseline vector components as a function of the Earth orientation (McCarthy, et al, 1979) as well as the familiar expressions
for the classical "variation of latitude" and UT1-UT0 which depend
on the orientation of the Earth.

The definition of the reference vectors can be accomplished in an internal adjustment to reduce the internal errors of a solution involving many possible observing instruments. The solution could be constrained to have no systematic deviation from the past IPMS and BIH
solutions.

In practice, we use a local Earth-fixed reference frame at the location of the observer (longitude λ and latitude ϕ). Let this frame
with unit base vectors j_1, j_2, j_3 be oriented so that j_3 is parallel
to e_3, j_1 is in the plane of the local meridian, and j_2 is oriented
to the west. Note that this is a left-handed system. If the components of the reference vector in this system at T_0 are r_1, r_2, r_3, and
if a number of such reference directions are to be used to determine
Earth orientation parameters, the system may be solved using a least-
-squares solution:

$$\begin{bmatrix} a_1 & a_4 & a_7 \\ a_2 & a_5 & a_8 \\ a_3 & a_6 & a_9 \end{bmatrix} \begin{bmatrix} x \\ y \\ UT1-UTC \end{bmatrix} = \begin{bmatrix} c_1 \\ c_2 \\ c_3 \end{bmatrix},$$

$a_1 = \Sigma r_3^2 + \Sigma r_1^2 \cos^2\lambda + \Sigma r_2^2 \sin^2\lambda - 2\Sigma r_1 r_2 \sin\lambda \cos\lambda,$

$a_2 = a_4 = \Sigma r_1^2 \sin\lambda \cos\lambda + \Sigma r_1 r_2 \cos^2\lambda - \Sigma r_1 r_2 \sin^2\lambda - \Sigma r_2^2 \sin\lambda \cos\lambda,$

$a_3 = a_7 = -\Sigma r_2 r_3 \cos\lambda - \Sigma r_1 r_3 \sin\lambda,$

$a_5 = \Sigma r_3^2 + \Sigma r_1^2 \sin^2\lambda + \Sigma r_2^2 \cos^2\lambda + 2\Sigma r_1 r_2 \sin\lambda \cos\lambda,$

$a_6 = a_8 = \Sigma r_3 r_1 \cos\lambda - \Sigma r_2 r_3 \sin\lambda,$

$a_9 = \Sigma r_1^2 + \Sigma r_2^2,$

$c_1 = \Sigma \delta r_1 r_3 \cos\lambda - \Sigma \delta r_2 r_3 \sin\lambda - \Sigma \delta r_3 r_1 \cos\lambda + \Sigma \delta r_3 r_2 \sin\lambda,$

$c_2 = \Sigma \delta r_1 r_3 \sin\lambda + \Sigma \delta r_2 r_3 \cos\lambda - \Sigma \delta r_3 r_1 \sin\lambda - \Sigma \delta r_3 r_2 \sin\lambda,$

$c_3 = \Sigma \delta r_2 r_1 - \Sigma \delta r_1 r_2.$

Preliminary results found from the application of this procedure to
data from two photographic zenith tubes, Doppler satellite data and
connected element interferometer observations show that the daily
values of polar coordinates can be determined with an accuracy of
±0".012 while daily values of UT1-UTC can be obtained with an accuracy
of ±0$.$0016.

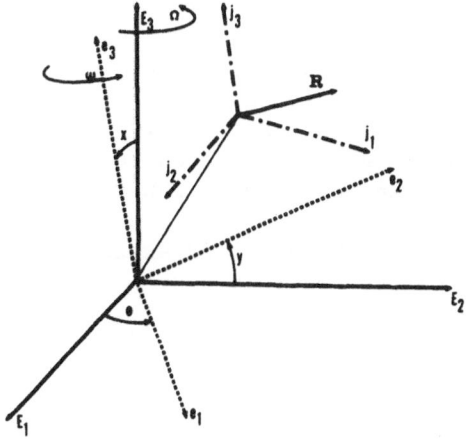

Fig. 3. Relationship among terrestrial reference frames.

3.2. RELATIONSHIP TO GEODETIC SYSTEMS

For the purpose of determining the changing orientation of the Earth
it is not strictly necessary to base a terrestrial system on a geodetic system of locations. The Conventional Earth System outlined
above can be related to Earth-centered geodetic systems if the location of the center of mass of the Earth with respect to a sufficient number of observing stations were determined. This is not necessary to describe the motion of the Earth about its center of mass,
however. The Conventional Earth System is not a geodetic system just
as the CIO and the BIH zero meridian do not presently form a geodetic
reference system suitable for station location. It does not provide
geodetic positions but directions to be used only to define the orientation of the Earth and Earth-centered geodetic systems.

3.3. SYSTEMATIC ERRORS IN $R(T_o)$

Since observations of $\delta r_i(T)$ are dependent on local conditions and observing procedures, most current observations contain systematic differences among themselves. The BIH and the IPMS allow for these errors
by applying empirical corrections to contributed data. In the development of a Conventional Earth System the systematic differences could
be evaluated to provide the best estimates of the $R_1(T_o)$ available at
that time. Systematic errors, both periodic and aperiodic, in the observations will continue to require the application of error models to
the observations. These, however, should be treated as corrections to
the observations and not to the $R(T_o)$. Periodically, estimates of
the systematic differences should be evaluated to determine improvements to the individual $R(T_o)$ in order to refine the definition and
maintain the system just as is done with the Fundamental Star System.
This may necessitate the introduction of models which describe the
relative motion of the reference directions to allow for true geophysical motion (i.e. station proper motions).

4. CONCLUSION

In view of the need for an improved terrestrial reference system suitable for use in determining the orientation of the Earth with respect
to a space-fixed frame, it appears that the presently adopted formal
definition of the CIO and the generally accepted BIH zero meridian are
not adequate. It is suggested that a more general definition capable
of incorporating observations from a number of techniques be formally
accepted. The Conventional Earth System presented here appears to meet
these needs and experimental use of this system shows it is a viable
solution to this problem.

REFERENCES

Fedorov, E. P.: 1979, "On the Coordinate Systems Used in the Study of Polar Motion", in Time and the Earth's Rotation, D. McCarthy and J. D. H. Pilkington, eds., D. Reidel Pub. Co., Dordrecht, pp. 89–101.

Fricke, W.: 1975, "Definition of the Celestial Reference Coordinate System in Fundamental Catalogs", in On Reference Coordinate Systems for Earth Dynamics, B. Kolaczek and G. Weiffenbach, eds., Warsaw Technical University, Warsaw, pp. 201–222.

Grafarend, E. W., Mueller, I. I., Papo, H. G., Richter, B.: 1979 "Concepts for Reference Frames in Geodesy and Geodynamics: The Reference Directions", Bull. Geodesique, 53, pp. 195–213.

IAU Commission 4 Report: 1980, Transactions of IAU, 17B, in press.

Kinoshita, H., Nakajima, K., Kubo, Y., Nakagawa, I., Sasao, T., and Yokoyama, K.: 1979, "Note on Nutation in Ephemerides", Pub. Int. Latitude Obs. Mizusawa, XII, pp. 71–108.

Kovalevsky, J.: 1979, "The Reference Systems" in Time and the Earth's Rotation, D. McCarthy and J. D. H. Pilkington, eds., D. Reidel Pub. Co., Dordrecht, pp. 151–163.

McCarthy, D. D., Klepczynski, W. J., Kaplan, G. H., Josties, F. J., Branham, R. L., Westerhout, G., Johnston, K. J., and Spencer, J. H., "Variation of Earth Orientation Parameters from Changes in the Orientation of the 35-km Baseline of the Green Bank Interferometer", in BIH Annual Report for 1979.

Woolard, E. W.: 1953, Astron. Papers Prepared for the Use of the American Ephemeris and Nautical Almanac, XV, part 1.

ON THE ESTABLISHMENT OF TERRESTRIAL COORDINATE SYSTEM BY CLASSICAL TECHNIQUES

Ya. S. Yatskiv

Main Astronomical Observatory,
Ukrainian Academy of Sciences,
Kiev, U.S.S.R.

ABSTRACT

The definition and practical realization of the astronomical system of terrestrial coordinates are discussed. In particular, the separation of secular polar motion and nonpolar drift of observatories are touched upon as far as those phenomena are crucially intertwined with the realization of a terrestrial coordinate system.

1. INTRODUCTION

Many different definitions and practical realizations of the terrestrial coordinate systems were discussed in the First Colloquium "On Reference Coordinate System for Earth Dynamics" held in Torun in 1974. To avoid repetition, this paper will be limited to a brief summary of the current state of establishing the astronomical system of terrestrial coordinates by classical techniques. The term "classical techniques" is used to refer to optical observations of stars by means of zenith telescopes, photographic zenith tubes, etc. The separation of the secular polar motion and the nonpolar drift of observatories; and the method for maintaining the conventional coordinate system, will also be discussed.

2. ON THE DEFINITION OF THE ASTRONOMICAL SYSTEM OF TERRESTRIAL COORDINATES

All classical techniques measure the orientation of reference directions on the Earth, i.e., plumb lines of observatories or directions to visible reference marks, with respect to the directions from the observer to stars, assuming the latter are known in a nonrotating celestial coordinate system, $(X) = (X_1, X_2, X_3)^T$.

E. M. Gaposchkin and B. Kołaczek (eds.), Reference Coordinate Systems for Earth Dynamics, 155–164.
Copyright © 1981 by D. Reidel Publishing Company.

Since distances are not measured these techniques enable only the
directions of the axes of a terrestrial coordinate system to be
determined.

For geometrical representation of the results by classical
techniques one takes the auxiliary unit sphere and draws from an
arbitrary centre the unit vectors, e_i, parallel to the plumb lines
of observatories. A vector e_i defines the position of the zenith,
Z_i, on the sphere. The usual procedure for expressing position
on the sphere is a specification of the directions e_i in terms of
arcs on the sphere that measure angular distances from selected
cardinal directions or circles. The astronomical latitude ϕ fixes
the position of a point in the local meridian plane, which passes
through the direction e_i and is parallel to the axis of the
Earth's rotation. ϕ is reckoned from the astronomical equator.
The position of the plane itself is specified by the angle that
it makes with the meridian plane through an arbitrarily chosen
reference point on the sphere. According to a resolution of the
Washington Conference in 1884 the meridian plane through the
Airy Transit Circle of the Greenwich Observatory was for many
years adopted as the initial reference plane.

This formal geometrical definition of the astronomical system of
terrestrial coordinates does not depend upon any interpretation
of directly observed phenomena which determine the positions of
the cardinal point and circle on the sphere. Really, the basis
of any system of terrestrial coordinates, $(x) = (x_1, x_2, x_3)^T$, is
essentially empirical. The system (x) is attached to the observed vectors e_i, or to corresponding zenith positions, z_i. The
minimum number of constraints sufficient to define the system
uniquely is three. Therefore, if we have more than two vectors
e_i, this attachment cannot be arbitrarily realized. To tie down
the three rotational degrees of freedom of the system, a certain
condition must be imposed, e.g., the minimization of the squares
of the zenith displacements from their initial positions in the
system (x), (Fedorov et al., 1972). Such a system is called the
conventional terrestrial coordinate system (Fedorov, 1979).
Where high precision of latitude and time determinations is
required it is necessary to specify explicitly the directions of
the third axis of the system (x), and the reference point for
reckoning longitude. According to the resolutions of the 13th
General Assembly of the IAU and the 14th General Assembly of the
IUGG the x_3-axis points to the CIO, which is defined by the conventional values of latitudes, ϕ_{oi}, of five ILS stations. The
longitude is now referred to the so-called "mean observatory"
as defined by the BIH. It was defined at an initial epoch as
the point where the meridian through Greenwich and the CIO cuts
the equator of the CIO. There was no authoritative approval of
such a fiducial point for longitude.

The rotation of the Earth is represented by the motion of the system (x) with respect to the nonrotating celestial system (X). This relative motion is not exactly predictable because of irregularities in the Earth's rotation due to unknown exitations. For observational purposes it is convenient to introduce an intermediate system $(\xi) = (\xi_1, \xi_2, \xi_3)^T$ whose rotation approximates as closely as possible that of the system (x) and at the same time is precisely predictable. Such a system is called the ephemeris terrestrial system (Fedorov, 1979). The selection of the ξ_3-axis of the ephemeris terrestrial system has been discussed by the IAU Working Group on Nutation in connection with the specification of the new set of nutational coefficients (Seidelmann et al., 1979). The proposed ξ_3-axis points to the "Celestial Ephemeris Pole", i.e., to the point that has no nearly diurnal motion with respect to either the nonrotating, (X), or terrestrial, (x), coordinate system. Two other axes rotate about the ξ_3-axis with "ephemeris angular velocity" specified by conventional formulae for the angle between the ξ_3-axis and the first axis of the non-rotating coordinate system, (X). In current practice, the latter is directed to the mean equinox. Recently, Guinot (1979) has proposed a new choice of fiducial point on the equator, the so-called Non-Rotating Reference Origin. The difference between these two reference points is given by the precession in right ascension. Classical astronomical latitude and time observations enable the directions of the axes of the system (ξ) to be determined with respect to the system (x).

3.   REALIZATION OF ASTRONOMICAL SYSTEM OF TERRESTRIAL COORDINATES

The astronomical system of terrestrial coordinates should be adequate for representation of the positions of zeniths on the auxiliary sphere and rotation of the Earth as a whole. The realization of any reference system consists of two stages:

1)   Determination of initial longitudes and latitudes of a number of observatories based on observations over some convenient time interval;

2)   Choice of the procedure for maintaining the system.

Different systems of terrestrial coordinates of particular interest in the study of the Earth's rotation have been realized. Some of them are in widespread current use in astronomy and geodesy:

MPO - The mean pole origin defined by the mean latitudes of observatories at any moment. For determining the mean latitude different methods may be used, such as the well-known Orlov's method.

CIO - The Conventional International Origin.

BIH - The reference system adopted by the BIH and referred
to the epoch of 1968.

BIH 1979 - The same as BIH 1968 except for conventional corrections added to the coordinates of the pole and to UT1.

In addition to above astronomical systems of terrestrial coordinates the following ones are discussed:

ST(SU) 1965 - The reference system adopted by Gubanov and
Yagudin (1979) for redetermining the UT scale
of the USSR for 1955-1974.

UT(SU) 1975 - The new reference system adopted by the USSR
Time Service in 1975 (Belotserkovskij and
Kaufman, 1979).

Some information concerning the observational data and the primary
reference system used for realization of the above reference
systems are given in Table 1.

The BIH 1968 system was realized by requiring coincidence of the
BIH and CIO poles in 1968 as well as continuity of UT1 with
previous values. After the initial adjustment of 1968, the BIH
system was kept independent of the ILS results. On the contrary,
ST(SU) 1965 system and the UT(SU) 1975 system have been adjusted
to that of the BIH only in the average the time intervals of
1962.0 - 1975.0 and 1957.0 - 1971.0, respectively. The merit
of any reference system depends not only upon the accuracy of
classical techniques, but also upon precise specification of
fundamental constants, computational procedures, etc. The comparison of polar coordinates and UT1 obtained in different
systems can yield information on the accuracy of the systems under
consideration.

Table 1.   Astronomical systems of terrestrial coordinates

Name of System	Number of Observa- tories	Number of Instruments		Interval of Observations	Primary Reference System
		Time	Latitude		
MPO	all	−	all	more than 1.6 yr	instrumental
CIO	5	−	5	1900.0-1906.0	instrumental
BIH 1968	51	48	39	1964.0-1967.0 1967.0-1968.0	CIO
ST(SU) 1965	15	29	−	1955.0-1975.0	BIH 1968
UT(SU) 1975	5	8	−	1957.0-1971.0 1974.0-1975.0	BIH 1968

4. SECULAR POLAR MOTION AND NONPOLAR DRIFT OF OBSERVATORIES

The separation of secular polar motion and displacements of individual observatories are crucially intertwined with the realization of a terrestrial coordinate system. The existence of the secular and long-period variations of the mean longitudes and latitudes of observatories has been proved by many authors. On the other hand, many determinations of the motion of mean pole of the epoch of observation with respect to the CIO, the so-called secular polar motion, have been carried out using the data of the five ILS stations. The extensive discussion on this matter has led to the conclusion, that at present there is no way to resolve the problem of how much is secular polar motion and how much is due to other secular effects providing that the number of stations is limited, for example, to five. However, different analyses were exersised to show qualitatively that the secular trend of the CIO relative to the mean pole, CIO-MPO, was, in the most part, due to the local nonpolar effects of the ILS stations. If the secular term of polar motion does not exist, we should clear up whether there is some long-period motion of the pole. Yatskiv et al. (1979) showed that the long-period components in polar motion which Markowitz (1970) has claimed to reveal are due to local, nonpolar variations of the latitudes of some stations, in particular, the Ukiah station. If the number of participating stations is large enough, as compared with the ILS, the effect of nonpolar drift of some stations on the relative motion of the reference systems would be balanced out. As an example, consider the variation of differences between the yearly mean values of coordinates of pole given by the ILS and the BIH for the period 1962-1979 (BIH Annual Report for 1979). These differences are compared with the variation of the mean latitude of Ukiah ($\overline{\psi}_u^x = -0.2583\psi_u$, $\overline{\psi}_u^y = +0.2559\psi_u$) as well as with the variation of the angle (S_{UM}) between the verticals of Ukiah and mean observatory (Pulkovo, Kazań, Poltava, Kitab). The values of -0.2583 and +0.2559 are the coefficients of the formulae used by the IPMS for calculating the coordinates of pole, x and y, from results of the five ILS stations.

There is a significant similarity between these three curves. As the variations of S_{UM} do not depend on polar motion, one can conclude that the long-period motion of the CIO with respect to the BIH 1968 system is caused mainly by the local nonpolar effect of the Ukiah station. These considerations lead to the conclusion that either the x_3-axis of the conventional terrestrial system, (x), should be directed toward the mean pole of the epoch of observation instead of the CIO, or a kinematic model of the nonpolar drifts of stations, the coordinates of which define the system, (x), should be introduced.

Comparison of the polar coordinates obtained in different systems
is capable of giving information on the relative motion of the
third axis of these systems, and remains the only means for
estimating the accuracy of the adopted conventional terrestrial
coordinate system. Figure 1 shows the difference, BIH-DMA, between the BIH 1968 system and the reference system adopted by
the DMTC for the reduction of the Doppler observations of artificial satellites. The trends of these systems with respect to
the MPO are also depicted. There is significant correlation
between the values, DMA-MPO, and the values, BIH-MPO, that gives
some indication on the existence of the secular motion of the
pole. However, one has to take into acount that the differences
between the BIH 1968 system and the DMA system are of the same
order as linear trends of BIH-MPO and DMA-MPO. Under the suppo-
sition, that there is no correlation between reference systems,
BIH, DMA and ILS, the estimation of the standard errors of the
yearly mean values of the polar coordinates have been determined:

x - coordinate: $\sigma DMA = \pm 0".004$, $\sigma BIH = \pm 0".003$, $\sigma ILS = \pm 0".011$
y - coordinate: $\sigma DMA = \pm 0".006$, $\sigma BIH = \pm 0".004$, $\sigma ILS = \pm 0".008$

When calculating these estimates, the value of ILS-BIH for 1976
have been rejected as outlier.

5. ON THE PROCEDURE FOR MAINTAINING THE CONVENTIONAL TERRESTRIAL COORDINATE SYSTEM

In current practice a minimum displacement principle, in least-
squares sense, is used to fix the reference system to a set of
observatories distributed over the Earth's crust (Fedorov, 1975b).
Let ρi be the displacement of the zenith Z_i in the conventional
terrestrial coordinate system, (x). According to a minimum displacement principle,

$$\sum_{i=4}^{n} \rho i^2 = \text{min,} \tag{1}$$

where n is number of observatories. Taking the displacement
along the parallel and meridian, the condition (1) becomes

$$\sum_{i=1}^{n} \{ (\ell_i - \ell_{oi})^2 \cos^2 b_{oi} + (b_i - b_{oi})^2 \} = \text{min,} \tag{2}$$

where ℓ_i, b_i are the longitude and latitude of the i-th observatory in the system, (x); ℓ_{oi}, b_{oi} are the values of ℓ_i and b_i
for the initial epoch T_o.

This statistical approach could be realized in the following
manner. At the start of perhaps a number of Chandlerian periods
the average values, $(\lambda_{oi}, \phi_{oi})$, of the observed astronomical
longitudes and latitudes, (λ_i, ϕ_i), referred to the ephemeris

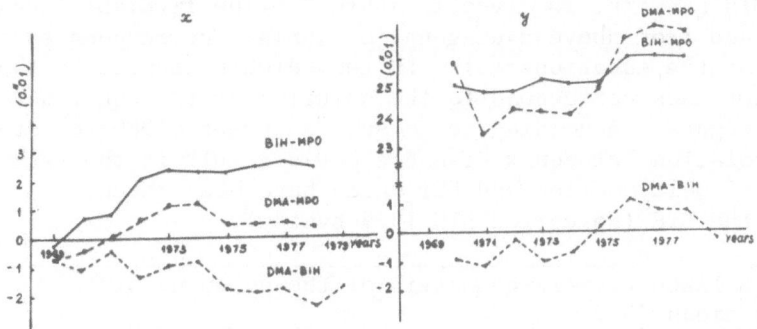

Figure 1.   Variations of the yearly mean differences: BIH-MPO
            (solid line), DMA-BIH (dashed line with solid circles)
            and DMA-MPO (dashed line with open circles).

system, (ξ), is used for calculating UT1 and polar coordinates,
(x,y), from the system of equation (3):

$$\phi_i - \phi_{oi} - (x \cos \lambda_{oi} + y \sin \lambda_{oi} + z) = v_i \qquad (3)$$

$$UTO_i - UTC - (-x \tan \phi_{oi} \sin \lambda_{oi} + y \tan \phi_{oi} \cos \lambda_{oi} + t) = v_i',$$

where UTO_i is the Greenwich mean time derived from observations
of star transits at the i-th observatory using the initial
longitude λ_{oi}; UTC is the universal coordinated time broadcast
by means of time signals; t is the value of UT1-UTC; z is nonpolar
latitude variation, which is common for all participating
observatories; v_i, v_i' are residual which is assumed to be independent random variables. Then the process is repeated for calculating the more accurate estimates of mean longitude and latitude
under the condition that UT1 and x, y are given. As a result,
the values λ_{oi}, ϕ_{oi} are obtained, which one can adopt as conventional coordinates of the i-th observatory for the initial
epoch T_o, i.e., ℓ_{oi}, b_{oi}. At some convenient time interval like
a one or two-year period, the estimates of the systematic errors
of individual instruments along with UT1 and polar motion are
determined from the solution of (3). The former as well as the
information on the observation accuracy are used to weight the
equations (3) when solving for subsequent time interval.

The difficulty of maintaining the conventional system, (x), is
caused, first of all, by the fact that poor network configuration
prevents the independent estimates of the unknowns x, y, z, t

from the solution of equations (3). For example, the covariance
matrix calculated for the worldwide net of about eighty observatories, the BIH network, is given in Table 2. The figures placed
to the right and from above a diagonal of Table 2 correspond to
the solution of the equations (3) with the weights adopted by the
BIH. The other ones correspond to the solution of the equations
(3) without weights. According to Korsuń and Emetz (1980) existence of correlation between x, t and z could result in the seasonal effect of polar motion and UT1 which have been recently
corrected by the BIH (so-called BIH 1979 system).

Table 2.   Normalized covariance matrix of the unknowns of
           equations (3).

	x	y	z	t
x		-0.14	-0.26	0.07
y	-0.05		0.04	0.04
z	-0.53	-0.11		0.02
t	-0.14	0.22	0.10	

The difficulty of maintaining the conventional coordinate system
is increased by the fact that the observatories are located on
different crustal plates that are in relative motion. Therefore,
the above assumption on the nature of residuals, v_i, v'_i, is not
valid and one should use some statistical model of these residuals
to solve the equations (3). Let us suppose

$$v_i = v_{s,i} + v_{r,i} , \qquad (4)$$

where $v_{s,i}$ is nonpolar systematic displacement of the i-th observ-
tory; $v_{r,i}$ is independent random variable. For this model, the
estimates of the unknown, x, y and UT1 of the equations (3)
would be consistent, if the vectorial mean value, $v_{s,i}$, taken
over the entire set of observatories, is zero, namely,

$$\sum_{i=1}^{n} v_{s,i}^{m} \cos \lambda_{oi} = 0, \quad \sum_{i=1}^{n} v_{s,i}^{m} \sin \lambda_{oi} = 0, \quad \sum_{i=1}^{n} v_{s,i}^{p} = 0 ,$$

where $v_{s,i}^{m}$, $v_{s,i}^{p}$ are the components of nonpolar displacement, $v_{s,i}$,
along the meridian and parallel respectively. The deviation from
this condition results in the relative motion of the different
astronomical systems of terrestrial coordinate. This consideration leads to a conclusion that the conventional terrestrial
coordinate system should be attached, in the statistical sense,
to the observatories at relatively stable sites having on the
average no mutual equatorial displacement of the zeniths and no

systematic meridional components of the zenith displacements. The realization of such a reference system could be achieved by means of the comparison of the different systems of terrestrial coordinates with each other and with the MPO in the manner that is similar to the compilation of general catalogues in fundamental astrometry. After realization of such a reference system, a special procedure has to be used to minimize the effects of the changes in the number of participating observatories and the observational programs. This procedure might be similar to that employed by the BIH, though some improvement is possible (Yatskiv et al., 1979)

REFERENCES

Belotserkovskij, D.Yu. and Kaufman, M.B.: 1979, in D.D. McCarthy and J.D.H. Pilkington (eds.) "IAU Symposium No. 82, Time and the Earth's Rotation", D. Reidel, Dordrecht, pp. 23-27.

Fedorov, E.P., Korsuń, A.A., Major, S.P., Panchenko, N.I., Taradij, W.K., Yatskiv, Ya.S.: 1972, "Motion of the Earth's Pole for the Years 1890.0-1969.0, Naukova Dumka, Kiev.

Fedorov, E.P.: 1975a, Astrometrija i Astrofizika, Kiev, No. 27, pp. 3-13.

Fedorov, E.P.: 1975b, in B. Kolaczek and G. Weiffenbach (eds.), "IAU Colloquium No. 26, On Reference Coordinate Systems for Earth Dynamics", Warsaw, pp. 63-77.

Fedorov, E.P.: 1979, in D.D. McCarthy and J.D.H. Pilkington (eds.) "IAU Symposium No. 82, Time and the Earth's Rotation", D. Reidel, Dordrecht, pp. 89-101.

Gubanov, V.S. and Yaguidin, L.I.: 1979, in D.D. McCarthy and J.D.H. Pilkington (eds.), "IAU Symposium No. 82, Time and the Earth's Rotation", D. Reidel, Dordrecht, pp. 47-51.

Guinot, B.: 1979, in D.D. McCarthy and J.D.H. Pilkington (eds.), "IAU Symposium No. 82, Time and the Earth's Rotation", D. Reidel, Dordrecht, pp. 7-18.

Kolaczek, B.: 1979, in D.D. McCarthy and J.D.H. Pilkington (eds.), "IAU Symposium No. 82, Time and the Earth's Rotation", D. Reidel, Dordrecht, pp. 125-128.

Korsuń, A.A. and Emetz, A.I.: 1980, in Proceedings of the 4th International Symposium, "Geodesy and Physics of the Earth", Karl-Marx-Stadt (in press).

Mandelbrot, B.B. and McCamy, K.: 1970, "On the Secular Polar Motion and the Chandler Wobble", Geophys. J. R. Astr. Soc., 21, pp. 217-232.

Markovitz, N.: 1970, in L. Mansihna, D.E. Smylie, A.E. Beck (eds.), "Earthquake Displacement Fields and the Rotation of the Earth", Springer-Verlag, New York, pp. 69-81.

Mironov, N.T.: 1973, Astron. Circular, No. 769, pp. 7-8.

Orlov, A.Ya.: 1941, Bull. Sternberg Institut, Moskva, No. 8.

Seidelmann, P.K. et al.: 1979, Final Report of the IAU Working
    Group on Nutation, 17th General Assembly of the IAU, Montreal.
Yatskiv, Ya.S., Korsuń, A.A., Mironov, N.T.: 1979, in D.D.
    McCarthy and J.D.H. Pilkington (eds.), "IAU Symposium No. 82,
    Time and the Earth's Rotation", D. Reidel, Dordrecht,
    pp. 29-39.

ON REFERENCE COORDINATE SYSTEMS USED IN POLAR MOTION DETERMINATIONS

B. Kołaczek
Planetary Geodesy Department
Space Research Centre of the Polish Academy of Sciences
Warsaw, POLAND

G. Teleki
Astronomical Observatory
Belgrade, YUGOSLAVIA

ABSTRACT

A short review of the reference pole presently used in polar motion determinations by classical astrometric methods is followed by a discussion of the systematic differences between systems of polar coordinates and the influence of the mean latitude of stations on pole position. The importance of homogenous processing astrometric data is stressed.

INTRODUCTION

Different reference poles have been defined for use in determination of polar motion and variations of Earth's rotation velocity by astrometric methods. These observations have been obtained since the beginning of regular activity of the International Latitude Service (ILS) in 1899 and of the Bureau International de l'Heure (BIH) and, later, the International Polar Motion Service (IPMS), as shown in Table 1 (BIH, 1968, Cecchini, 1970, IPMS, 1962). At present the Conventional International Origin (CIO) and reference poles of BIH, 1968, and IPMS systems are used.

CIO was defined by adoption of the conventional latitudes of 5 ILS stations and introduced into practice on January 1, 1968, according to the resolutions of General Assemblies of the IAU and IUGG (IAU, 1967, IUGG, 1967).

E. M. Gaposchkin and B. Kolaczek (eds.), Reference Coordinate Systems for Earth Dynamics, 165–173.
Copyright © 1981 by D. Reidel Publishing Company.

BIH, 1968, system was defined by determination of latitudes and
longitudes of the instruments participating in time and latitude
determination during the years 1966.50 - 1967.45 and located at
50 stations (BIH, 1968). The following instruments participated in
these observations of latitude: VZT-19(22), A-11(12), PZT-9(14);
and of time: IP-16(13), A-15(14), PZT-9(14), IPP-8(12). Numbers
in parenthesis denote numbers of instruments cooperating with BIH
in 1979. A system of weights characterizing the long period
stability of instruments was adopted by BIH. The coincidence
of the reference pole of the BIH, 1968, system with CIO and continuity of UT1 in 1968.0 were assured. In 1979 the improved
system BIH, 1979, was adopted by introducing the conventional annual and semi-annual corrections to x and UT1 obtained from comparisons of BIH astrometric results with DMA and EROLD results
(BIH, 1979, Feissel, 1980).

Table 1. Reference Poles Used in Determination of Polar Motion

Service	Period	Reference Pole
ILS	1899 - 1949 1949.0 - 1968.0 1968.0 -	Mean poles of different epochs The new system, 1900 - 1905* CIO
BIH	1955 - 1958 1958.65 - 1959.15 1959.2 - 1968.0 1968.0 -	Cecchini pole** Transformation for Cecchini pole to the mean pole of the date Mean pole of the date CIO
IPMS	1962.0 - 1968.0 1968.0 -	The new system, 1900 - 1905* CIO

*Mean position of the true celestial pole in 1900 - 1905 (IUGG,
  1960).
**Mean terrestrial pole of 1949 - 1958 (Cecchini, 1970).

Coordinates of the pole (x,y) and UTO in the BIH, 1968, system
are computed by known equations:

$$x \cos L_{0,i} + y \sin L_{0,i} + z = \theta_i - \phi_{0,i}$$
$$x \tan \phi_{0,i} \sin L_{0,i} + y \tan \phi_{0,i} \cos L_{0,i} + t = UTO - TAI$$

(1)

The residuals of latitude and universal time R and S of individual
series, referred to the BIH, 1968, system are added respectively
in order to preserve the system. Here $\phi_{0,i}$ and $L_{0,i}$ denote the
latitude and longitude of a station adopted by BIH.

The residual R and S, are expressed by BIH as follows:

$$R_j = a_j + b_j\sin2\pi\theta + c_j\cos2\pi\theta + d_j\sin4\pi\theta + e_j\cos4\pi\theta$$

$$S_j = a'_j + b'_j\sin2\pi\theta + c'_j\cos2\pi\theta + d'_j\sin4\pi\theta + e'_j\cos4\pi\theta$$

(2)

and are determined every year for every instrument by least squares solutions of residuals during the four preceding years. This has been the procedure since 1975 (Feissel, 1980). The values of coefficients of the equation (2) are denoted by A,B,C, etc. and published in BIH Annual Reports.

IPMS determines polar motions with respect to the CIO continuing computations of pole coordinates in the ILS system. The IPMS determines pole coordinates in two other systems, IPMS$_L$ and IPMS$_{L+T}$, based respectively on latitude data from about 70 instruments and the combined latitude and time data of about 120 instruments. Mean latitudes and longitudes of all instruments are determined with respect to CIO and Zero Point of the BIH, 1968, system, respectively, using data from several years. These are occasionally changed due to change of instruments or programs. Pole coordinates are computed in the IPMS systems using equation (1) without any additional systematic corrections. In the case of differences of UTO-TAI the empirical term τ is added.

The present accuracies of pole coordinates of BIH and IPMS systems are of order of 3 to 4 and 10 to 15 milliseconds of arc, respectively. Systematic differences between pole coordinates of these systems are two to three times larger.

DISCUSSION OF THE SYSTEMATIC DIFFERENCES BETWEEN POLE COORDINATES OF DIFFERENT SYSTEMS

Differences of polar coordinates computed in different systems were developed by BIH using formula (2). Variations of the coefficients A of the differences of ILS-BIH and IPMS-BIH are plotted in Figures 1 and 2 together with coordinates of the barycenters of ILS and IPMS polar orbit in respect to CIO (BIH, 1979, IPMS, 1977). Correlation of these two curves in the case of ILS system shows that there are no secular variations of relative positions of the BIH reference pole and that of CIO. Large variation of the coefficient A and of positions of the barycenters of ILS polar orbit are connected with latitude variations of stations and not with secular polar motion because a coefficient A contains only systematic differences between systems. In the case of IPMS there is some discrepancy between variations of IPMS barycenters and variations of the coefficient A. This can be seen in the relative motion of CIO with respect to reference poles of the BIH and IPMS systems. It should be mentioned that motion of the mean

pole computed by the Orlov filter for ILS (A. A. Korsuń, 1980)
is in agreement with the variation of the coefficient A presented here.

The variations of coefficients of individual stations that coop-
erate with BIH in the last 12 years are of order of 0.1 seconds
of arc for the 12 latitude stations and of order of 0.01 - 0.02
seconds of arc for the 20 stations. This is roughly the BIH
weights. Variations of the A coefficients of stations correlate
with their mean latitude variations computed by the Orlov's
filter (Figure 3).

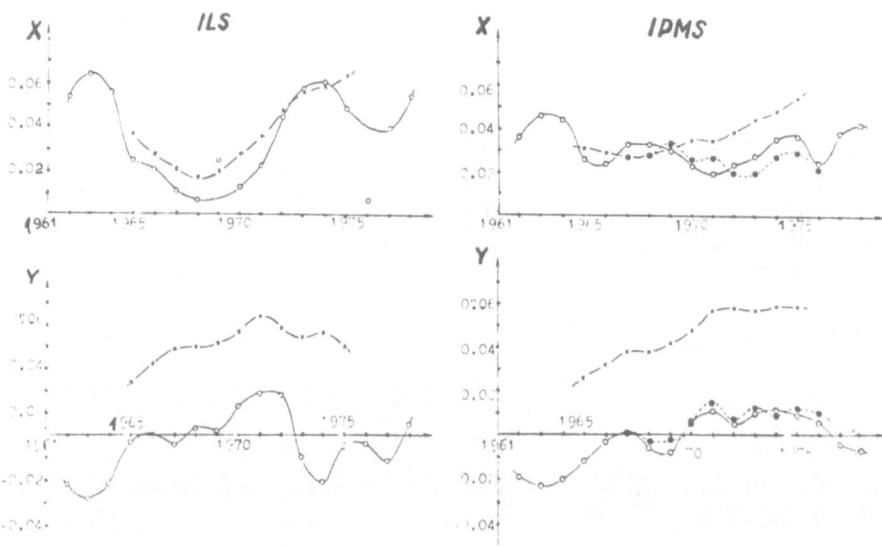

Figure 1. ILS Barycenters x-x and   Figure 2. ILS Barycenters
           Coefficients A_{BIH}o-o   x-x and Coeffi-
                                                    cients A_{BIH}o-o

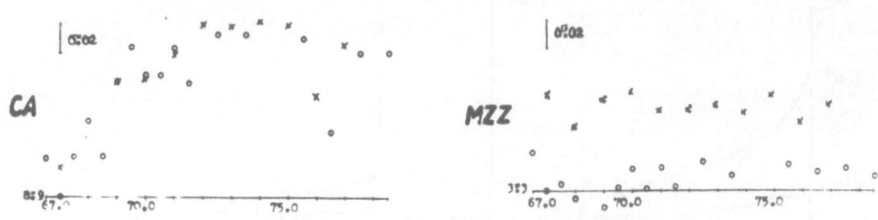

Figure 3. Orlov's Mean Latitudes (x-x) and Coefficients A_{BIH}
(o-o) of Carloforte and Mizusawa.

The influence of mean latitude variation on ILS polar orbit was
analysed by computation of ILS polodia for the period of 1968-
1974 using the differences of latitudes (θ) and mean latitudes
(θ_m) and comparing these results with polodia computed from
latitude data themselves. Computations were also made for different subsets of 5 to 20 stations (given in the Table 2).

Table 2. List of Stations Considered in the Analysis of the
Effect of Large Variations of a Mean Latitude on
Derived Polar Coordinates

I. Stations with small variations of mean latitudes in 1968-1974.		II. Stations with large variations of mean latitudes in 1968-1974.	
Mt. Stromlo	Greenwich	Richmond	Turku
Blagovestchensk	Paris	Washington	Neuchatel
Irkoutsk	Pulkovo	Gaithersburg	Hamburg
Warsaw	Poltava	Mizusawa	Uccle
Pecny	Tokio	Belgrade	Kitab

Mean square error of polar coordinates x,y and the error of a
single observation ($\delta\varepsilon$) computed from the $\theta-\theta_m$ data are much
smaller than for the θ data themselves (Figure 4). The systematic differences $(O-C)_{IPMS}$ are much smaller for $\theta-\theta_m$ data than
for θ data in the case of small number of stations and stations
with large mean latitude variations (Figure 5).

There is noticeable difference between ILS polodia computed from
$\theta-\theta_m$ and θ data in the period of 1962-1968 (Figure 6). ILS
polodia computed from $\theta-\theta_m$ data is much closer to IPMS polodia
computed by IPMS in this period.

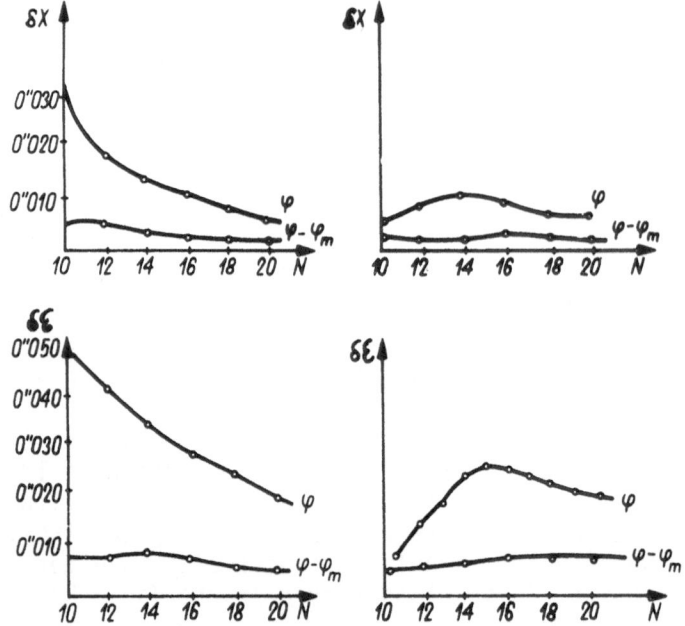

Figure 4.   Mean square errors of x and of a single observation:
a) the II set of stations plus n stations of the
   set I (Table 2).
b) the I set of stations plus n stations of the
   set II (Table 2).

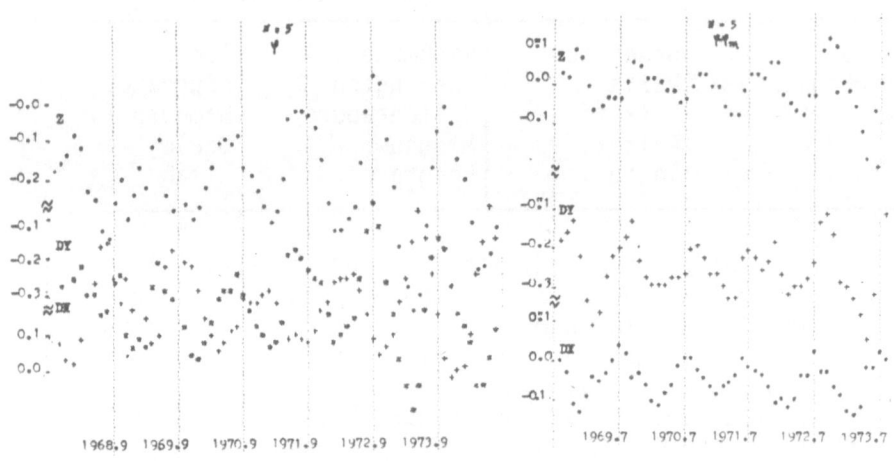

Figure 5a	Figure 5b
Variations of (O-C) computed for x,y and term z from θ data of five stations.	Variations of (O-C) computed for x,y and term z from θ-θ$_m$ data of five stations.

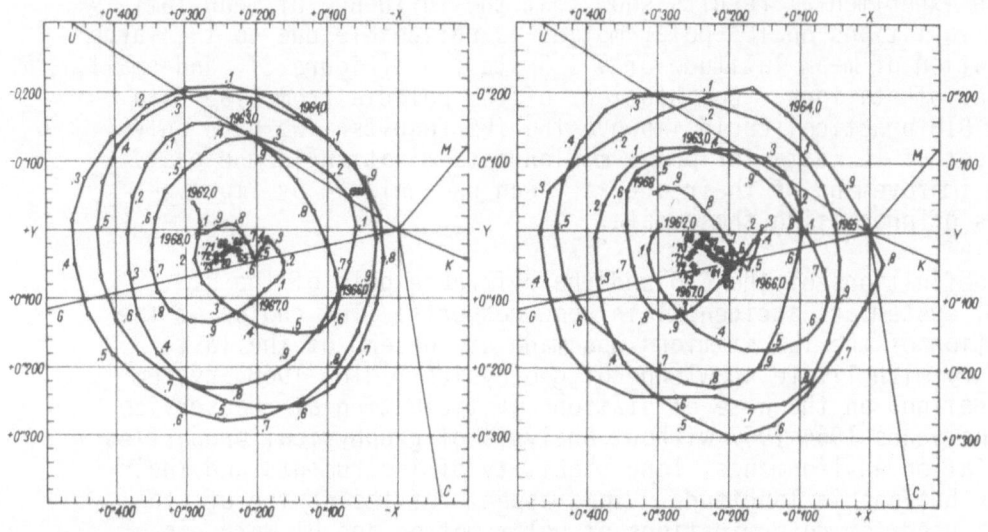

Figure 6a. IPMS Polodia in   Figure 6b. ILS Polodia in
1962-1968.   1962-1968

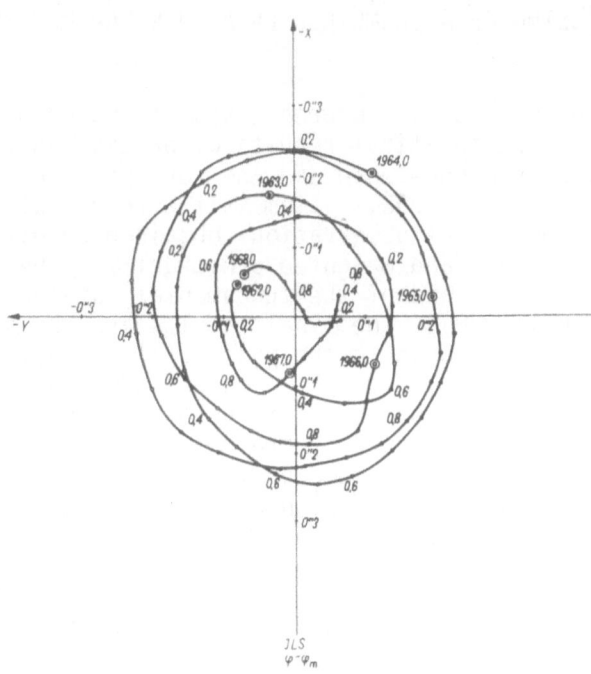

Figure 6c. ILS Polodia computed from θ-θ$_m$
data in 1962-1968.

These experimental results show that the influence of mean latitude variations on ILS polar motion is noticeable due to the large
variation of mean latitude of 5 ILS stations (Figure 3), and small
number of stations. Computations of ILS polodia from $\theta-\theta_m$ data
(the BIH practice) could improve the ILS results. We have only
one set of ILS data for polar motion determination in the past.
Some improvement of their accuracy can be achieved by improvements of processing these data.

The definitions of the CIO and the reference pole of the BIH,
1968, system are accidental in some respects. The choice of the
location of the ILS stations was made at the end of the last
century with little knowledge of geophysics. BIH, 1968, system
was defined on the base of stations participating at the service
in the years 1964-1967 without analysis of geophysical properties
of station environments, long stability of instruments and their
distributions in longitude. Some aspects of the choice of stations used for determinations of polar motion and UT were discussed by Djurovič (1978) and Kołaczek (1978). Now we should
consider the establishment of the best astrometric system taking
into account present geophysical knowledge and the experience of
the BIH and IPMS services.

REMARKS ON PROCESSING OF ASTROMETRIC DATA IN POLAR MOTION
DETERMINATIONS

The celestial and terrestrial reference systems based on astrometric observations are the final results of processing these
measurements. The reference pole and the zero point defined
statistically as the BIH, 1968, system is the result of the adjustment of many individual observations obtained by different
methods at many stations and weighted arbitrarily. The same
situation exists in the case of the fundamental catalogues FK4 or
FK5 (Tucker, Teleki, 1978). The final results are the sum of
real physical parameters used in the processing of that data but
not enough is yet known. The results are dependent on applied
methods of data reduction and on the transformations to the common
systems. The final results usually are not free from influences
of systematic errors which are complex functions of time and other
factors, e.g., temperature, and which cannot be eliminated by
purely statistical methods. Applying statistical principles, we
must seek to increase the accuracy of individual observations.
At present, a common program ought to be obligatory for each
participating station. Such a program would describe separately
each kind of instrument, would establish uniform observational
methods and treatment of instruments and pavilions, would specify
determination of critical instrument constants, would detail
measurements of important geophysical and atmospheric parameters,
and would establish uniform processing of observations.

At present, methods for the basic processing of observational data differ from station to station and are usually not described completely. Observational data are smoothed in different ways, creating arbitrary deformation of data. In transformation of data to one common system an arbitrary system of weights and some empirical corrections are adopted without understanding the physical character of all interfering phenomena. In requiring more accurate data to calculate more realistic parameters of physical phenomena it is necessary to assure homogeneity of methods of data processing. It is constantly accentuated but not fulfilled (e.g., Kołaczek, Weiffenbach, 1974).

New observing techniques promising much higher accuracy of determinations of polar motion and UT open the new epoch in studies of the Earth's rotation. At the threshold of this epoch the Merit Program is being organized in order to compare possibilities of all available techniques and to outline the future plans of regular determination of polar motion and UT. At this moment it is necessary to reopen discussion concerning the practice of the classical astrometric methods in order to improve their accuracy by improving the processing of these data. More attention to all perturbing physical phenomena ought to be paid in astrometric methods as they are in modern techniques.

REFERENCES

BIH: 1968-1979, BIH Annual Report for the years 1968-1979.
Cecchini, G.: 1970, Report presented at XIV IAU General Assembly, Brighton.
Djurovič, D.: 1978, Proceedings of the IAU Symposium No. 82, pp. 75-78.
Feissel, M.: 1980, Bull. Geodesique, Vol. 54, pp. 87-102.
IAU: 1967, IAU Transactions, Vol. XIII.
IUGG: 1967, Comptes Rendus de la XIV Assembleé Generale de l'U.G.G.I.
IPMS: 1974-1977, Annual Report of IPMS for the year 1974-1977.
Kołaczek, B.: 1978, Proceedings of the IAU Symposium No. 82, pp. 125-128.
Kołaczek, B., Weiffenbach, G.: 1974, Proceedings of the IAU Colloquium No. 26.
Korsuń, A. A., Emez, A. I.: 1980, 4th Intern. Symposium, "Geodesy and Physics of the Earth", Karl-Marx-Stadt, May, 1980 (in press).
Tucker, R. H., Teleki, G.: 1974, Proceedings of the IAU Colloquium No. 48, pp. 545-556.

NOTE: This work is supported in part by the Smithsonian Institution Foreign Currency Program Grant, No. FR-6-50015.

THE SHORT PERIOD LATITUDE VARIATIONS DERIVED FROM RECALCULATED PAST ILS OBSERVATIONS

Seiji Manabe
International Latitude Observatory of Mizusawa
Mizusawa, Iwate, 023 Japan

ABSTRACT

Ten frequency components of short period latitude variations and corrections for the declination at an epoch and the proper motion in declination are estimated simultaneously. The short period variations are nearly diurnal and semi-diurnal and are supposed to be due to nutations and Earth tides. Analysis is by the least-squares chain method with weights for individual observations, recently developed by Manabe et al. (1977). Weights are assigned on the assumption that the observational errors follow a normal distribution, whose dispersion changes night by night. Among all possible 1024 combinations of the short period variations, the best is chosen on the basis of the minimum AIC procedure. The data used are the results of the homogeneous recalculation of the ILS observations, 1899-1978. This is considered to be the optimal astrometric data set for investigating nutations, because of its long time-span, homogeneity and size. Declinations are estimated for each of the seven observational periods with mean errors smaller than $0\overset{''}{.}01$. The mean errors of the proper motions are of the order $0\overset{''}{.}002/y$. Among the short period variations, significant estimates are obtained for the terms corresponding to the lunar fortnightly, solar semi-annual and lunar nodal nutations, and the M_2 tide. The mean errors are shown to be greatly reduced and the entire data set is analyzed. An effort is made to distinguish the preferred theoretical model.

INTRODUCTION

It is a challenging problem for astrometry to decide which theoretical model of nutations is most consistent with observations. For example an accuracy of $0\overset{''}{.}001$ is required to distinguish between Molodensky's (1961) and Wahr's (1979) models. Manabe et al. (1979) estimated the nutations and Earth tides from the recalculated ILS data (Yumi and Yokoyama 1980) in each of seven periods. They obtained nutation parameters which are generally in agreement with theoretical predictions. However, the mean errors of their estimates are not small enough to distinguish the preferred model. Therefore, it is useful to

175

E. M. Gaposchkin and B. Kołaczek (eds.), Reference Coordinate Systems for Earth Dynamics, 175–179.
Copyright © 1981 by D. Reidel Publishing Company.

estimate the nutations using the whole 80-year series of ILS data. Furthermore, it is very important to combine the data in different periods and analyze them simultaneously in order to realize a uniform system of declinations and to provide a homogeneous series of polar motion data. The most serious obstacle to doing this is an irregularity in the scale values of micrometer screws. In the present analysis, after eliminating long period latitude variations and daily variations of scale values from observational equations by using the least-squares chain method (Manabe et al. 1977), we estimate constant and secular parts of declination corrections and the short period variations of latitude. The estimates thus obtained are free from systematic errors due to irregular variations of scale values. The frequency components of the short period variations that are taken into account are the fortnighly, semi-annual, annual and principal nutations and the M_2 and S_2 tides. In the following notations are the same as those found in Manabe et al. (1977).

METHOD OF ANALYSIS

The VZT observable is a difference D of readings of a micrometer. Apart from accidental errors, D depends systematically on latitude variations, declination errors and errors in the adopted scale values. These parameters are so small that their products can be neglected. Then, after making various corrections such as differential refractions by using the adopted scale values m_i at t_i for a star s_i, the observational equations for the r-th night are given by

$$E(\Delta D_r) = X_r \beta + L_r (P_r \, \Delta m_r)^T \tag{1}$$

where $\beta = (\Delta \delta^T \, \Delta u'^T \, A^T)^T$, Δm_r is a daily correction of the adopted scale values, $\Delta D_r = D_r - \text{diag}[m_1 \ldots m_{n_r}]\zeta_r$ is an O-C vector and ζ_r is a vector of zenith distances calculated on the basis of catalogue places and the IAU nutation table. Matrices X_r and L_r are $X_r = \text{diag}[m_1 \ldots m_{n_r}] \, (R_r \, T_r \, K_r)$ and $L_r = (-e_{n_r} \, \zeta_r)$ with n_r-vector $e_{n_r} = (1 \ldots 1)^T$. According to the least-squares chain method, the last term in equation (1) can be eliminated by using a matrix $W_r = (W_{ij}^r)$ with

$$W_{ij}^r = w_i \delta_{ij} - w_i w_j \left\{ m_i m_j [w\zeta\zeta] - (m_i \zeta_j + m_j \zeta_i) \, [wm\zeta] + \zeta_i \zeta_j [wmm] \right\}$$

$$/ \, \left\{ [wmm] \, [w\zeta\zeta] - [wm\zeta]^2 \right\} \quad , \tag{2}$$

where w_i is a weight and [ ] is Gauss' summation symbol. Let us denote the resultant normal equation by $S\hat{\beta}=Q$, where $S = \sum\limits_{r} X_r^T W_r X_r$

and $Q = \sum_{\tilde{r}} X_r^T W_r \Delta D_r$. If we neglect irregular and temperature dependent variations of m_i which are of relative order 10^{-4}, and small periodic variations of ζ_i due to the nutations, S satisfies a relation

$$S \, H = 0 \quad , \quad \text{with } H = \begin{bmatrix} e_n & 0 & a \\ 0 & e_n & b \\ 0 & 0 & 0 \end{bmatrix} \quad ,$$

where n-vectors a and b are constant and secular parts of ζ such that $\zeta_i = a_{si} + (\bar{t}_r - t_0)b_{si}$. Therefore, the rank of S is deficient by three and three additional conditions are required on β. We adopt conditions

$$H^T \beta = 0 \text{ or } e_n^T \Delta\delta = 0 \quad , \quad e_n^T \Delta u' = 0 \text{ and } a^T \Delta\delta + b^T \Delta u' = 0 \quad (3)$$

at an epoch of correction. It can be shown that these conditions are equivalent to setting $|\beta| = \min$.

The equation $S\hat{\beta}=Q$ has been solved with conditions (3) for all possible 1024 combinations of the short period variations. The best combination of the short period variations has been chosen with the aid of the minimum AIC procedure (Akaike 1973).

RESULTS

The table shows the minimum AIC estimates of the principal terms of the short period variations. The last three columns show the ratio γ of the estimated and the theoretical nutation amplitudes to those of a rigid Earth (Kinoshita 1977) for the diurnal components, and the tidal factor $\Lambda = 1 + k - \ell$ for the semi-diurnal components.

It is clear that the 2L-α term agrees well with the theoretical values. But the mean error is too large to distinguish the better theoretical value. The agreement of the 2L'-α term with the theoretical values appears poor. However, if the contribution of the O_1 tide to this component ($-0.0016 \sin(2L'-\alpha)$ at $39°8'$ N with $\Lambda = 1.2$) is subtracted from the estimate, we obtain much better agreement. The most serious discrepancies between the present estimates and the theoretical values are in the $\pm \, \Omega-\alpha$ terms. The differences are much larger than the mean errors, which are small enough to distinguish between the two models. This may be due to some unknown systematic errors with long time-scale. In fact if we use the data for 1899-1955 only, we obtain $\gamma = 0.99687 \pm 17$ for $\Omega-\alpha$ and $\gamma = 1.00380 \pm 119$ for $-\Omega-\alpha$. The differences between these values and those in the table far exceed the mean errors. The γ's of other components also change, but the differences can be regarded as caused by statistical fluctuations.

Table.   Minimum AIC estimates of the short period variations at
       1950.0.

	Amp.	Phase	γ or Λ	Wahr	M-II
$2L'-\alpha$	0″.0923	0°.87	1.0081	1.0283	1.022
	± 10	59	105		
$-2L'-\alpha$	0.0036	32.36	1.0439		
	8	13.16	2398		
$2L-\alpha$	0.5476	0.04	1.0332	1.0344	1.0315
	21	22	389		
$-2L-\alpha$	0.0253	-19.87	1.1201		
	20	4.39	8728		
$\ell-\alpha$	0.0430	170.18	1.7196		
	50	40.41	2004		
$-\ell-\alpha$	0.0496	-3.35	1.9856	1.2246	1.2247
	39	4.47	1576		
$\Omega-\alpha$	8.0282	-0.01	0.9973	0.99640	0.99667
	10	1	1		
$-\Omega-\alpha$	1.1791	0.33	1.0015	1.00314	1.00290
	9	5	8		
$2L'-2\alpha$	0.0069	78.89	0.8961		
	5	4.17	649		
$2L-2\alpha$	0.0067	77.78	1.8611		
	40	73.65	1.1111		
Periods 1 - 5					
$\Omega-\alpha$	8.0247	-0.02	0.9969		
	14	1	2		
$-\Omega-\alpha$	1.1818	-0.21	1.0038		
	14	7	12		

    As for the declination corrections, the mean errors at 1950
range from 0″.007 for the stars observed throughout the whole
period to 0″.2 for the stars observed during 1899-1906.   The range
of the mean errors of the proper motion corrections is from
0″.00017 /y to 0″.0054/y.   It is remarkable that the mean errors
for stars observed only for six years are much reduced in the
present calculation from those in the past calculation (Manabe
et al. 1979) which also uses only six years of data.

REFERENCES

Akaike, H. 1973, 2nd International Symposium on Information Theory, eds B. N. Petrov and F. Csáki, 257.
Kinoshita, H. 1977, Celest. Mech., 19, 215.
Manabe, S., Sasao, T. and Sakai, S. 1977, Publ. Int. Latit. Obs. Mizusawa, 11, 23.
Manabe, S., Sakai, S. and Sasao, T. 1979, Publ. Int. Latit. Obs. Mizusawa, 13, 53.
Molodensky, M. S. 1961, Commun. Obs. Roy. Belgique, 3e Série, 10, Fasc. 2.
Wahr, J. M. 1979, Ph.D. Thesis, Univ. Colorado.
Yumi, S. and Yokoyama, K. 1980, Results of the International Latitude Service in a Homogeneous System 1899.9 — 1979.0, Mizusawa, Japan.

DETERMINATION OF 'DECADE' FLUCTUATIONS
IN THE EARTH'S ROTATION 1620-1978

L V Morrison   F R Stephenson
Royal Greenwich Observatory   University of Liverpool
UK   UK

ABSTRACT: The 'decade' fluctuations in the Earth's rotation in the
period 1620-1978 are derived from astronomical observations. The
torques on the mantle which produce these fluctuations attain a magnitude of 10^{18}Nm (10^{25} dyn cm) around 1900.

1. INTRODUCTION

The terrestrial frame of reference fixed in the mantle varies from
a uniformly rotating frame because of tidal interactions with the Moon
and Sun and changes in the distribution of angular momentum between the
core, mantle and atmosphere. From an analysis of the apparent fluctuations in the position of the Moon relative to the stars, Brouwer (1952)
found variations in the rate of rotation of the mantle occurring over
decades which are two orders of magnitude greater than those due to
tidal forces. Munk and MacDonald (1960) argued that the changes are too
great to be explained by the re-distribution of angular momentum between
the atmosphere and mantle, or by changes in the moment of inertia of the
mantle. They, therefore, concluded that the changes were caused by the
interchange of angular momentum between the core and mantle.

In this paper we report our results obtained by extending the
number and range of observations available to Brouwer, particularly in
the 17th century. Our results should provide a tighter constraint on
the proposed mechanisms for the interchange of angular momentum between
the core and mantle.

2. OBSERVATIONS AND REDUCTION

The main source of observations for this analysis are the timings
of occultations of stars by the Moon taken from the catalogues of
Morrison and Stephenson (1981) for the years 1623-1860, Morrison and
Lukac (1980) for 1861-1942 and Morrison (1978a) for 1943-1971. Timings
of contact 4 of solar eclipses were used to supplement the occultation
observations before 1670 (Morrison and Stephenson, 1981).

E. M. Gaposchkin and B. Kołaczek (eds.), Reference Coordinate Systems for Earth Dynamics, 181–185.
Copyright © 1981 by D. Reidel Publishing Company.

The reduction of timings of occultations and eclipses leads to the determination of the difference ΔT between the uniform time-scale derived from the time-argument of the lunar ephemeris (dynamical time, TD) and the time-scale derived from the rotational period of the Earth (universal time, UT). Beginning with 1955 the values of ΔT are given by TAI−UT1+32^{s}184, where TAI is the international atomic time-scale TAI+32^{s}184, by adding the following empirical correction to the time-argument and hence the values of ΔT: −1^{s}821(−1.54+2.33T−1.78T^2), where T is measured in centuries from 1900.0 (see Morrison, 1979a). The greatest uncertainty in this empirical correction is ±1.0 in the coefficient of T^2 which arises from the tidal acceleration. This produces a possible systematic error of ± 1^{s}8(T−0.55)^2 in the values of ΔT before 1955. Each determination of ΔT is also subject to random errors in timing, the catalogue position of the star, the limb-profile heights of the Moon, and the lunar ephemeris. These produce the standard deviations (σ) listed in Table I for the values of ΔT derived from single observations. The errors in ΔT derived from TAI−UT1+32^{s}184 beginning in 1955 are negligible in the present context. The numbers of observations within 3σ of the mean in each period are also listed in Table I.

Table I

Standard deviation of ΔT and number of observations within 3σ

Period	σ	$N < 3\sigma$	Period	σ	$N < 3\sigma$
1620-1669	1m	94	1820-1860	1.5s	1265
1670-1699	15s	65	1861-1942	1.3s	24800
1700-1759	5s	169	1943-1954	1.0s	~10000
1760-1819	2s	313			

In the period 1620-1860 the individual values of ΔT were smoothed by cubic splines having 13 knots spaced at proportionately smaller intervals of time according to the increase in the number of observations. From 1861 onwards annual means were calculated and smoothed using a 5-point convolute (see Morrison, 1979b). The resultant smoothed curve for ΔT is shown in the upper section of Fig 1. This is an improvement over the curve deduced by Brouwer in that we have extended the range and number of observations and have applied corrections for the limb-profile heights of the Moon in reducing the occultations. The erratic changes in Brouwer's results on a time-scale of less than 5 years do not appear in our results.

3. ANALYSIS

3.1 Difference in time-scales, ΔT, and the rotational displacement angle $\Delta\theta$.

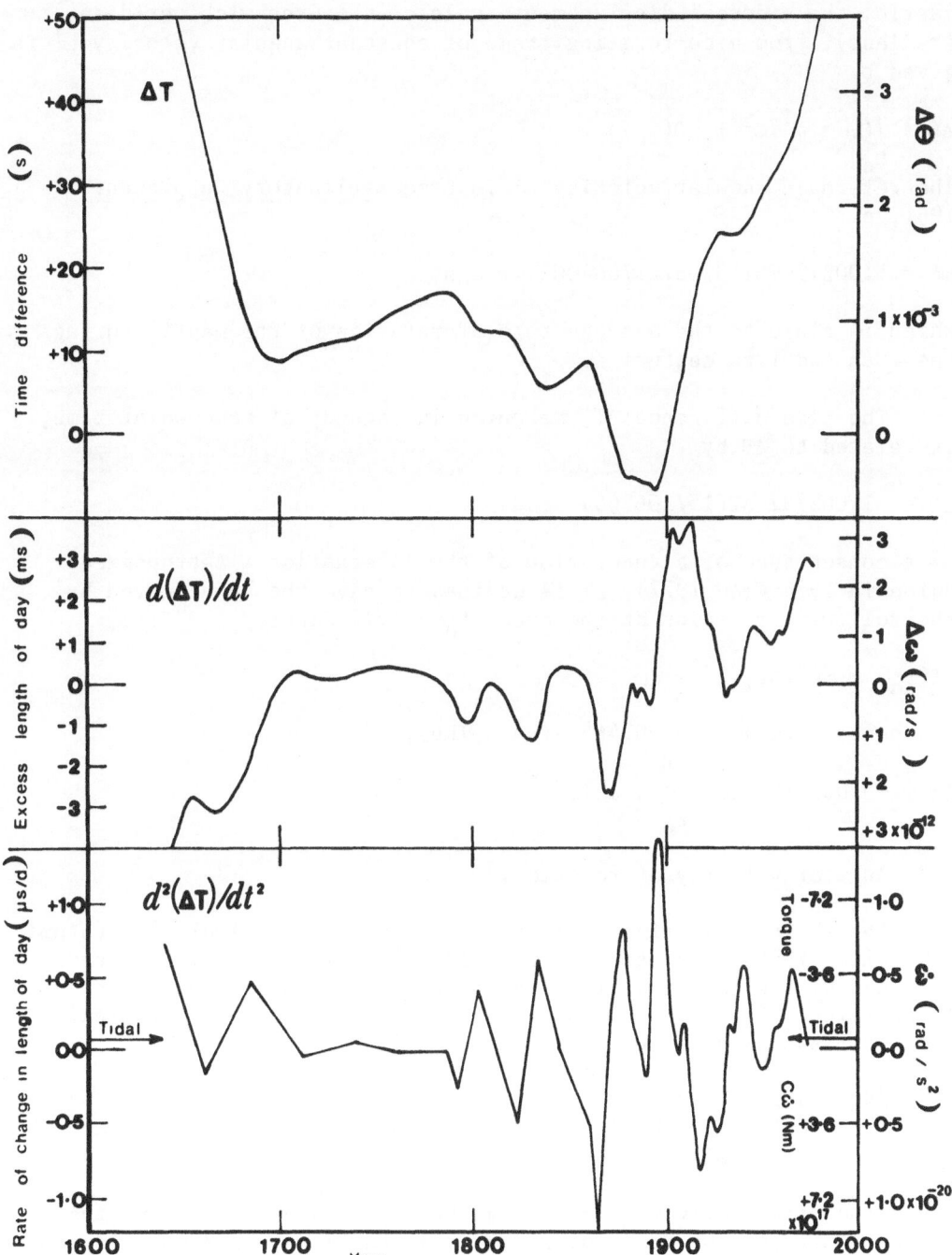

Fig. 1.   Smoothed values of ΔT, and its first and second
derivative, derived mainly from timings of lunar occultations.
The bottom section shows the deduced magnitude and temporal
behaviour of the torque operating on the mantle.

If $\omega = \omega(t)$ is the actual sidereal rate of rotation of the mantle, the sidereal displacement angle of the Greenwich meridian, $\Delta\theta$ (radians), from a co-rotating frame of constant angular velocity ω^1 is given by

$$\Delta\theta = \int_{t_o}^{t} (\omega - \omega^1)dt + \Delta\theta(t_o).$$

The reference angular velocity ω^1 is (see Explanatory Supplement 1961, p.76)

$$\omega^1 = 1.00273\ 78119\ 06(2\pi/86400)\ \ \text{rad/s},$$

which is close to the average rate of rotation of the mantle during the 18th and 19th centuries.

The time-difference ΔT, measured in seconds of mean solar time, is related to $\Delta\theta$ by

$$\Delta\theta = -1.002737\ \Delta T(15/206265)\ \ \text{rad}.$$

As a consequence of a resolution of the International Astronomical Union in 1976 (IAU 1977), ΔT is defined to have the value given by the following relation at the epoch t_o = 1977 January 1 0^h TAI,

$$\Delta T(t_o) = +47\overset{s}{.}52,$$

and hence $\Delta\theta(t_o) = -0.003465$ rad. Thus,

$$\Delta\theta = -0.003465\ \text{rad} + \int_{t_o}^{t} (\omega - \omega^1)dt.$$

3.2 Angular velocity of rotation.

The first derivative of ΔT with respect to dynamical time (atomic time after 1954) is plotted in the middle section of Fig. 1. It was obtained by taking the derivative of the spline curves up to 1860 and by applying a 5-point convolute to the annual values of ΔT thereafter (see Morrison 1979b). We have

$$d(\Delta T)/dt = d(TD-UT)/dt$$
$$= 86400\ \text{SI seconds per SI day} - \text{number of mean solar seconds per SI day.}$$

Thus, the first derivative of ΔT can be regarded as expressing the excess length of the mean solar day over the SI day of 86400 SI seconds, and is conveniently measured in units of milliseconds. The right ordinate in Fig 1 gives

$$d(\Delta\theta)/dt = \omega - \omega^1 = \Delta\omega\ \ \text{rad/s},$$

where the value of the reference velocity ω^1 is given in section 3.1.

3.3 Angular acceleration of rotation and torque.

The second derivative of ΔT with respect to dynamical (atomic) time is plotted in the lower section of Fig. 1. Before 1861 the second derivative consists of 13 straight lines between the knots of the cubic splines. From 1861 onwards the smoothed curve was derived by applying an 11-point convolute to the annual values of ΔT (see Morrison 1979b). The second derivative measures the rate of change in the length of the mean solar day and is conveniently measured in microseconds per day. Multiplying the angular acceleration by the principal moment of inertia of the mantle, $C = 7.2 \times 10^{37}$ kg m^2, gives the magnitude of the torque acting on the mantle (in units of Newton metres $= 10^7$ dyne cm). The results before 1800 are unreliable due to the sparsity of observations. From Fig. 1 it is seen that around 1900 the torques reach an amplitude of 10^{18} Nm in about 5 years. The tidal acceleration due to the Moon, Sun and atmosphere is $-(68\pm5) \times 10^{-23}$ rad/s^2 (Morrison, 1978b), which produces a torque of $-(0.49\pm0.04) \times 10^{17}$ Nm. The magnitude and temporal behaviour of the non-tidal torques impose fairly severe constraints on the possible mechanisms for core-mantle coupling (see Lambeck 1980).

REFERENCES

Brouwer, D., 1952. Astr. J., 57, 125
Explanatory Supplement to the Astronomical Ephemeris and the American
    Ephemeris and Nautical Almanac, 1974, HMSO, London.
IAU, 1977. Trans. Int. Astr. Un., 16B, 56.
Lambeck, K., 1980. The Earth's Variable Rotation, Cambridge
    University Press.
Morrison, L. V., 1978a. R. Greenwich Obs. Bull. 183.
Morrison, L. V., 1978b. in Tidal Friction and the Earth's Rotation,
    eds. Brosche and Sundermann, pp 22-27, Springer-Verlag, Berlin.
Morrison, L. V., 1979a. Mon. Not. R. astr. Soc., 187, 41.
Morrison, L. V., 1979b. Geophys. J. R. astr. Soc., 58, 349
Morrison, L. V., and Lukac, M. R., 1980. R. Greenwich Obs. Bull.
    (in preparation)
Morrison, L. V., and Stephenson, F. R., 1981. R. Greenwich Obs. Bull.
    (in preparation)
Munk, W. H., and Mac Donald, G. J. F., 1960. The Rotation of the Earth,
    Cambridge University Press.

ON THE ANNUAL EARTH'S ROTATION

Angelo Poma and Edoardo Proverbio
Stazione Astronomica Internazionale, Cagliari, Italy and
Istituto di Astronomia dell'Università, Cagliari, Italy

ABSTRACT

A number of works have presented evidence that the seasonal terms of the Earth's rotation are variables from year to year. In this paper, introducing a parameter associated to a mean seasonal kinetic energy we show evidence of about 4 year and 7 yr periodicity. This result is briefly discussed.

The importance of the study of the Earth's annual motion in astronomy and geophysics has been outlined by Lambeck (1980) in his recent book: The Earth's variable rotation. The problem can also be considered important for the definition of a reference system. Because of fluctuations in the Earth's rotation, in fact, the astronomical scale of Universal Time is not uniform. To reduce it, at least in part, UT2 was introduced to take into account the apparently regular annual and semiannual variations. But, owing to the evidence that also seasonal variations are variable, the suppression of UT2 has been recently proposed (Guinot, 1979) at the IAU S. Fernando Symposium n. 82.

Attempts to explain and analyze the variation of seasonal fluctuations were made by several authors. Since seasonal terms are generally attributed in the largest part to the zonal wind circulation, also their variability is assumed to being originated by the same causes. The results of the approaches are not, however, entirely satisfactory. Lambeck and Cazenave (1973) found an increasing amplitude of the semi-annual term and a decreasing amplitude of the annual term. Okazaki (1975) made in greater detail an analysis of the amplitude changes of the annual component and suggested that it varies with a repeating period of 6 yr correlated with the westerly zonal winds at the 500-mb in the zone 35°-55° N. Variations of the amplitudes and phases of the components with periods of 1, 0.5 and 0.33 yr have been also investigated by Korsun and Sidorenkov (1974) for the interval 1956.5-1972.8 . They found several periods and explained them as being the result of the amplitude

187

E. M. Gaposchkin and B. Kołaczek (eds.), Reference Coordinate Systems for Earth Dynamics, 187–190.
Copyright © 1981 by D. Reidel Publishing Company.

modulations of the seasonal waves; in particular a 6-7 yr period was emphasized and attributed to "polar tides".

Sidorenkov (1979) himself suggested an "intermisphere engine" as the responsible mechanism of the seasonal variation. In this case the form of the function describing the seasonal variation should be different from the classical Stoyko formula. Starting from this point of view we have shown that the mean seasonal kinetic energy has_ a fairly regular variation (Poma and Proverbio, 1980). We shall discuss this point in the next section.

Finally a complete study of the Earth's motion should also include the annual polar motion but,because of its proximity to the Chandler resonance, such an,investigation is a more difficult task.

2. THE SEASONAL FUNCTION

Let us describe the seasonal variations of the Earth's rotation as a quite general periodic function having fundamental period T equal to one year and the mean annual value equal to zero. Then we can represent this "seasonal function" S as a trigonometric series

$$S = \sum_k (a_k \cos 2k\pi t + b_k \sin 2k\pi t) \tag{1}$$

where t is a fraction of a year and k = 1,2,3,...

According to the observational evidence the Fourier coefficients of the harmonics of order k > 3 can be neglected, with a high degree of approximation. Let us introduce the mean amplitude A defined by

$$A^2 = P = 2T^{-1} \int_0^T S^2 \, dt \tag{2}$$

and related to the Fourier coefficients a_k and b_k by the Parseval equation

$$P = \sum_k (a_k^2 + b_k^2) \tag{3}$$

We assume the quantity P, which is proportional to a mean kinetic energy, as a parameter representative of the seasonal variation of the rate of the Earth's rotation.

In order to evaluate this parameter we have used the values of the rotational velocity m_3, derived from the 5-day data of UT1 - IAT, published by the BIH, for the period 1962.0 - 1979.0. The values of the fun-

ction S have been obtained from those of m3 after filtering the long
term variations by the use of running means (Korsun and Sidorenkov, 1974).
Another method which led to equivalent results was used and described
by us in the above mentioned paper.

Finally, by using as input data the values of S thus obtained, both
sides of equation (3) have been calculated over an annual period sli-
ding by quarterly intervals. The integral of the (2) has been numerically computed and the coefficients a_k and b_k (k = 1,2,3) have been determined by the least squares method starting from equations of conditions
(1). The results for A^2 and p agree fairly well with one another;
a_k and b_k of the annual, semi-annual and quarterly components are also
in discrete agreement with those found by Korsun and Sidorenkov (1974).
This should confirm the validity of the method and the reliability of
results. The values of P expressed in (ms/day)^2 are plotted in Fig. 1.
The annual values for the period 1956-1961 have been carried out using
relatively few data and must be considered less precise and homogenous.
It can be clearly seen from the diagram that P varies apparently in a
regular manner. It must be first emphasized that the variation of P appears to be more regular than the above mentioned fluctuations obtained,
analysing separately the amplitude of the annual or semi-annual components, by the above mentioned authors.

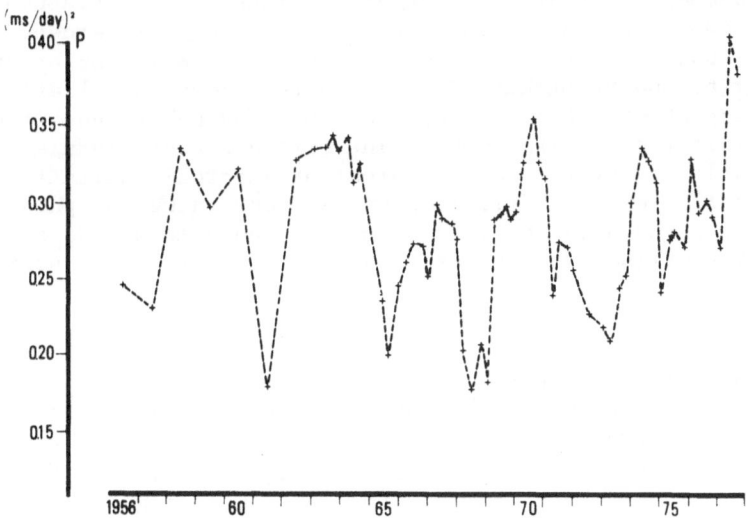

Figure 1.   Variation of P, square of the mean seasonal amplitude

These results seem suggestive of a modulation of the seasonal variation in the rate of rotation with periodicity of about 7 yr and, less
definite, of about 3,5 year. If this latter is removed an interesting
result is obtained. The residuals are very well fitted by a sine curve
of period 7 yr and amplitude 0.022 (ms/day)^2. (Fig. 2)

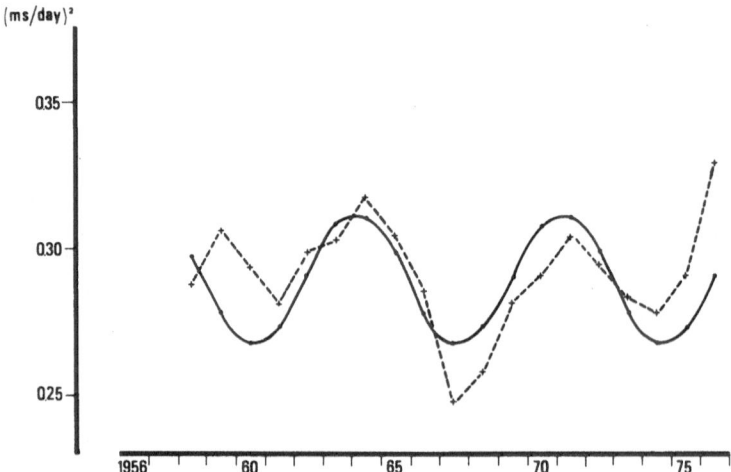

Figure 2. 4-yr running means of P (dashed line); the sine curve, with 7-yr period, has been computed by the least square method.

CONCLUSIONS

The existence of periodical fluctuations in the seasonal variation of l.o.d. of a period of about 3.5 and 7 yr are suggested by our results.

The assumption that the modulations found in the seasonal variations of l.o.d. could have different causes cannot be disregarded. The perio-dicity of about 7.5-yr discovered in the spectrum in the atmospheric mass distribution by Dickman (1977) and the global oscillation of the ocean having about a 3.5 year period showed by Kakuta and Onodera (1972) could explain the results found. The origin of both modulations might also be explained by the same external or internal excitation function. So there still remain difficult problems concerning the physics and energy flux engaged in the same modulation phenomena. We are indebted to to Dr. L.V. Morrison for the critical reaching of the manuscript.

REFERENCES

Dickman, S.R.,1977, Geophys. J.R. astr. Soc., 51, 229
Guinot, B., 1979, Proc. of the IAU Symposium n. 82, p. 16
Kakuta, C., Onodera, E., 1972, Publ. Int. Obs. Mizusawa 8,2
Korsun, A.A., Sidorenkov, N.S., 1974, Sov. Astron., Vol. 18, No. 3
Lambeck, K., The Earth's Variable Rotation, Cambridge Univ. Press,1980
Lambeck,K., Cazenave, A., 1973, Geophys. J.R. Astron. Soc., 32, 79
Okazaki, S., 1975, Publ. Astron. Soc. Japan, 27, 367
Poma, A., Proverbio, E., Proc. 4th Intern. Symp. "Geodesy and Physics of the Earth", Karl-Marx-Stadt, 1980 in press
Sidorenkov, N.S., 1979, Proc. of the IAU Symposium n. 82, pp. 61-64

POSITIONING WITH THE GLOBAL POSITIONING SYSTEM

Clyde C. Goad
National Geodetic Survey
National Ocean Survey
National Oceanic and Atmospheric Administration
Rockville, MD 20852
USA

ABSTRACT

This year (1980) the U. S. Department of Defense has scheduled to
have in operation six satellites of the Global Positioning System
(GPS), which will provide timing and three dimensional position
recovery potential to North America during certain segments of a
day. By the mid-eighties, continuous timing and three dimensional
recovery from 18 GPS satellites are planned. Although the GPS is
designed for fast position recovery to the 10-meter level, extended data collection periods could yield subdecimeter relative
positioning on a routine basis.

INTRODUCTION

One of the most promising systems for highly accurate position
determination in the next decade is the NAVSTAR Global Positioning
System (GPS) being developed by the U. S. Department of Defense.
This system, which has been described extensively in recent
literature (e.g., Parkinson, 1979), will satisfy many present
navigation requirements and new demands for geodetic control, in
a very cost-effective manner.

P-CODE

The P-code (a 0 degree or 180 degree pseudorandom phase shift
sequence) is used to modulate the 1575.42 MHz (L1) and 1227.60 MHz
(L2) carrier frequencies emitted by the satellites at a 10.23 MHz
rate. A dense spectrum of sidebands is produced because the code
does not repeat itself for a week. The spectrum is spread over
roughly a 20 MHz bandwidth centered on the carrier frequency.

191

E. M. Gaposchkin and B. Kołaczek (eds.), Reference Coordinate Systems for Earth Dynamics, 191–194.
Copyright © 1981 by D. Reidel Publishing Company.

In standard GPS navigation receivers, the incoming signals are
cross-correlated with similar reference signals generated by the
receiver. The time delay between the reference signal and the
satellite signal is determined by finding the time delay which
maximizes the cross-correlation. The time delay, when multiplied
by the speed of light in a vacuum, is called the pseudorange.
The difference between the L1 and L2 pseudoranges is used to calculate the ionospheric contribution to the higher frequency (L1)
pseudorange. By combining the L1 pseudo-random range information
for four or more satellites the receiver position with respect to
the GPS reference system and the offset of the receiver clock from
GPS system time can be determined.

RECONSTRUCTED CARRIER PHASE METHOD

The difference between the incoming satellite signal and the
properly time-delayed receiver reference signal is essentially a
sine wave beat frequency. After removing the data message, a
low-frequency signal generated in the receiver is phase-locked to
the beat frequency to improve the signal-to-noise ratio. This
signal is called the reconstructed carrier, even though it is at
a much lower frequency than the original carrier. Whenever the
distance from the satellite to the receiver changes by one wavelength of the carrier frequency, the phase changes by one cycle
in the reconstructed carrier. The phase also depends on differences in initial phase and frequency between the receiver and
satellite clocks.

The mathematical description of the reconstructed carrier phase
technique is given in Bossler, et al. (1980). It is seen that
the phase difference between two stations from a satellite will
give the range difference with an ambiguity equal to some multiple of the carrier wavelength along with any clock mismatch between the two ground stations. Again differencing this phase
difference at the two ground stations with the same measurement
but to another satellite (a second difference) will eliminate
the mismatch from the receivers' clocks. The ambiguity will still
remain, however. The method of resolving these ambiguities is to
solve for measurement biases along with the vector components from
one station to the other. The biases will be poorly determined if
only a few minutes of observations are available, but the uncertainty decreases as the observation time increases. However, because of the need to correct for the ionosphere, the effective ambiguity length for this method is 10.7 cm, instead of the 19.0 cm
wavelength for the L1 frequency.

It appears likely that the main accuracy limitation in using the
reconstructed carrier phase method will come from uncertainty in
the tropospheric propagation corrections. The portion resulting
from the dry part of the atmosphere can be well modelled from

surface measurements of the pressure. However, the portion of
the correction due to water vapor distribution is more difficult.
If the most accurate values for relative positions are desired,
it will probably be necessary to incorporate water vapor radiometers at each receiver site for base lines greater than roughly
3 km in length.

A simulated test situation for the P-code used in Bossler, et al.
(1980) was repeated using the double difference phase technique.
The solution state consisted of the differences in absolute receiver coordinates and measurement biases representing the ambiguities. The data rate was assumed to be only one phase measurement per receiver pair per satellite every 10 minutes to avoid
overly optimistic results from oversampling. Measurement noise
was assumed to be random and to have an amplitude of 2 cm to reflect the 1 cm level of residual error in the tropospheric refraction correction for each of the four measurements used in
forming the double difference.

The results indicate that resolution of the ambiguities is possible, even allowing for the shorter effective ambiguity length mentioned earlier. A detailed description and results may be found
in Bossler, et al. (1980).

It is then legitimate to carry out a new simulation in which biases are not solved for as long as the ambiguities can be clearly
resolved. The results of such a simulation are as follows:

$$\text{sigma (dx)} = 0.6 \text{ cm}$$

$$\text{sigma (dy)} = 0.5 \text{ cm}$$

$$\text{sigma (dz)} = 1.3 \text{ cm}$$

These results indicate that the ultimate resolution potential
using 2 hours of tracking data may be at the 1 or 2 cm level.
This is much better than the 10 cm level originally desired.

CONCLUSIONS

It is felt that using the reconstructed carrier phase method will
be useful in exploiting the potential of the NAVSTAR Global
Positioning System. Obviously there is a great deal of additional research needed to achieve this goal. Studies to date indicate that a highly portable receiver with an antenna similar in
size to those used with modern Doppler receivers can be developed.
Depending on funding, prototype receivers may be available by
1982. If prototype receivers perform according to expectations,
operational copies could be available in the 1984-1985 time
frame.

REFERENCES

Bossler, J. D., C. C. Goad, and P. L. Bender, 1980, Using the
    Global Positioning System (GPS) for geodetic positioning,
    accepted for publication in Bulletin Geodesique.

Parkinson, B. W., 1979, Global positioning system (NAVSTAR),
    Bulletin Geodesique, 53, 89-108.

RELATIONSHIP BETWEEN TERRESTRIAL AND SATELLITE DOPPLER SYSTEMS

Gérard Lachapelle
Sheltech Canada
400-4 Ave S.W., P.O. Box 100
Calgary, T2P 2H5, Canada

Jan Kouba
Earth Physics Branch
1 Observatory Crescent
Ottawa, K1A 0Y3, Canada

ABSTRACT

   Classical horizontal geodetic networks are commonly combined with
space observations, mostly satellite Doppler, in order to optimize the
accuracy of geodetic control points and, thus, satisfy as many types of
users as possible. Since satellite Doppler observations refer to a
fully defined three-dimensional reference system and terrestrial observations, through the presence of Laplace stations (astronomical longitude and azimuth), contribute also to the pole and longitude orientati-
ons, it is imperative to ensure the highest possible degree of compatibility between the astronomical and satellite Doppler systems to maintain optimization of the accuracy of control points. Since gravity and
geopotential (in the form of spherical harmonics) data are usually combined to evaluate geoid undulations and deflections of the vertical
which are in turn used to reduce terrestrial angular and range observations, it is equally imperative to ensure that the satellite Doppler
system and that of the geopotential solution are truly geocentric and
thus compatible with the gravity data which should refer to a single
equipotential surface. In order to estimate the degree of compatibility in terms of longitude orientation between satellite Doppler and
geodetic astronomical systems as realized by current observations,
astrogeodetic (based on CIO pole, BIH longitudes, and NWL9D satellite
Doppler system) and gravimetric deflections of the vertical were compared at several hundred stations of the Canadian geodetic framework
and U.S. transcontinental traverse. It was found that, when using the
U.S. data subset only, incompatibility between the zero geodetic meridian plane of the NWL9D system and the zero astronomic meridian plane
of the BIH was of the order of $0\overset{''}{.}8$, which is in good agreement with
previous results. However, inter-comparisons between various North
American subsets revealed inconsistencies between areas of up to $0\overset{''}{.}8$
(between Canadian and U.S. geodetic astronomical longitude observations).
These results are based on the assumption that gravimetric deflections
are bias free. The geocentricity of the NWL9D system with respect to
other systems such as the Goddard Earth Models and SAO Standard Earths
is also analyzed by comparing satellite Doppler derived geoid undulati-

E. M. Gaposchkin and B. Kołaczek (eds.), Reference Coordinate Systems for Earth Dynamics, 195–203.

ons with GEM and SAO SE undulations. An incompatibility of 4 m in Z (axis) exists between the origin of the NWL9D system and that of the other systems.

1. INTRODUCTION

Classical horizontal geodetic networks consist of terrestrial angle and distance measurements supplemented by astronomical observations which provide geodetic azimuth orientation through the well known Laplace equation

$$\alpha = A - (\lambda_a - \lambda_g) \sin \phi_g$$

where α is the resulting geodetic azimuth calculated from observed astronomical azimuth A and longitude λ_a, and derived geodetic longitude λ_g and latitude ϕ_g. The establishement of an earth fixed conventional geodetic reference system was traditionally done by specifying the geodetic coordinates of a point on the earth surface and the geodetic azimuth (through the use of astronomical observations) of a line from this point. This ensured the geodetic system to be compatible with the astronomical system (Star catalogue, pole, and zero meridian) used to reduce astronomical observations. Additional Laplace azimuth observations were made at other points to provide additional strength to the geodetic networks. This was the case with the North American Datum 1927 (NAD27) (Can. Inst. of Surv. 1974).

The advent of space techniques and, in particular, of satellite Doppler positioning, has provided geodesists with the capability of determining earth fixed geocentric three-dimensional positions with a sub-metre accuracy. Satellite Doppler data will be merged with classical horizontal geodetic networks to define the North American Datum 1983 (NAD83) (U.S. Dept of Commerce 1978). Several hundred satellite Doppler stations have already been established in Canada (Boal & Kouba 1978) and the U.S. (Strange & Hothem 1976) for this purpose. These North American satellite Doppler stations are estimated to be accurate and consistent at the 1 m level (Kouba & Hothem 1978). The station positions refer to the NWL9D (or, equivalently, NWL9Z) reference system (Anderle 1974, 1976) which was intended to be geocentric and to have its z axis coinciding with that implied by the Conventional International Origin (CIO pole). The longitude origin was specified to be consistent with the zero meridian of the Bureau International de l'Heure (BIH) within 1". In view of the merging of satellite and terrestrial data, it is necessary that terrestrial astronomical observations (used in Laplace azimuths) and satellite positions refer to the same astronomical pole and zero meridian, e.g., CIO pole and BIH meridian, otherwise network distortions will occur. It has already been established that the z axis of the NWL9D system is consistent with CIO pole within $0\overset{''}{.}05$ (e.g., Hothem 1979). In view of this and of the fact that it would be difficult to study both pole and longitude origin from North American data only due to coupling between the pole x coordinate and $\partial\lambda$ (longitude origin), the present investigation deals with the longitude origin (Section 2) and geocentricity (Section 3).

It was decided among North American countries that NAD83 will be geocentric. Geocentricity will be realized through NWL9D satellite Doppler positions using an adequate datum definition (e.g., Kouba 1978). It is therefore important that the NWL9D satellite coordinate system be "as geocentric" as possible or that at least its position with respect to the true geocentre be as accurately known as possible in order to apply appropriate transformations. In addition, the NWL9D system should be compatible (in terms of geocentricity) with the geopotential model to be used to provide low harmonic components of geoid undulations and deflections of the vertical to be used to reduce terrestrial angle and distance measurements to the reference ellipsoid. The geocentricity of the NWL9D system was tested against other "geocentric" systems and results are reported in Section 3.

2. ZERO MERIDIAN OF NWL9D VERSUS BIH

The longitudes of NWL9D can be compared with those of BIH using either Very Long Baseline Interferometry (VLBI) or astrogravimetric techniques. Results using the VLBI technique, which is based on direct comparisons of satellite Doppler and VLBI data, are very consistent and indicate that NWL9D longitudes (East) should be increased by 0."8 ($\overline{+}$0."1) to make the zero geodetic meridian plane of the NWL9D system parallel with the zero astronomical meridian plane of the BIH (Hothem 1979; Langley et al 1979). However, results using the astrogravimetric technique and data in North America are not consistent and suggest significant biases in the geodetic astronomical longitude data.

The astrogravimetric technique, which is described in detail in Kouba & Lachapelle (1979), consists of comparing (absolute) geocentric gravimetric prime vertical components η_g (of the deflection of the vertical) defined as angular differences between normals to the ellipsoid and to the geoid with astrogeodetic prime vertical components η_a defined as

$$\eta_a = (\lambda_a - \lambda_g) \cos \phi$$

where λ_a is the astronomical longitude referred to CIO pole and BIH zero meridian and λ_g the geodetic longitude based on the NWL9D system. Unless both BIH and NWL9D zero meridian planes are parallel, η_a will be biased and this bias will be the difference between η_g and η_a since both quantities should theoretically be the same. The additional presence of random errors in both η_a and η_g necessitates the use of several well distributed data points. It is also possible to determine additional biases such as offset parameters Δx, Δy, and Δz and pole coordinate differences δx and δy using meridian components ξ_g and ξ_a of the deflection of the vertical in addition to η_g and η_a in a least squares solution. However, Δx, Δy, and Δz are better determined through the use of

TABLE 1

COMPARISON OF NWL9D AND BIH LONGITUDES IN NORTH
AMERICA USING THE ASTROGRAVIMETRIC TECHNIQUE

Solution No.	Description	δy^*	$\delta\lambda^{*\dagger}$	$\hat{\sigma}_o$
1	597 Canadian points East of long. 248°E	.01∓.08	.34∓.14	1.10
2	142 Canadian points Laplace stations $\phi<60°$, $\lambda>248°E$	-.14∓.14	-.01∓.21	.83
3	628 U.S. points $\lambda>248°E$	-.30∓.04	.76∓.05	.65
4	47 Canadian Laplaces $49°<\phi<53°$ $248°E<\lambda<280°E$	-.27∓.19	.16∓.30	.88
5	27 Canadian Laplaces as sol. No.4 with year $\geq$ 1968	-.29∓.22	.34∓.34	.75
6	101 U.S. points $46°<\phi<49°$ $248°<\lambda<280°E$	-.19∓.09	.46∓.13	.55

* In arcsecs
† With respect to NWL9D zero meridian plane

a worldwide data set (Cf. Section 3). In the present case, they were
constrained to zero which implies that the geodetic coordinates λ_a and
ϕ_a used to derive η_a and ξ_a refer to a geocentric system, i.e., NWL9D
is geocentric. The non-geocentricity of NWL9D in the z axis reported
in Section 3 will affect mostly δy but not $\delta\lambda$. δx is fixed to zero in
view of the coupling effect mentioned earlier.

The η_g and ξ_g components for several hundred points in Canada and
the U.S. for which η_a and ξ_a were available were predicted using a combination of geopotential coefficients (In this case, GEM10B - See Lerch
& Wagner 1978) and surface gravity data (Lachapelle 1978) according to
the method described in Lachapelle (1977). Canadian astronomical data
used to derive η_a and ξ_a was partly reduced to conform with CIO pole
and BIH zero meridian as described by Vamosi (1977). Canadian geodetic
coordinates were obtained from the October 1977 test adjustment of the
terrestrial and satellite Doppler data (Beattie et al 1978) and are
related to the NWL9D system through a longitude correction of 0".65
(Kouba 1978). U.S. η_a and ξ_a were provided by the National Geodetic
Survey; the geodetic coordinates were obtained from an unconstrained

adjustment of the transcontinental traverse and then converted to NWL9D
(Gergen 1979); astronomical longitudes were referred to the BIH zero
meridian.

Results are listed in Table 1. The $\delta\lambda$ values represent the amount
(in arcsecs) which should be added to NWL9D longitudes to make these
compatible with BIH longitudes. Differences between the various solutions of $\delta\lambda$ are due to regional biases in either or both gravimetric
and astrogeodetic deflections of the vertical. Solution No. 3 is in
agreement with VLBI results quoted earlier and with astrogravimetric
results of White & Huber (1979) which are based on a limited U.S. data
sample. Solution No. 2, which is in excellent agreement with an azimuth
misorientation of 0".5 (which corresponds to the original -0".65 correction applied to the NWL9D longitudes for the October 1977 adjustment -
See Kouba 1978) between terrestrial and combined terrestrial-satellite
Doppler data solutions of the October 1977 adjustment (Beattie et al
1978), implies that Canadian geodetic astronomical and NWL9D longitudes
are in agreement; this contradicts VLBI results and astrogravimetric
results in the U.S. (Solution No. 3). However, Solutions No. 4, 5 and
6, which are made of data subsets from Solutions No. 2 and 3, show that
results for $\delta\lambda$ vary significantly. If we assume that gravimetric components η_g and ξ_g are bias free, this indicates that North American geodetic astronomical longitude observations are not only incompatible
between Canada and the U.S. but are also affected by significant regional
biases. This could affect the new NAD83 unless adequate precautions are
taken. The biases are larger than anticipated previously and are not
fully understood at present.

Inconsistencies are also noted between the various solutions for
δy. This could be due to either incompatibilities between astronomical
latitude observations/reductions or/and biases in the gravimetric meridian components ξ_g of the deflection of the vertical. The negative sign
trend for δy is consistent with the z axis offset of the NWL9D system
from the geocentre reported in Section 3. A 4 m offset would amount to
-0".09 in δy. The values of $\hat{\sigma}_0$ in Table 1 indicate that for all six
solutions but No. 1 the estimated variances of either or both gravimetric
and astrogeodetic deflection components were pessimistic. The estimated
variances of the gravimetric deflection components were calculated
according to Lachapelle (1977) and ranged from 1" to 2". The estimated
variances of the astrogeodetic components were set to 0".7 (ξ_a) and
1"·cosϕ (η_a) respectively.

3. GEOCENTRICITY OF NWL9D

This was analysed by comparing satellite Doppler (NWL9D) derived
geoid undulations N_D (N_D is h minus H where h is the satellite Doppler
derived ellipsoid height and H, the sea level or orthometric height)
with undulations derived from various geopotential models of the gravity
field. Since all systems should, in principle, be geocentric, the first-
degree harmonic of the geoid undulation,

$$N_1(\phi,\lambda) = \Delta x \cos\phi \cos\lambda + \Delta y \cos\phi \sin\lambda + \Delta z \sin\phi,$$

should be zero in both cases and a comparison solution in Δx, Δy and Δz between both sets of undulations should give zero for these translation parameters. A well distributed, worldwide set of undulations is required to obtain a meaningful solution. Such a solution, which also included a fourth parameter, namely N_0 the zero-degree harmonic of the geoid undulation (which provides information about the semi-major axis of the mean earth ellipsoid), was carried out by the U.S. Defense Mapping Agency Hydrographic/Topographic Center (DMAHTC) (Grappo 1979) using 290 globally balanced satellite Doppler stations. Geopotential models used for these comparisons were GEM10 (Lerch et al 1977) and GEM10B (Lerch & Wagner 1978). DMAHTC (Grappo, personal communication, July 1980) recently expanded comparisons to include GEM9 (Lerch et al 1977), Smithsonian Astrophysical Observatory (SAO) Standard Earth (SE) III (Gaposchkin 1973) SAO SE IV.3 (Gaposchkin 1976), SAO Global Gravity Field and WGS72(12,12). These results, summarized in Table 2, are more conclusive since GEM, SAO and WGS72 geopotential models are practically mutually independant. All solutions exhibit a fairly consistent Δz value of 4 m which is the z coordinate of the "geocentre" of NWL9D with respect to the "geocentre" of the geopotential models. This result is in agreement with that of (Hothem 1979) obtained from a direct comparison of satellite Doppler (NWL9D) and satellite laser ranging stations in the U.S. The high degree of consistency (in terms of geocentricity) between SAO and GEM geopotential models was also well demonstrated by Schaab & Groten (1979) using 10° grid data sets for comparison.

TABLE 2

GEOCENTRICITY OF NWL9D VERSUS GEM, SAO SE AND WGS72 MODELS

(Using 290 globally balanced Doppler stations)

Geopotential Model	Δx [*]	Δy [*]	Δz [*]	a [*]
GEM9	0.9	−0.1	3.9	6378138.5
GEM10	0.7	−0.2	4.0	6378135.7
GEM10B	0.6	0.3	4.3	6378136.6
SAO SE III	0.6	−0.4	2.7	6378138.4
SAO SE IV.3	0.5	0.2	3.4	6378138.6
SAO GRAV. MODEL	1.2	0.1	3.3	6378138.2
WGS72(12,12)	0.8	−0.5	4.8	6378139.3

* In metres

The Δz value of 4.8 m obtained when using WGS72(12,12) geopotential model is interesting since WGS72 and NWL10E (which is used to calculate orbits for the NWL9D system) geopotential models are expected to be correlated and, thus, to have the same "geocentre". However, the above result is also consistent with Anderle (1980) who reports a Δz of 2.4 m when comparing GEOS-3 altimetric data (using orbits in NWL9D) with the NWL10E geopotential model. Also, Malyevac & Colquitt (1980) finds no significant difference between their SEASAT-1 orbit solutions using NWL10E and GEM10 geopotential models respectively. Yet, J.G. Marsh (Personal communication, March 1980) of NASA/Goddard Space Flight Center reports a 5 m difference between satellite positions using GSFC and NWSC (Naval Surface Weapons Center) orbits respectively. These findings suggest incompatibilities in station computation but compatibility (in terms of geocentricity) of the NWL10E geopotential model (which is used for the orbit computations in the NWL9D system) with GEM and SAO models. Since these models are independant and consistent (in terms of geocentricity), they can be assumed to be truly geocentric at the 1 m accuracy level. This implies that the NWL9D system is off the geocentre by about 4 m in z. Results reported by West (1980) are in disagreement with the above since comparisons between SEASAT-1 altimetric data (using orbits in NWL9D) and GEM10B and WGS72 geopotential models give Δz values between 0.0 and 0.3 m.

The results listed in Table 2 for the semi-major axis (a) of the mean earth ellipsoid will not be discussed in detail here. However, it is recalled that they are in agreement with the value of 6378137 m adapted by the International Association of Geodesy for Geodetic Reference System 1980 at its Canberra 1979 General Assembly following recommendations by Moritz (1979). The 4 m offset in the z axis of the NWL9D system is the cause for the best fitting semi-major axis of the mean earth ellipsoid obtained from North American data only (Lachapelle 1979; Grappo 1979) to be systematically 3 m lower than the above value of 6378137 m based on worldwide data.

4. CONCLUSIONS

The NWL9D system, which is used to calculate satellite Doppler positions worldwide, appears to be compatible with CIO pole at the 0."05 accuracy level. Its longitudes (East) should be increased by 0."8 (∓0."1) to make its zero geodetic meridian plane parallel with the zero astronomic meridian plane of the BIH. z cooordinates should be increased by 4 m (∓1m) to make its centre coincide with the geocentre.

ACKNOWLEDGEMENT

The first author initiated this research while still at the Geodetic Survey of Canada, Ottawa. Mr. G.A. Grappo, DMAHTC, Washington, D.C., and Dr. J.G. Marsh, NASA/GSFC, Greenbelt, Md, have kindly provided us with unpublished data.

REFERENCES

ANDERLE, R.J. (1974) Role of Artificial Earth Satellites in Redefinition of the North American Datum. The Can. Surv., Vol. 28, No.5, pp. 590-597.

ANDERLE, R.J. (1976) Point Positioning Concept Using Precise Ephemeris. Proc. Satellite Doppler Positioning Intern. Geod. Symp., pp. 47-75.

ANDERLE, R.J. (1980) Accuracy of Mean Earth Ellipsoid Based on Doppler, Laser, and Altimeter Observations. Naval Surface Weapons Center Techn. Rep. 80-4.

BEATTIE, D.S., J.A.R. BLAIS, and M.C. PINCH (1978) Test Adjustment of the Canadian Primary Horizontal Network. Proc. Second Intern. Symp. on Problems Related to the Redefinition of North American Geodetic Networks, NOAA, U.S. Dept of Commerce.

BOAL, J.D., and J. KOUBA (1978) Adjustment and Analysis of the Satellite Doppler Network in Canada. Ibidem.

CANADIAN INSTITUTE OF SURVEYING (1974) Abbreviated Proceedings of Interr Symp. on Problems Related to the Redefinition of North American Geodetic Networks. The Can. Surv., Vol. 28, No.5.

GAPOSCHKIN, E.M. (1973) Smithsonian Standard Earth (III). Smithsonian Astrophysical Observatory Spec. Rep. No. 353.

GAPOSCHKIN, E.M. (1976) Gravity Field Determination Using Laser Obser-vations. Center for Astrophysics Preprint Series No. 548.

GERGEN, J.G. (1979) The Relationship of Doppler Satellite Positions to the U.S. Transcontinental Traverse. Proc. Second Intern. Geodetic Symp. on Satellite Doppler Positioning, Austin.

GRAPPO, G.A. (1979) Determination of the Earth's Mean Equatorial Radius and Center of Gravity From Doppler-Derived and Gravimetric Geoid Heights. Presented at XVII General Assembly of the Intern. Union of Geodesy and Geophysics, Canberra.

HOTHEM, L.D. (1979) Determination of Accuracy, Orientation and Scale of Satellite Doppler Point-Positioning Coordinates. Proc. Second Intern. Geodetic Symp. on Satellite Doppler Positioning, Austin.

KOUBA, J. (1978) Datum Considerations for Test Adjustments of Canadian Primary Horizontal Networks. Proc. Second Intern. Symp. on Problems Related to the Redefinition of North American Geodetic Networks, NOAA, U.S. Dept of Commerce.

KOUBA, J., and L.D. HOTHEM (1978) Compatibility of Canadian and U.S. Doppler Station Network. Ibidem.

KOUBA, J., and G. LACHAPELLE (1979) Orientation of Doppler and Astro-nomical Observations in Canada and the United States Using Gravi-metric Deflections of the Vertical. Collected Papers of the Geo-detic Survey of Canada, pp. 55-74.

LACHAPELLE, G. (1977) Estimation of Disturbing Potential Components
    Using a Combined Integral Formulae and Collocation Approach.
    Manuscripta Geodaetica, Vol. 2, pp. 233-262.

LACHAPELLE, G. (1978) Evaluation of 1°x 1° Mean Free Air Gravity Ano-
    malies in North America. Collected Papers of the Geodetic Survey
    of Canada, pp. 183-213.

LACHAPELLE, G. (1979) Comparison of Doppler-Derived and Gravimetric
    Geoid Undulations in North America. Proc. Second Intern. Geodetic
    Symp. on Satellite Doppler Positioning, Austin.

LANGLEY, R.B., W.H. CANNON, W.T. PETRACHENKO, and J. KOUBA (1979)
    LBI and Satellite Doppler: Baseline Comparisons. Ibidem.

LERCH, F.J., S.M. KLOSKO, and R.E. LAUBSCHER (1977) Gravity Model
    Improvement Using Geos-3 (GEM9 & 10). Presented at Spring Meeting
    of American Geophysical Union, Washington, D.C.

LERCH, F.J., and C.A. WAGNER (1978) Gravity Model Improvement Using
    Geos-3 Altimeter (GEM10A & GEM10B). Presented at Spring Meeting
    of American Geophysical Union, Miami.

MALYEVAC, C.W., and E.S. COLQUITT (1980) NSWC Doppler Computed
    SEASAT-1 Orbits. Internal Report, Naval Surface Weapons Center.

MORITZ, H. (1979) Report of Special Study Group No. 5.39 of IAG on
    Fundamental Geodetic Constants. Presented at XVII General Assembly
    of Intern. Union of Geodesy and Geophysics, Canberra.

SCHAAB, H., and E. GROTEN (1979) Comparison of Geocentric Origins of
    Global Systems from Uniformly Distributed Data. Bull. Geod. 53,
    pp. 11-17.

STRANGE, W.E., and L.D. HOTHEM (1976) The National Geodetic Survey
    Doppler Satellite Positioning Program. Proc. Satellite Doppler
    Positioning Intern. Geodetic Symp., pp. 207-227.

U.S. DEPT OF COMMERCE (1978) Proceedings of Second Intern. Symp. on
    Problems Related to the Redefinition of North American Geodetic
    Networks. Washington, D.C.

VAMOSI, S. (1977) Reduction of Astronomical Latitudes, Longitudes and
    Azimuths 1910-1975 to the FK4 System and to the CIO Pole. Collec-
    ted Papers of the Geodetic Survey of Canada, pp. 251-284.

WEST, G.B. (1980) SEASAT-1 Satellite Altimeter Observations in the
    Determination of a Mean Ellipsoid. Presented at Spring Meeting of
    American Geophysical Union, Toronto.

WHITE, H.L., and D.N. HUBER (1979) Longitude Orientation of the Doppler
    Reference System as Determined from Astronomic and Gravity Obser-
    vations. Proc. Second Intern. Geodetic Symp. on Satellite Doppler
    Positioning, Austin.

SOME CONSIDERATIONS IN THE USE OF VERY-LONG-BASELINE-
INTERFEROMETRY TO ESTABLISH REFERENCE COORDINATE SYSTEMS
FOR GEODYNAMICS

Douglas S. Robertson
National Geodetic Survey, National Ocean Survey,
National Oceanic and Atmospheric Administration
Rockville, Maryland 20852 U.S.A.

ABSTRACT

Present knowledge of the number, distribution, proper motion and
structures of extragalactic radio sources indicates that there should be
no problem in defining a celestial reference frame with stabilities of a
few milliseconds of arc over time spans of the order of a decade. One of
the limiting factors appears to be the structure of the sources. By
measuring and monitoring these structures, the stability could probably be
improved by as much as one or two orders of magnitude. Even without this
improvement, a network of properly distributed fixed observatories making
regular interferometric observations of these radio sources could be used
to define a terrestrial coordinate system that could be maintained at the
few centimeter level over indefinitely long time periods. Such a stable
terrestrial reference system would be useful for a host of modern geodetic
and geodynamic applications, including, in particular, studies of the time
varying deformations and relative motions of lithospheric plates. The
National Geodetic Survey has already begun work on a three station base
network of permanent observatories under project POLARIS as a first step
toward implementing the new celestial and terrestrial reference frames.
It is hoped that others will join in the effort and make the new reference
frames a reality by the middle of this decade.

1. Introduction

From the very inception of both astrometry and geodesy, a central
problem has been the search for a suitable celestial reference frame from
which the motions of the Earth could be measured. Central to this search
has been the identification of suitable fiducial points. Galactic stars
were used as fiducial points as early as the dawn of history. By the late
nineteenth century evidence for both random and systematic motions of
these objects had begun to accumulate. Early in the twentieth century the
discovery of the extra-galactic nature of spiral nebulae raised the
possibility that celestial objects could be located at sufficient
distances that their proper motions would be negligible. However, the
attempt to use galaxies as fiducial points was hampered by their large

E. M. Gaposchkin and B. Kołaczek (eds.), Reference Coordinate Systems for Earth Dynamics, 205–216.
Copyright © 1981 by D. Reidel Publishing Company.

angular extent and diffuse nature (Sandig, 1974). In 1963 the discovery
of the extra-galactic nature of quasars and related radio sources opened
the possibility that such compact extra-galactic objects could be used as
fiducial points. The development of Very-Long-Baseline Interferometry
(VLBI) a few years later not only introduced a technique which enabled
observers to measure the positions of such radio sources with
unprecedented precision, but also demonstrated that the radio sources were
of extraordinarily small angular extent, of the order of a millisecond of
arc. At the same time it was quickly demonstrated that the radio sources
were not point-like on a scale of milliseconds of arc, but rather
contained structure which could be quite complex in form and vary with
time. These structures are small, and their effects on the VLBI
observations are small (typically a few centimeters). Nevertheless, they
must be dealt with if the full precision of the VLBI techniques is to be
exploited. These structures are also a function of the observing
frequency. Unless otherwise noted, the discussion in this paper will
focus on structures at X-band (8 GHz), a commonly employed frequency for
geodetic applications.

The study of extra-galactic radio sources is in its infancy. There
is at present no consensus among researchers on basic questions such as
how many radio sources exist, where they are located, and what physical
mechanisms are responsible for their observed structures and features.
Preliminary studies of radio source catalogs suggest that there are
probably thousands X-band radio sources detectable with present VLBI
equipment (Shaffer, 1980), of which only about a hundred have actually
been observed. Of that hundred, only about thirty have been studied in
enough detail to determine anything about their structure at the
millisecond of arc level. Even this amount of information is sufficient
to determine that these sources can be used to form a celestial reference
frame that is an order of magnitude more stable than the present stellar
reference frame, and there are good reasons to hope that improvements of
two or more additional orders of magnitude may eventually be achievable.
In this paper I discuss some of the observations of radio source
structures and their significance for the definition of a celestial
reference coordinate system, and also briefly discuss the methods by which
such a coordinate system can be related to other useful celestial and
terrestrial coordinate systems.

2. Radio Source Characteristics

2.1 Introduction

In interpreting source structure observations it must be kept in mind
that in many cases the measurements are extremely limited. A single
measurement of the amplitude and phase of the VLBI fringes from one
baseline can be used to infer at most a single component of the
two-dimensional Fourier transform of the brightness distribution of the
source (see, e.g., Cohen, 1973). Therefore, observations over many
baselines are required to determine the brightness distribution of the
source reliably. Furthermore, the phase information in the VLBI fringes

is normally corrupted by systematic errors; as a result many of the determinations of source structure are made using fringe amplitude data alone. The source structure determinations are therefore often based on data which, by themselves, are inadequate to define unambiguously the structures of the radio sources being observed. The interpretations are then based on underlying assumptions or models which, while plausible, may not accurately reflect the real structure of the sources. In other words, many of the structure models contained in the literature are merely consistent with the observed data, rather than required by those data. Recently the use of differenced phase observations (the so-called "closure phase") has enabled experimenters to recover a portion of the phase information of the VLBI fringes, and thereby greatly increase the observational constraints on the structure of the sources (Rogers et al., 1974). This technique, when used with a large number of different baselines, should enable experimenters to determine quite reliable source structure maps in the future. At present, we have sufficient evidence to conclude, first, that structure at the millisecond of arc level does exist, and, second, that the source structure can vary significantly on a time scale of months.

2.1.1 Structure Characteristics

A wide variety of structures has already been identified, including simple points, points with halos, double and multiple points, jets, etc. (See, e.g., Wittels et al., 1975, Shaffer et al., 1975, Schilizzi et al., 1975, and Kellerman et al., 1977.) The effect of a typical millisecond-level structure on the VLBI observables was considered by Cotton (1980). According to his calculations, this effect should have a maximum amplitude of about 5 cm. Further, his results indicate that this maximum amplitude is reached only at points close to nulls in the fringe visibility function. By avoiding such nulls it should be possible to keep the effects of millisecond-level structures well below the level of 5 cm. Although this has not been rigorously established for all possible structures, it seems unlikely that the case considered by Cotton is atypical. Therefore, it seems likely that with only the most rudimentary allowances for structure effects (e.g., avoiding data close to nulls in the fringe visibility function) it should be possible to limit the systematic errors in baseline estimation caused by source structures to a level substantially less than a decimeter.

It is only when we wish to improve dramatically on this level of accuracy that the necessity for dealing with source structures arises. Two strategies for dealing with source structures suggest themselves. The simplest course would be to make an exhaustive search for a set of radio sources, well distributed in the sky, which have no detectable structure at the level of accuracy desired. If such a set of sources could be found, then presumably errors resulting from unmodeled structure effects would cease to exist. A more general course of action would be to determine the structure of a set of sources to the desired precision by means of VLBI observations. Once the structure is known, then corrections to the VLBI observations can be made to the desired degree of accuracy.

2.1.2 Time Variations of Radio Source Structure

The source structure problems introduced in section 2.1 are further complicated by the fact that the structure may vary with time. In fact, variability appears to be the rule among radio sources rather than the exception. Large numbers of radio sources exhibit considerable variation not only in total brightness (Dent and Kapitzky, 1976), but also in detailed structure. The observed changes could be caused by either physical motions of compact components of the sources, or changes in brightness of relatively stationary components, or both. It is not always easy or even possible to distinguish between these possibilities. In one sense it doesn't matter which possibility proves to be correct, because both phenomena would require corrections to the VLBI observations of about the same magnitude. However, as we shall see, it should be easier to deal with the case of stationary components of varying brightness than the case of sources whose components are physically moving.

The magnitude of the problems caused by source structure variations is indicated in Table 1. Here I have listed the apparent relative velocities of compact components of radio sources as inferred from various VLBI observations, tabulated in order of increasing red-shift (Z) of the radio source. In most cases the interpretation of the observed changes in the source structures as motions of discrete components is somewhat controversial (see, e.g., Cohen et al., 1977). The differences between the tabulated velocities for a given source could be a result of different observing techniques, different model assumptions, actual variations in the radio sources themselves, or all of the above. In spite of the possible unreliability of specific inferred velocities, these velocities can be used to estimate the scale of the corrections required. The magnitudes of the velocities in table 1 range from several milliseconds of arc per year, in the case of 3C120, down to a small fraction of a millisecond per year. In other words, in the absence of explicit corrections for these effects the source structure variations could lead to systematic errors in VLBI position determinations at the decimeter level on a time scale of years to decades.

The best method for dealing with effects of variations in source structure will depend on reliable determinations of the exact nature of the variations. The obvious and simplest strategy would be to select sources whose structure is observed to not vary. One serious problem with this idea is the possibility that the sources so selected might be simply in a temporarily quiescent phase, and on a time scale of years might begin to exhibit serious structural changes. Unfortunately there is at present no adequate way to guarantee that this will not occur. None of these structure variations have been monitored for more than a decade. A longer observing span will be required before we can discuss their long-term behavior with confidence. An idea which bears investigation is the possibility that stable sources could be found among radio sources with high red-shifts, on the assumption that their (presumed) greater distance will mitigate the effects of actual variations within the source. (Note that the red-shift of each of the objects listed in Table 1 is fairly low.)

Table 1. Observed velocities of radio source components, in milliseconds of arc per year.

SOURCE	Z	VELOCITY	REFERENCE
3C84	0.018	0.8 ± 0.3 ms/yr	Kellerman et al., 1971
		0.12	Preuss et al., 1979
3C120	0.032	1.5	Shaffer et al., 1972
		1.1	Shaffer et al., 1972
		4.2	Shapiro et al., 1973
		5.0	Kellerman et al., 1973
		1.0	Kellerman et al., 1973
		1.2	Kellerman et al., 1973
		0.6	Wittels et al., 1975
		1.8	Cohen et al., 1977
		2.9	Cohen et al., 1977
		1.51 ± 0.13	Seielstad et al., 1979
		3.12 ± 0.34	Seielstad et al., 1979
3C273B	0.158	0.99	Cohen et al., 1971
		0.47	Cohen et al., 1971
		0.9	Schilizzi et al., 1975
		0.32	Cohen et al., 1977
		0.41	Seielstad et al., 1979
3C279	0.538	0.43 ± 0.1	Whitney et al., 1971
		0.72	Cohen et al., 1971
		0.66	Cohen et al., 1971
		0.26	Cohen et al., 1971
		1. (?)	Kellerman et al., 1974
		0.27	Cohen et al., 1977
		0.5 ± 0.1	Cotton et al., 1979
3C345	0.595	0.2	Cohen et al., 1976
		0.09	Wittels et al., 1976A
		0.09 ± 0.03	Wittels et al., 1976B
		0.17	Cohen et al., 1977
		0.16 ± 0.01	Seielstad et al., 1979

It may in fact not be possible to find a set of sources that are reliably free of structure variations. In that case the observing strategy to be followed would depend heavily on the nature of the structure variations. The simplest case would be one in which the structure variations were caused by changes in the radio brightness of relatively stationary components of the source. In this case it would be necessary first to map the relative locations of the components (with VLBI observations), and then monitor the intensity variations and make the

corrections discussed in section 2.1. The accuracy achievable in this
case would be comparable to that discussed in section 2.1. Indeed, the
only difference between this case and the case of nonvarying structure is
the time-varying nature of the corrections and the concomitant necessity
of monitoring the changes in the radio source. The case in which structure
variations are caused by actual physical motions of components of the
radio source is considerably more complicated. It may be possible in this
situation to identify a portion of the structure (e.g., the center of
expansion of an expanding source) whose motion is, if not zero, at least
demonstrably less than the motion of the faster components. If this can
be done, then the slower moving point can be used as a reference point to
which the changing source maps can be referred, thereby reducing or
eliminating the problem.

The worst case that we need to consider is the case that all radio
sources will be found to have moving components, and stationary component
cannot be identified. It would then be necessary to rely on averaging
techniques, based on the assumption that the average motion of all of the
radio sources has no bias, to determine the coordinate system. This
technique would then resemble the technique used in classical astrometry
for dealing with the proper motions of stars, but with two important
differences: the motions are orders of magnitude smaller, and biases are
unlikely to occur in the radio source motions of the sort that are
introduced in stellar motions by the rotation of the Milky Way galaxy.

2.2 Relative Proper Motion Measurements

Up to now we have been dealing only with observations of single
sources. Another important set of constraints on the behavior of radio
sources comes from differential measurements of closely spaced pairs of
radio sources. The significance of the close spacing of the sources in
the sky is that many of the systematic errors affecting VLBI observations
(e.g., atmospheric effects) can be very nearly cancelled out, allowing the
exploitation of the full precision of the VLBI fringe phase observables.
In a case recently reported in the literature (Shapiro et al., 1979), the
relative coordinates of 3C345 and NRAO512 were measured with an
uncertainty of about 0.3 millisecond of arc, and an upper bound of 0.5
millisecond of arc per year was placed on the relative proper motion of
the centers of brightness of the two sources. Notice that this upper
bound is larger than the apparent motion of components within 3C345
itself. The importance of this result is two fold: first, it demonstrates
a technique for determining the relative coordinates of radio sources with
sub-millisecond of arc precision; second, it suggests that proper motions
of radio sources should not introduce systematic errors in the
determinations of terrestrial coordinates at the few centimeter level for
at least several years. Furthermore, since this particular determination
of relative proper motion is merely an upper bound, we can hope that the
actual motion (if any) and its effects are considerably smaller.

2.3 Conclusions

The effects of radio source structures and time variations appear to pose no difficulty for the definition of coordinate systems with precisions at about the level of a few milliseconds and time scales of less than a decade. In order to progress much beyond these limits, it will become necessary to deal with source structure effects. The level of difficulty involved in this task will depend on whether or not a sufficient number of well-behaved radio sources can be found. If ten to twenty point-like, time invariant sources can be found, then the task will be easy. It should be noted that there are selection effects operating in many of the source structure determinations completed to date. Many studies of radio source structure have been done by scientists whose main interest has been the study of the behavior of the radio sources themselves. Therefore, the work has quite naturally been biased toward the most complex and highly time variable (and therefore the most intrinsically interesting) radio sources. Another important selection effect results from the limited sensitivity of many early VLBI systems. This limited sensitivity forced the observers to concentrate on the brightest radio sources in the sky (e.g., 3C84); in many cases these bright sources have had highly complex structure. As a result the literature on the subject of source structure may very well not present a representative sample of radio source behavior. In order to define a celestial coordinate system which can be used to make terrestrial measurements whose accuracy is much greater than 5 cm, a detailed search is needed to discover and monitor the sources most suitable for defining a reference coordinate system.

3. Relating a Coordinate System Based on Radio Sources to Other Useful Coordinate Systems

In order to utilize a coordinate system based on radio source locations, we will have to relate that coordinate system to other useful celestial and terrestrial coordinate systems. The following is a brief sketch of some of the methods that could be employed for some of the more commonly used coordinate systems.

3.1 Terrestrial Coordinate Systems

3.1.1 Geographic Coordinate Systems

If we define a geographic coordinate system in terms of the locations of fixed radio observatories, then VLBI observations can be employed to relate such a coordinate system directly to a coordinate system based on radio source locations. If the Earth were perfectly rigid this would suffice to form a basic reference network to which other types of geodetic measurements could be referred. For the real Earth the radio observatories are attached to crustal plates, which not only move relative to each other, but may not be perfectly rigid themselves. The motions of the plates relative to each other are believed to range from a few centimeters per year up to around 15 centimeters per year. Just as in the

case of radio source structure problems, the simplest method for dealing
with such tectonic motions may prove to be a policy of simple avoidance.
It should be possible to find stable areas in the interior of continental
plates where the relative motions of points on the plates are
substantially smaller than the relative motion between plates. (If it were
not possible to find such areas, that is, if points within plates move
relative to each other with velocities comparable to the velocities of the
relative plate motions, then the entire concept of crustal plates would
cease to have much meaning.) We should therefore be able to situate radio
telescopes in areas where their relative motions are expected to be of the
order of centimeters per decade or less, and we could then use such a
network for local geodetic control. More importantly, we could use the
network as a reference grid from which measurements into less stable areas
could be made. In the same fashion, measurements made between similar
grids located on different plates could be used to define the magnitude
and time scale of interplate motions.

3.1.2 Spin axis and Equator

By monitoring the orientation of an array of fixed radio telescopes
as the Earth completes a diurnal rotation, it is possible to estimate the
orientation of the Earth's spin axis with respect to the geographic
coordinate system defined by the location of the telescopes ("polar
motion" or "wobble") as well as its orientation with respect to the
celestial coordinate system ("nutation"). At the same time, it is
possible to monitor the Earth's rotation about its spin axis ("UT1"). The
capability of VLBI observations to link a celestial coordinate system
directly to the two most important terrestrial coordinate systems forms
the basis of the National Geodetic Survey's project POLARIS, which will be
discussed in section 4.

3.2 Celestial Coordinate Systems

Curiously, it is more difficult to connect a celestial coordinate
system based on radio sources to other celestial coordinate systems than
it is to connect it to terrestrial coordinate systems. Nevertheless there
are several promising methods for making such connections, which should be
pursued.

3.2.1 Stellar Coordinate Systems

There are two general methods for relating radio source coordinates
to stellar coordinates. Both methods require the existence of sources of
radiation that are sufficiently bright to be detected in both the optical
range and the radio range of the spectrum. A large number of radio
sources have optical counterparts whose coordinates could be measured with
optical techniques. Unfortunately the optical counterparts tend to be
exceedingly faint (magnitude 14 or smaller), too faint to be observed with
classical astrometric instruments. Another problem is that the
uncertainty of the classical observations is several orders of magnitude
larger than that of the VLBI observations. Both of these problems could

be alleviated by using the U.S. National Aeronautics and Space
Administration's (NASA's) forthcoming space telescope for optical
astrometric work, a possibility which is currently under study (Shelus,
1980). A second method for relating the two coordinate systems entails
finding ordinary stars which radiate significant amounts of energy in the
radio portion of the spectrum. Unfortunately nearly all stars are
exceedingly faint in this portion of the spectrum. To date, only a single
star in the FK4 catalog, Algol, has been observed with radio
interferometry, and that one only during infrequent radio "flare" periods
(Clark et al., 1976). Both of these methods for relating the coordinate
systems depend heavily on the assumption that the location of the origin
of the radio radiation is coincident with the origin of the optical
radiation. Much time and further study will be required to shed light on
the validity of this assumption.

3.2.2 Ecliptic Coordinate Systems

Identifying the location of the ecliptic plane and other Solar System
coordinates relative to the location of radio sources is made difficult by
the paucity of sources of radio energy in the Solar System. One method
which could be of use involves monitoring the occultation of
extra-galactic radio sources by the Moon. At present the accuracy of this
method is limited mainly by our knowledge of the details of the lunar
limb. Another useful method involves the observation of artificial radio
sources (spacecraft) with VLBI techniques. Some promising results have
been obtained from observations of the VIKING spacecraft in orbit around
the planet Mars (Ratner, 1980). This method is limited mainly by the
difficulty and expense of acquiring suitable spacecraft.

4. Operational Plans for Defining Radio Source Based Coordinate Systems

The following is a brief outline of some of the present operational
plans for defining and making use of a celestial coordinate system based
on radio sources.

4.1 Project POLARIS

The U.S. National Oceanic and Atmospheric Administration's National
Geodetic Survey has begun to develop a network of three VLBI stations
within the continental United States under project POLARIS. The project
is described more fully in Carter (1980). The primary objective of this
project will be the determination of polar motion and universal time with
a precision of ten centimeters or better on a time scale of several
observations per week. A secondary objective of this network will be the
determination of fundamental reference coordinate systems for geodetic and
astrometric work. The basic network was selected to provide as large a
triangle as possible within the United States without involving
tectonically unstable areas such as the far western portions of the United
States. The network will involve stations in Massachusetts, western Texas
and southern Florida. Testing of portions of the network will begin in
September, 1980, in conjunction with the IAU/IUGG project MERIT. Full

operational status of the three U.S. stations should be achieved by 1983.
Since the usefulness of such a network would be greatly enhanced by
cooperative observations from additional stations, particularly stations
located outside the North American plate, the National Geodetic Survey
will invite and would like to encourage such cooperation. In addition,
the possibilities for employing the transportable VLBI receivers currentl
under development by NASA in conjunction with the POLARIS network are
being actively considered. Such receivers would be useful in helping to
resolve a host of modern geodetic and geodynamic problems, including such
things as time varying deformations and relative motions of lithospheric
plates. The detailed operational plans for the POLARIS network are in a
rudimentary state of development; many important questions including the
amount of observing time that can be dedicated to research into the natur
and behavior of radio sources, or the amount of time dedicated to crustal
deformation studies remain to be resolved on the basis of the maximum
available scientific return, consistent, of course, with the basic
objectives of the project.

4.2 NASA/DSN Plans

NASA's Deep Space Network intends to operate its three 210 ft
antennas as a VLBI network in order to provide polar motion and UT1
determinations in support of spacecraft tracking operations. The plans
for this operation are outlined in a paper by Fanselow et al., (1980)
presented at this conference.

4.2 Other

There are several other groups of scientists doing research in VLBI,
including groups at NASA, JPL, M.I.T., Haystack Observatory, Cal Tech,
NRAO, also groups in Canada, Europe, Japan, and China. Their
activities are too numerous and varied to be cataloged here, and while
most of these groups do not plan to make regularly scheduled observations
for the purpose of determining reference coordinate systems, their work
will nevertheless make an important contribution to this effort,
particularly in the areas of source structure determinations, source
catalog observations, and studies of tectonic activity.

5. Conclusions

The goal of using VLBI observations to establish a reference
coordinate system for geodynamics with a precision at the sub-decimeter
level seems to be within our grasp. Serious possibilities exist for
improving this precision level by an order of magnitude or more. A great
deal of work is currently underway to achieve these goals, and a large
number of researchers around the world are working on aspects of this
problem. The full realization of the goals of determining and making use
of such coordinate systems will be achieved only with a high degree of
international cooperation of the sort which this conference can help to
foster. Given such cooperation, the future is likely to see a pattern of
important and interesting new developments and discoveries.

REFERENCES

Carter, W. E.: 1980, NASA Conf. Pub. 2115, pp. 455–460.
Clark, T. A., L. K. Hutton, C. Ma, I. I. Shapiro, J. J. Wittels, D. S. Robertson, H. F. Hinteregger, C. A. Knight, A. E. E. Rogers, A. R. Whitney, A. E. Niell, G. M. Resch, and W. J. Webster: 1976, Ap. J. Letters, 206, 1107.
Cohen, M. H.: 1973, Proc. IEEE, 61, pp. 1195.
Cohen, M. H., W. Cannon, G. H. Purcell, D. B. Shaffer, J. J. Broderick, K. I. Kellerman, and D. L. Jauncey: 1971, Ap. J., 170, pp. 207–217.
Cohen, M. H., A. T. Moffet, J. D. Romney, R. T. Schilizzi, G. A. Seielstad, K. I. Kellerman, G. H. Purcell, D. B. Shaffer, I. I. K. Pauliny-Toth, E. Preuss, A. Witzel, and R. Rinehart: 1976, Ap. J., 206, L1–L3.
Cohen, M. H., K. I. Kellerman, D. B. Shaffer, R. P. Linfield, A. T. Moffet, J. D. Romney, G. A. Seielstad, I. I. K. Pauliny-Toth, E. Preuss, A. Witzel, R. T. Schilizzi, and B. J. Geldzahler: 1977, Nature, 268, pp. 405–409.
Cotton, W. D.: 1980, NASA Conf. Pub., 2115, pp. 193–197.
Cotton, W. D., C. C. Counselman III, R. B. Geller, I. I. Shapiro, J. J. Wittels, H. F. Hinteregger, C. A. Knight, A. E. E. Rogers, A. R. Whitney, and T. A. Clark: 1979, Ap. J., 229, pp. L115–L117.
Dent, W. A., and Kapitzky, J. E.: 1976, Astron. J., 81, pp. 1053–1068.
Fanselow, J. L., O. J. Sovers, J. B. Thomas, F. R. Bletzaker, T. J. Kearns, G. H. Purcell, Jr., D. H. Rogstad, L. J. Skjerve, and L.E. Young: 1980, this volume.
Kellerman, K. I., D. L. Jauncey, M. H. Cohen, D. B. Shaffer, B. G. Clark, J. Broderick, B. Ronnang, O. E. H. Rydbeck, L. Matveyenko, I. Moiseyev, V. V. Vitkevitch, B. F. C. Cooper, and R. Batchelor: 1971, Ap. J., 169, pp. 1–24.
Kellerman, K. I., B. G. Clark, D. L. Jauncey, J. J. Broderick, D. B. Shaffer, M. H. Cohen, and A. E. Niell: 1973, Ap. J., 183, L51–L55.
Kellerman, K. I., B. G. Clark, D. B. Shaffer, M. H. Cohen, D. L. Jauncey, J. J. Broderick, and A. E. Niell: 1974, Ap. J., 189, pp. L19–L22.
Kellerman, K. I., D. B. Shaffer, G. H. Purcell, I. I. K. Pauliny-Toth, E. Preuss, A. Witzel, D. Graham, R. T. Schilizzi, M. H. Cohen, A. T. Moffet, J. D. Romney, and A. E. Niell: 1977, Ap. J., 211, pp. 658–668.
Preuss, E., K. I. Kellerman, I. I. K. Pauliny-Toth, A. Witzel, and D. B. Shaffer: 1979, Astron. Astrophys., 79, pp. 268–273.
Ratner, M.: 1980, private communication.
Rogers, A. E. E., H. F. Hinteregger, A. R. Whitney, C. C. Counselman, I. I. Shapiro, J. J. Wittels, W. K. Klemperer, W. W. Warnock, T. A. Clark, L. K. Hutton, G. E. Marandino, B. O. Ronnang, O. E. H. Rydbeck, and A. E. Niell: 1974, Ap. J., 193, pp. 293–301.

Sandig, H. U.: 1974, Proc. IAU Colloq. No. 26, pp. 241-246.

Schilizzi, R. T., M. H. Cohen, J. D. Romney, D. B. Shaffer, K. I.
    Kellerman, G. W. Swenson, J. L. Yen, and R. Rinehart: 1975,
    Ap. J., 201, pp. 263-274.

Seielstad, G. A., M. H. Cohen, R. P. Linfield, A. T. Moffet, J.
    D. Romney, R. T. Schilizzi, and D. B. Shaffer: 1979, Ap. J.,
    229, pp. 53-72.

Shaffer, D. B., M. H. Cohen, D. L. Jauncey, and K. I. Kellerman:
    1972, Ap. J., 173, pp. L147-L150.

Shaffer, D. B., M. H. Cohen, J. D. Romney, R. T. Schilizzi, K. I.
    Kellerman, G. W. Swenson, J. L. Yen, and R. Rinehart: 1975,
    Ap. J., 201, pp. 256-262.

Shaffer, D. B.: 1980, private communication.

Shapiro, I. I., H. F. Hinteregger, C. A. Knight, J. J. Punsky,
    D. S. Robertson, A. E. E. Rogers, A. R. Whitney, T. A. Clark,
    G. E. Marandino, R. M. Goldstein, and D. J. Spitzmesser: 1973,
    Ap. J., 183, L47-L50.

Shapiro, I. I., J. J. Wittels, C. C. Counselman III, D. S.
    Robertson, A. R. Whitney, H. F. Hinteregger, C. A. Knight,
    A. E. E. Rogers, T. A. Clark, L. K. Hutton, and A. E. Niell:
    1979, Astron. J., 84, pp. 1459-1469.

Shelus, P.: 1980, private communication.

Whitney, A. R., I. I. Shapiro, A. E. E. Rogers, D. S. Robertson,
    C. A. Knight, T. A. Clark, R. M. Goldstein, G. E. Marandino,
    and N. R. Vandenberg: 1971, Science, 173, pp. 225-230.

Wittels, J. J., C. A. Knight, I. I. Shapiro, H. F. Hinteregger,
    A. E. E. Rogers, A. R. Whitney, T. A. Clark, L. K. Hutton,
    G. E. Marandino, A. E. Niell, B. O. Ronnang, O. E. H. Rydbeck,
    W. K. Klemperer, and W. W. Warnock: 1975, Ap. J. 196, pp. 13-
    39.

Wittels, J. J., W. D. Cotton, C. C. Counselman, I. I. Shapiro,
    H. F. Hinteregger, C. A. Knight, A. E. E. Rogers, A. R.
    Whitney, T. A. Clark, L. K. Hutton, B. O. Ronnang, O. E. H.
    Rydbeck, and A. E. Niell: 1976a, Ap. J., 206, pp. L75-L78.

Wittels, J. J., I. I. Shapiro, W. D. Cotton, C. C. Counselman,
    H. F. Hinteregger, C. A. Knight, A. E. E. Rogers, A. R.
    Whitney, T. A. Clark, L. K. Hutton, A. E. Niell, B. O.
    Ronnang, and O. E. H. Rydbeck: 1976b, Astron. J., 81, pp. 933-
    945.

USE OF GRAVITY MEASUREMENTS IN DEFINING AND REALIZING REFERENCE SYSTEMS FOR GEODYNAMICS

J. D. Boulanger, N. N. Pariisky
Academy of Sciences of the USSR
Institute of Physics of the Earth, Moscow
L. P. Pellinen
The Central Research Institute of Geodesy
Air Survey and Cartography, Moscow

ABSTRACT

Single measurements of gravity cannot give sufficient information about the position of measuring points with respect to some terrestrial reference system. Only a set of gravimetric stations all over the Earth combined with a determination of their coordinates allows one to determine (from the solution of Molodensky's problem) the heights of these stations with respect to a level ellipsoid with center at the geocenter. Given in addition their heights above some reference ellipsoid, whose position in the Earth's body is fixed through a set of reference points on its surface, the position of the geocenter in the same reference system may be obtained.

Wider opportunities are opened through repeated gravity measurements. They are bound up with the fact that an essential part of nontidal variations of gravity is caused by the height variations of gravity stations. In this report the relation between gravity and height variations is considered, taking into account the latest results of gravity variation investigations.

RELATION BETWEEN GRAVITY VARIATIONS AND HEIGHT VARIATIONS OF GRAVITY STATIONS

Neglecting the centrifugal force variations, for the time being let us represent, according to [1], the variation dg of gravity anomalies on the Earth surface by expansion in spherical harmonics. For the spherical approximation we have, for harmonics of the n'th degree

$$dg_n = (n - 1) \frac{dV_n}{R} - 2 \frac{\gamma}{R} dH_n^\gamma \quad .$$

(1)

Here dV is the variation of the gravity potential at the Earth's surface, dH^γ is the normal height variation (equal to $dH - d\zeta$ where dH is the geodetic height variation, $d\zeta$ is the quasigeoidal height variation), R

217

is the mean Earth radius, and γ is the mean gravity at the Earth's surface.

Zero and first degree harmonics are worth special notice. If effects associated with possible time-dependent variations of the universal gravitational constant (less than 0.1 μGal per year) are neglected, then $dV_0 = 0$ and we obtain

$$dg_0 = -2^{\gamma}/R \ dH_0 \quad . \tag{2}$$

We assume that dg_0 and the observed variations of the angular velocity a) of the Earth's rotation are both caused by the variations of its mean density (or rather radial deformation of certain spherical layers). The known values $d\omega/\omega$ may then be used to estimate the order of possible variations of the radius of the Earth and surface gravity values. According to [2,3], observed irregular variations $d\omega/\omega$ amount to 0.5×10^{-8} for periods of 1 to 3 years, corresponding to the values $dH_0 = 1.8$ cm and $dg_0 = 5.7$ μGal when uniform expansion (or contraction) takes place. When the deformation occurs only in upper layers of the Earth's crust these values will be twice as large.

Using (1) we have for the first degree harmonics

$$dg_1 = -2^{\gamma}/R \ (dH_1 - d\zeta_1) \quad . \tag{3}$$

If some initial reference system has been fixed the dH_1 term would characterize the general displacement of the stations of repeated gravity measurements and the $d\zeta_1$ term that of the geocenter. The result obtained is quite obvious: repeated gravity measurements yield information only about relative displacements of points on the Earth's surface and the geocenter (see also [4,5]).

Hence it is important that repeated gravity measurements be carried out at those stations that realize the global geocentric reference system. Achievement of an accuracy of 10 μGal in absolute or adequate relative (also useful in this case) measurements would allow determination of relative displacements of these stations and the geocenter with an accuracy of 3 cm, which is not yet attainable by other methods. The results obtained will be used not so much for determination of geocenter displacements relative to the Earth's crust as for studying the stability of the reference system fixed by the stations on the Earth's surface.

The work of S. M. Molodensky [1] confirms that elastic Earth deformations in general result in strongly correlated temporal variations of gravity potential and normal heights. He uses the expression for dg_n in the form

$$dg_n = -2^{\gamma}/R \; dH_n^{\gamma} \; (1 - \Xi_n) \tag{4}$$

and gives an estimate of the parameter Ξ_n that depends on the depth ℓ and the character of the deformation source causing the gravity variations. As shown in [1], $|\Xi_n| < 0.1$ for dilatation stresses, for depth ℓ less than 1930 km. For shear stresses $|\Xi_n| < 0.3$. The fact that the second term in (1) is clearly predominant shows the possibility of estimating the order of gravity variations from vertical crust movement data. The latter amount to some mm per year, and corresponding annual variations of gravity are 1-2 μGal.

Planetary gravity variations corresponding to sources located near the core-mantle boundary have some specific features. According to [1] variations of this kind need not be accompanied by height variations at all.

An estimate of possible planetary variations dg_2 is given in [2,3] under the assumption that the values $d\omega$ observed are responsible. It is supposed that the deformation axis conincides with the rotational axis of the Earth so that dg_2 comprises only the second zonal harmonic. The ratio of dg/g to $d\omega/\omega$ has its maximum value at the pole, at a depth ℓ = 120 km. When $d\omega/\omega$ = 0.5 × 10^{-8}, the variations dg can be as much as 29 μGal. The change of dg-variations between the equator and latitudes 60° amounts to 33 μGal.

GRAVITY VARIATIONS CAUSED BY CENTRIFUGAL FORCE CHANGES

The most important effect is caused by polar motion. According to [6,7] the variations expected do not exceed 5 μGal and are largest in higher latitudes. They can be computed from polar motion data within accuracy limits not exceeding some tenths of μGal.

Irregularities in the Earth's rotation give rise to neglected gravity variations of about 0.03 μGal, if $d\omega/\omega$ = 0.5 × 10^{-8}.

ATMOSPHERIC EFFECTS AND INFLUENCE OF WATER REDISTRIBUTION IN THE EARTH'S CRUST

Air and water masses are the most mobile components of our planet and the effect of their movement on gravity variations has been pointed out by many authors.

Atmospheric corrections amount to some tens of 1 μGal but they may be easily accounted for from air-pressure distribution data on the Earth's surface. Only the indirect effect of Earth's crust deformations caused by air-pressure variations remains unclear.

Accounting for gravity variations caused by sub-soil water level variations is the most difficult problem of high-recision gravimetric measurements. From Vikhirev's [8] estimates the effects of natural sub-soil water level variations may be observed practically everywhere. The highest amplitudes of gravity variations caused by variations of the mean annual sub-soil water level amount to 70 μGal. The duration of such variations may be 10 years or more. They are superimposed on annual variations that cause dg-variations of as much as 100 μGal.

The effects of water quantity changes at great depths are expected to be signigicant in extensive artesian basins and may even surpass those of sub-soil water level variations.

Certain problems arise in coastal areas because of water-level variations and corresponding loading on the lithosphere.

Taking into account that high-precision gravity stations are not numerous, the sites for these stations should be chosen on bedrock outcrops. But in such places uplifts of the Earth's crust can often be observed. This is why conclusions about gravity variations based on local observations may turn out to be unrepresentative when studying global geodynamic processes and establishing an Earth reference system. In order to diminish the effect of local gravity variations other measurements should be taken at the same time as hydrogeological investigations: for example, repeated relative gravity measurements between the fundamental stations and closely located satellite stations.

RECENT INFORMATION ON VARIATIONS OF ABSOLUTE GRAVITY

In the late sixties a breakthrough took place in instrumental gravimetry. New instruments — absolute ballistic gravity meters — were developed. At present these instruments have higher accuracy than those for relative gravity measurements, and allow determination of global gravity variations even when only a limited number of stations are used. The stationary absolute gravity meter of Prof. Sakuma in Sèvres, France, and the transportable one of Prof. Faller in USA have been used to determine the origin and scale of the International Gravity Standard Net of 1971 (IGSN-71). Regular absolute gravity measurements were carried out in Sèvres from 1967 to 1973.

A transportable absolute instrument named GABL has been developed in the Institute of Automation and Electrometry of the Siberian Branch of Academy of Sciences of the USSR. The measurements with this instrument started in 1972. Since the instrument has been a somewhat improved and at present allows gravity measurements with a relative accuracy of about 6 to 8×10^{-9}. A little later a transportable gravity meter was developed in Italy. In 1977 the GABL-instrument was compared with the Italian absolute gravity meter and with that of Prof. Sakuma in Sèvres. This comparison has confirmed the high accuracy of all three instruments.

Let us review in detail two groups of measurements with the GABL-instrument.

Boulanger [9,10] has compared gravity values at European stations of the IGSN-71 Potsdam S-13 (two determinations), Helsinki and Sèvres A_3, obtained in 1976-78 either with the GABL-instrument or by relative methods in the system of the International gravimetric station at Ledovo (Moscow), with the results obtained in 1969-70 when establishing the IGSN-71 system. The mean increase in the value of gravity at these stations is 45 ± 2.7 µGal.

The change recorded seems to be explained as follows. According to Prof. Sakuma's data the least gravity value recorded in Sèvres was observed when establishing the IGSN-71. This value increased in 1972 by 45 µGal. When adjusting the IGNS-71 the g-value in Sèvres was given a very large weight and the gravity value obtained in 1969 was adopted as the origin of the system. The result was that gravity values at all the stations of IGSN-71 were too low.

In Figure 1 the results of repeated gravity determinations carried out with the GABL-instrument at the stations Novosibirsk, Ledovo and Potsdam are presented according to data obtained by Boulanger. For the sake of better visual demonstration all the meaurements are reduced to the gravity value at Ledovo. It can be seen that in 1975-78 the mean rates decrease in the gravity values at all three stations are practically identical and equal to 9.9 ± 1.3 µGal/year. In the first approximation the variations observed may be represented as quasiperiodical with a period of 5 years and an amplitude of about 20 µGal. Note that a similar gravity variation with about the same period and amplitude of about 25 µGal was observed in Sèvres from 1967 to 1973. The measurements with the GABL-instrument were carried out at three stations situated along a line over 5.000 km long, so this phenomenon may be said to extend over at least a considerable part of Eurasia.

For further investigations of the observed phenomenon it will be necessary to continue the observations at former stations and at the same time to extend them to other stations. It is obvious that the cause and character of gravity variations observed,

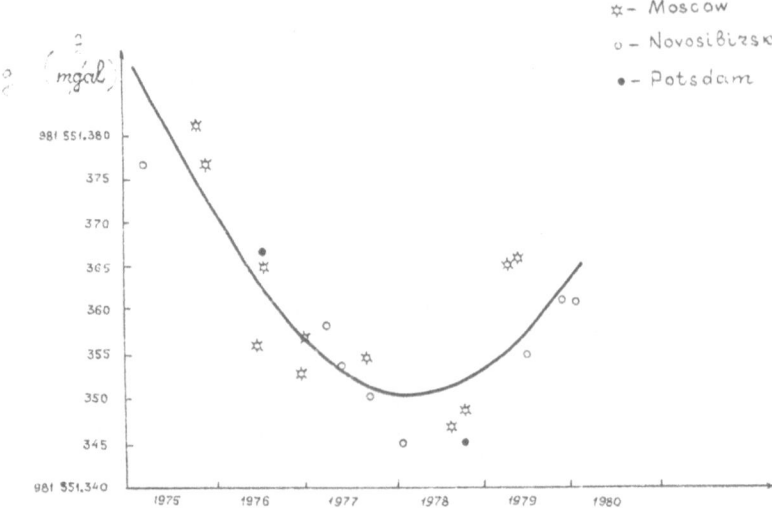

Figure 1.   Secular change in gravity referred to the gravity at
Ledovo.

as well as the effect of systematic instrumental errors, may be
decided with greater certainty after carrying out additional
regular comparisons of ballistic gravity meters of different kinds
and simultaneous absolute measurements at some stations located
far away from each other.

SOME CONSIDERATIONS CONCERNING GLOBAL INVESTIGATIONS OF NON-TIDAL
GRAVITY VARIATIONS

For global investigations of non-tidal gravity variations
aimed at realizing reference systems for geodynamics it is necessary to perform repeated measurements at a great number of stations
uniformly distributed on the Earth's surface. Such a net may
simultaneously decide whether irregular variations of the Earth's
rotation are connected with variations of the inertia tensor or
with internal motions.

Sites for these stations should be chosen to be stable hydro-
geologically, if possible, on crystalline bedrock not less than
100 km seacoasts and large lakes. To diminish the effects of
local vertical crustal movements and other factors resulting in
local gravity variations, it is desirable to establish in the
vicinity of each station special satellite-stations connected
with the primary ones by high-precision relative gravity measurements carried out at regular time intervals. It is also desirable
to establish a net of stations at which regular measurements by
stationary instruments can be carried out. The tidal parameters
should be determined at the stations of repeated measurements

using several instruments. Data describing the hydrologic regime,
atmospheric conditions and height variations with respect to sea-
level should also be recorded. The published results should
contain detailed information about all the reductions made, or
the data necessary for calculating such reductions.

The instrumental potentialities of gravity measurements are
far from being exhausted. Let us hope that these measurements
will in the future become one of the most effective means of investigating geodynamic phenomena and establishing global coordinate systems.

References

Молоденский С.М., 1980, "Изв. АН СССР. Физика
Земли",   4, 3-14.

Молоденский М.С., Молоденский С.М., Парийский Н.Н.,
1975, "Изв. АН СССР. Физика Земли",   6, 3-11.

Pariisky N.N., 1978, "Bull. di Geofis. Teor. ed. Appl.", 20,
No. 80, 413-418.

Mather R.S., 1973, "Bull. géod.", No. 108, 187-209.

Юркина М.И., 1978, "Геод. и картография", М.,
30-35.

Bursa M., 1972, "Stud. Geophys. et Geod.", 16, No. 2, 122-125.

Lambeck K., 1973, "Stud. Geophys. et Geod.", 17, No. 4, 269-271.

Вихирев Б.В., 1976, "Повторн. гравиметр. наблюдения
М., ВНИИГеофизика, 4-23.

Boulanger Yu.D., 1979, Publ. Finn. Geod. Inst., No. 89,
Helsinki, 20-26.

Буланже Ю.Д., 1980, "Геод. и картография,   5, 14-
20.

LOCAL GEOMETRIC AND GRAVIMETRIC DATUM AND ITS RELATION TO THE GLOBAL TERRESTRIAL REFERENCE SYSTEM

R. Kelm
Deutsches Geodätisches Forschungsinstitut, Abt. I
Marstallplatz 8, D-8000 München 22

ABSTRACT

In active tectonic zones relative earth surface displacements of mm-order and relative gravity disturbances of 0.1 mgal size may be expected between two survey epochs. To monitor and separate these small effects significantly special local free networks have to be designed which are measured by relative observations of highest accuracy (mm- and μgal-level). For the connection of the local nets to a global terrestrial reference system a suitable local datum for point coordinates (geometric datum) and gravity field parameters (gravimetric datum) has to be defined within the terrestrial reference system.

The definition of the local datum is given and its dependency on the relative observation type discussed. For the datum realisation mainly satellite techniques are proposed. The tolerance interval of the datum parameters with respect to the global terrestrial system are analysed. Adjustment results referred to the local datum can directly be transformed into the terrestrial system without loss of significance, if the datum does not exceed the tolerance intervals. As a consequence requirements for the fundamental stations establishing the terrestrial reference system are discussed.

1. INTRODUCTION

Each geodetic observable quantity depends on physical fields which are of dynamical nature with periodical and nonperiodical frequencies, as there are the gravity field, the atmospheric field and the strain field of the earth. Especially the strain and the gravity field are strongly correlated, because they are mainly excited by the same sources, e.g. by the change of mass distribution in the earth crust or by tidal effects. The more accurately the geodetic observations can be measured, the more sensitive they react on the frequency band of the physical fields and the more difficult it is to separate the systematic effects from each other by suitable estimation models. On the search for a separation tool

225

E. M. Gaposchkin and B. Kołaczek (eds.), Reference Coordinate Systems for Earth Dynamics, 225–228.
Copyright © 1981 by D. Reidel Publishing Company.

the fact among others is to be considered that physical fields behave
in local areas of the earth surface differently than in the global
frame. This is a problem of relating suitable estimates of local observations to coordinate estimates in a global terrestrial reference
system. The paper aims at characterising this problem in order to give
some impulses for further research. Therefore it is sufficient to treat
a very simplified case: We only introduce two characteristical types of
observations. For the determination of horizontal strain rates multi-
wave distance measurements with a priori accuracy up to 1 cm for that
distance, and for the vertical direction levelling-combined with relative gravity measurements leading to a distance accuracy up to 1 cm. The
task here is to analyse the possibility of connecting suitable three-
dimensional estimates of functions of distance, levelling and relative
gravity parameters related to a local reference system with coordinate
estimates of a global terrestrial reference system.

2. DEFINITIONS AND PROBLEMS

A Conventional Terrestrial reference System (CTS) is supposed to be
available as defined e.g., by H. Moritz (1979), with the mean rotation
axis Z and a Y-axis perpendicular to it and therefore situated in the
mean equatorial plane - only this two - dimensional case is regarded
further on in the simulation studies.

Some definitions are useful for the description of the task and resulting problems:
The geometric geodetic datum defines the relation between a local reference system and CTS and is given by geometric datum parameters (GDP):
GDP are not significantly estimable by any kind of local observations
and are therfore not dependent on estimable functions in the local reference system. In general, translation, rotation and scale parameters
belong to GDP. When e.g., only distances are observed, not more than a
scale parameter is determined by this observation type, and as GDP remain the translation and rotation parameters.

In the same way a gravimetric geodetic datum can be defined: It relates
a local gravity field to a global one within CTS, and its parameters
(here generally functions) cannot be determined by gravimetric-sensitive
local observations. Because the task here is mainly of geometrical nature,
this datum will not be dealt with further on in this paper.

In the CTS it is possible to estimate unbiased Cartesian coordinates
e.g., by dynamical satellite techniques. This is not possible in a
local reference frame when only local observations are available. Nevertheless it is useful to define coordinate estimations: Estimable coordinates are Cartesian coordinates within a local reference system which
are estimated by local observation quantities with consideration of
minimal constraints, the kind and number of which are given by the GDP.
For instance in the Y, Z-plane GDP are only two translation parameters
to a certain network point. Then there are two minimal constraints which

have to be chosen well-defined , but can be distributed arbitrarely;
e.g., the two coordinates of a point are taken as errorless in the
estimation procedure.

The datum realization is regarded as the procedure which delivers the
GDP by special observation techniques independent of the local observations.

With these definitions the problem here is described in four topics:
TOP 1: Which kind of GDP has to be defined in the local reference frame?
TOP 2: Which tolerance intervals for the GDP before the datum realization are allowed in order to transform the estimate coordinates into CTS
after datum realization without significant systematical error influences?
TOP 3: Which observation techniques are suitable for the datum realization?
TOP 4: Vice versa - which requirements can be proposed for the establishment of CTS?

3. ANALYSIS AND RESULTS

TOP 1: By local observations translation parameters to CTS cannot
significantly be determined. Distance observations only give information
about the scale in the local network. An open question is if the indirect
observation type of geopotential difference ΔV - derived from levelling
and relative gravity results - significantly fixes the rotation parameter. For the analysis ΔV is linearised around approximate coordinates
and gravity field functions given by geometric and gravimetric datum
parameters based on a mass point model for the local gravity field. The
resulting observation equation is in the Y, Z-plane

$$\Delta V = (b + \delta b) \, \delta y + (c + \delta c) \, \delta z \tag{1}$$

with

$\quad \Delta V$: observed minus approximate potential difference

$\quad \delta y, \, \delta z$: unknown differentials of the coordinate differences of
$\qquad\qquad$ two points

$\quad b, \, \delta b$: sensitivity matrix elements

δb is only effected by local gravity field disturbancies and is negligible for this analysis. b and c are functions of the geocentric radius
vector and do not differ much from each other. When (1) is transformed
by infinitesimal rotation, the rotation part of ΔV - $\delta \Delta V$ - as a function
of the infinitesimal rotation parameter $\delta \varphi$ is

$$\delta \Delta \sim (y \Delta z - z \Delta y) \, \delta \varphi \tag{2}$$

with

$\quad y, \, z$: coordinates of a point in Y, Z-plane

$\quad \Delta y, \, \Delta z$: coordinate differences between two points.

An analysis of (2) shows that

   sign yΔz = - sign zΔy

so that the <u>rotational parameter</u> is estimable in fact. So in this problem only translation parameters are GDP.
<u>TOP 2:</u> For the analysis the results of a numerical simulation study in the Y, Z-plane, as described in the primary task, is interpreted. Presumed that errorless observations are available and the local gravity field is estimated with significant accuracy the true potential difference is disturbed by about 0.01 m^2/sec^2 when the GDP (here only translation parameters) differ from the true values by 1 m. This means that a <u>systematical error</u> of 1 mm in the vertical direction occurs - that can be tolerated. So tolerance intervals of the GDP up to 1 m are allowed.
<u>TOP 3:</u> For the transformation of <u>estimable coordinates</u> of a local net to CTS the exact values of GDP (here the coordinates of the points to which the translation parameters are related) have to be determined by observation types which connect fundamental stations of CTS to that point. In near future this could be realized by laser satellite techniques expecting an accuracy on cm-level.
<u>TOP 4:</u> Only one idea should be given here: Let us review the initial idea presented in the introduction. For comparison of local and global effects - both sensitive to cm level - the accuracy of the estimated coordinates must reach this level, too. But the coordinates of CTS - fundamental stations are mean values - averaged over the time. For a sensitivity analysis it is required that all local and global estimated coordinates relate to the same epoch -. Therefore it is proposed that for each fundamental station motions as a function of t with respect to CTS have to be determined.

4. CONCLUSIONS

For the analysis of the local and global behaviour of the figure and the gravity field of the earth globally <u>and</u> locally observed geodetic quantities are to be used both consistently transformed into CTS. Besides further advantages it helps to separate highly correlated systematic effects caused by physical fields from each other. It is a problem of relating estimable coordinates of local areas - each independently surveyed and estimated in arbitrary epochs - to the CTS. The analysis and results given here - though only being preliminary and restricted to simplified case - nourish the hope that the problems can be solved to a significant accuracy in near future. But nevertheless much research work has to be done for the establishment of CTS and the connection theory between local networks and CTS.

REFERENCE

Moritz, H. (1979), "Concepts in Geodetic Reference Frames", <u>Dept. of Geod. Sci. Rep.</u> 294, Ohio State University, Columbus

ERRORS IN POLE COORDINATE OBTAINED BY THE DOPPLER SATELLITE POSITIONINGS

S. Takagi
International Latitude Observatory of Mizusawa
Japan

ABSTRACT

The pole coordinate obtained by the Doppler observation of Earth satellite shows good results, but there are some important difference from those of the astronomical observation, for example, that of the IPMS.

The Doppler observation of the Navy Navigation satellite has proved to be precise enough to supersede the astronomical observation in the geodynamics. (Anderle, 1965, 1966, 1970, 1971a, 1971b, 1971c, 1974). Yumi et al. (1979) showed that the local terms in the astronomical observation was not found in the Doppler latitude obtained at Mizusawa. This shows the superiority of the Doppler observation to the astronomical one.

However, our investigations show:
a.   Pole coordinate obtained with the Doppler observation has a large secular variation (about $0\overset{''}{.}02$ per year) with respect to that of the IPMS (Report of the IPMS at the IAU Meetings, 1979).

b.   Doppler latitudes derived from north and southbound passes of a satellite show systematic differences (Kakuta et al., 1979).

c.   Doppler latitude shows some unexpected deviation from the astronomical latitude (Takagi, 1974).

It will be necessary for us to make detailed theoretical analysis of errors in the Doppler data before we proceed to use the Doppler results in place of the astronomical ones in the studies of geodynamics. Lambeck (1971) investigated effects of pole motion in the orbital elements of an Earth satellite. Takagi (1975) summarized theoretical results which should be considered in deriving pole coordinates from the results of the Doppler observations of an Earth satellite. O'Tool constructed a precise system of reduction of the Doppler satellite observation (O'Tool, 1976).

E. M. Gaposchkin and B. Kołaczek (eds.), Reference Coordinate Systems for Earth Dynamics, 229–231.
Copyright © 1981 by D. Reidel Publishing Company.

The reduction method of the Doppler observation has been developed by several organizations for the geodetic use, (Anderle, 1974; Beuglass, 1975; Smith et al., 1976). The principles of these methods are not essentially different from each other. The pole motion derived from these methods has complicated correlations with errors in the orbital and instrumental parameters (Anderle, 1974; O'Tool, 1976).

The reduction method is: the data for twenty one TRANET stations are used to determine eleven parameters to be applied to the two day span; two bias parameters in each satellite pass and nine parameters. The nine parameters are six constants of integration, pole coordinates, one scaling factor for atmospheric drag effect, while two parameters are a scaling factor for tropospheric refraction (a scale factor of unity is used as an apriori observation of refraction correction with standard deviation of 0.1) and a frequency bias parameter for each pass.

The perturbations in the satellite orbit include the Earth's gravitational field (up to 19 degree and higher order terms, over the total number about 400), the solid Earth tides (Love's number = 0.26), the solar and lunar attraction, atmospheric drag and direct solar radiation. The satellite orbit is computed by twelfth-order numerical integration of equations of motion.

The next step is to compute coordinates of stations with respect to the two bias parameters for each pass.

The random error in the pole coordinate is estimated as 7.4 cm and 6.8 cm in x and y respectively. The standard deviations of the individual pole position with respect to the mean position for a five day interval are 66 cm and 50 cm in x and y on the average (Anderle, 1976).

| | Mean Difference | | | | S.D. of Difference | | | |
| | DMA-BIH | | DMA-IPMS | | DMA-BIH | | DMA-IPMS | |
	x	y	x	y	x	y	x	y
1973	-.3	- .9	-.2	-1.2	.6	.8	.5	.6
1974	-.1	-1.1	.0	- .9	.5	.5	.6	.4
1975	-.4	-1.1	.0	- .5	.6	.5	.3	.6

S.D. of the Doppler Observation at Mizusawa
(1976.0 - 1977.5)

	Longitude	Latitude	Height (meter)
Daily Mean	$0.''015$	$0.''018$	$0.^{m}147$
5 Days Mean	0.027	0.032	0.288

It is certain that we must make further investigations of the motion of the Earth satellite to get the true pole motion. The precision of the satellite observation is good enough to be used in the investigation of geodynamics and the further improvements of the theory of motion of the satellite affected by the geophysical phenomena, such as, pole motion, will be expected to promote the studies in the geodynamics.

The orbital elements obtained from the satellite observation are combined with the reference system which can not be revised from the observation itself, for the orbital elements to be improved appear in combination with parameters which are to be determined simultaneously. Accordingly, parameters, such as, precession and nutation, should be determined from another point of view. This fact means that the reference system determined from Doppler observations should be to be compared with the fixed reference system determined from the so-called astronomical observation, such as astrometry and VLBI.

REFERENCES

Anderle, R. J., 1965, J. Geophy. Res., 70, 2453-2458.
Anderle, R. J., 1966, in The Use of Artificial Satellite for Geodesy 2, G. Veis, editor, National Technical University, Athens.
Anderle, R. J., 1970, NWL - Technical Report, TR-2432.
Anderle, R. J., 1971a, NWL - Technical Report, TR-2508.
Anderle, R. J., 1971b, NWL - Technical Report, TR-2559.
Anderle, R. J., 1971c, NWL - Technical Report, TR-2889.
Anderle, R. J., 1974, J. Geophys. Res., 79, 5319.
Beuglas, L. K., 1975, NSWC/DL - Technical Report, TR-3173.
Kakuta, C. et al., 1979, J. Geod. Soc. Japan, 25, 194-208.
Kakuta, C., 1978, in The Use of Artificial Satellite for Geodesy and Geodynamics. G. Veis and E. Livieratos, eds., National Technical University, Athens.
Kozai, Y., 1975, Orbit of an Artificial Satellite. Special Publ. of NHK, Japan (Excellent summaries of his discussions on the motion of the artificial satellite and we referred to this book without notice).
Lambeck, K., 1971, Bull. Geod., 101, 263.
Lambeck, K., 1973, Celes. Mech., 7, 139.
O'Tool, J. W., 1976, Proc. Int. Geod. Symposium, Satellite Doppler Positionings, 2 (Las Cruces, USA). DMA editor.
Smith, R. W. et al., 1976, Technical Report No. DMATC 76-1.
Takagi, S., 1974, Proc. IAU Colloquium No. 26 (Torun).
Takagi, S., 1975, Publ. Int. Lat. Obs. Mizusawa, 10, 53-108.
Takagi, S., 1977, Publ. Int. Lat. Obs. Mizusawa, 11, 57-75.
Yumi, S. et al., 1979, Proc. Int. Geod. Symposium. Applied Research Laboratory, University of Texas, Austin.

MOTIONS OF ARTIFICIAL SATELLITES AND COORDINATE SYSTEMS

Yoshihide Kozai
Tokyo Astronomical Observatory, Mitaka, Tokyo 181, Japan

Abstract. In order to compute satellite motions with centimeter accuracy, the reference system, to which they are referred, should be carefully chosen. In fact there are many kinds of the reference systems. In this paper advantages and disadvantages of various reference systems are discussed.

There are many different reference systems, to which satellite positions are referred. Everybody prefers one system to the others, however, there is no system which everybody prefers the best. The mean equator and equinox at a certain epoch are adopted at some institutes as their reference frame for satellite positions, whereas at others the true equator and equinox of data are adopted. Instead of the true equinox the mean equinox of a fixed date can be chosen. It is also possible to adopt the mean equator and equinox at the beginning of a year and to change the system at the beginning of every year. More precisely, there are many choices for the z-axis; namely, the figure axis or the celestial pole axis or the instantaneous spinning axis. Furthermore we may ask what kind of the figure axis should be adopted; the true or mean figure axis. I do not want to conclude here which is the best system, however, will try to discuss how we should do to compute the satellite positions with centimeter accuracy for each case.

None of the reference systems is inertial as their origins are at the geocenter which is in accelerated motion. The effects of the motion of the geocenter to satellite motions are usually included in the lunisolar gravitational perturbations. In fact the disturbing function due to the sun and the moon for the satellite motions is derived by writing their equations of motion in an inertial coordinate system with its origin at the barycenter of the solar system which is assumed to be moving with an uniform velocity.

If there is no rotational motion of the coordinate axes of the reference system with respect to the inertial system, it is not

233

E. M. Gaposchkin and B. Kołaczek (eds.), Reference Coordinate Systems for Earth Dynamics, 233–237.
Copyright © 1981 by D. Reidel Publishing Company.

necessary to add any term to the disturbing function of the satellite
motions, since it is a quasi-inertial system (Moritz, 1979). The
coordinate system with the mean equator and equinox at a certain epoch
as the reference frame is such a system. It has an advantage as it is
easy to describe the equations of motion in this system. However, to
write the expression of the geopotential is not so easy as the coordinates of any point are time-dependent in the system even though the
solid earth is assumed. It is also possible to adopt a quasi-inertial
system to formulate the equations of motion and to introduce an
auxiliary reference to express the geopotential and the station coordinates. As the auxiliary reference frame usually the equator of date
is adopted.

When the equator and the equinox of date are adopted as the frame
of reference, the expression of the geopotential becomes simpler as
the coordinates of any earth fixed point are time-independent. It has
a disadvantage, however, since perturbations are produced as the
coordinate axes move. To discuss more precisely on this system it is
necessary to specify what axis is adopted as its z-axis. Indeed there
are many axes. The system, in which the equator of date and the mean
equinox at an epoch is adopted, is a kind of this system, as there is
no essential difference to treat the perturbations.

Roughly speaking, there are two choices for the z-axis, the figure
axis or the celestial pole axis. The figure axis is the axis of the
maximum momentum of inertia. Therefore, if it is adopted C_{21} and S_{21}
terms vanish in the geopotential. However, as the sun and the moon
are generating tides on the earth, the figure pole is moving around
its mean position by as much as 60 meters (Moritz, 1979). Therefore,
the figure axis is not fixed to the earth in any sense.

However, if the mean position of the figure pole which does not
move with respect to the earth is adopted as the direction of the z-
axis, the coordinates of any of the observing stations are expressed
as the constant values plus tidal motions and the geopotential is
expressed as the sum of the averaged part and the variable part due to
the tidal deformation which can be formulated or can be derived by
analyzing satellite motions. The rotation of the earth around the
mean figure axis is not so simple as that around the celestial pole
axis. However, as its rotation rate is constant with error less than
10^{-6} at most (the distance between the mean figure pole and the
celestial pole being 6 meters), the geopotential can be expressed with
10^{-10} accuracy by assuming that the rotation rate around the mean
figure axis is constant. Therefore, it seems that it is easy to
express both the station coordinates and the geopotential in this
reference system. However, the reference system moves by the precession and the forced and free nutation in the inertial system.

When the celestial pole axis is adopted, the station coordinates
are expressed by the sum of the constant part, the tidal effects and

the effects of the polar motion and the geopotential is the sum of the constant part, the tidal part and the time-dependent C_{21} and S_{21} terms due to the polar displacement. The system does not move due to the free nutation with respect to the inertial system, however, of course, moves due to the precession and the forced nutation. This system has an advantage as the nutation theory is referred to this axis according to an IAU resolution in 1979. It is practically convenient to adopt CIO as the direction to the z-axis.

When any non-uniformly moving reference coordinate system is adopted the equations of motion should include additional terms due to the centrifugal and the Coriolis forces, although acceleration of the motion of the reference system is very small as it usually moves due to the precession and the nutation. However, as the velocity, the energy and the angular momentum of the satellite with respect to the system are different from those with respect to the quasi-inertial system and vary with time, the osculating semi-major axis and eccentricity change with time even for the two-body problem. Of course, the other orbital elements also include their perturbations due to the motion of the system.

There is also a different kind of reference frame, such as one which has been adopted at the Smithsonian Astrophysical Observatory. It is basically a quasi-inertial system even though it has not been stated so before, since the basic equations of motion are those in the inertial system without any additional term. However, in order to make the expressions of the geopotential, the station coordinates and the perturbations much simpler an auxiliary reference for them has been introduced. It is referred to the equator of date and the mean equinox at 1950.0, with respect to which the inclination, the argument of perigee and the longitude of the ascending node of the satellite are given. However, the semi-major axis and the eccentricity are referred to the quasi-inertial system in the sense that the velocity, the energy and the angular momentum computed by formulae for the two-body problem with the osculating semimajor axis and eccentricity are those with respect to the quasi-inertial system, namely, that defined by the mean equator and equinox at 1950.0. The semimajor axis and the eccentricity, therefore, are not disturbed by the motion of the auxiliary system, that is, constant for the two-body problem, however, have some small indirect perturbations through other elements due to the motion of the equator. Therefore, one must be careful to analyze doppler data of the satellite in this system.

For the three angular elements with respect to the auxiliary system (Kozai and Kinoshita, 1973) proved that by adding $\partial i/\partial t$, $\partial \omega/\partial t$ and $\partial \Omega/\partial t$ to the right-hand sides of Lagrange's planetary equations for the inclination, the argument of perigee and the longitude of the ascending node, respectively, the equations hold for any moving auxiliary system. Similarly, any other type of equations of motion can be modified for such systems. For the other orbital elements the equations need not have any additional term as the definitions of the

semimajor axis, the eccentricity and the mean anomaly are the same as
those for the quasi-inertial system.

The partial derivatives are derived by using geometrical relations
between the moving and the fixed systems. In fact as the reference
system is moving the values of the three angular quantities, i, ω and
Ω take values different from the original ones even if there were not
perturbation at all. The partial derivatives are the time derivatives
of the angular quantities without taking into account any perturbation
in the orbital elements. Namely, the elements are constant in deriving
the derivatives.

In the previous paper (Kozai and Kinoshita, 1973) the following
expressions are derived for the partial derivatives:

$$\frac{\partial i}{\partial t} = - \frac{\partial}{\partial t} \{\sin \theta \cos (\alpha - \Omega)\} \quad ,$$

$$\frac{\partial \omega}{\partial t} = \csc i \frac{\partial}{\partial t} \{\sin \theta \sin (\alpha - \Omega)\} \quad , \tag{1}$$

$$\frac{\partial \Omega}{\partial t} = (1 - \cos \theta) \frac{d\alpha}{dt} - \cot i \frac{\partial}{\partial t} \{\sin \theta \sin (\alpha - \Omega)\} \quad ,$$

where θ and α are, respectively, the inclination and the longitude of
the ascending node of the moving reference plane with respect to the
fixed one and are explicitly time-dependent.

When the moving references are the equator and the equinox of date
and the fixed ones are the mean equator and equinox at an epoch, $\sin \theta$
$\sin \alpha$, and $\sin \theta \cos \alpha$ are expressed as,

$$\sin \theta \sin \alpha = \sin \varepsilon_1 \sin \psi \quad ,$$

$$\sin \theta \cos \alpha = \sin (\varepsilon_0 - \varepsilon_1) + 2 \sin^2(\psi/2) \sin \varepsilon_1 \sin \varepsilon_0 \quad , \tag{2}$$

where ψ is the arc between the ascending nodes of the ecliptic at the
epoch referred to the equators and ε_0 and ε_1 are their inclinations.
And ε_1 and ψ change due to the precession and the nutation and for
some cases the free nutation. The expression (2) can be approximated
by,

$$\sin \theta \sin \alpha = (0.3979 + \varepsilon_1 - \varepsilon_0) \sin \psi \quad ,$$

$$\sin \theta \cos \alpha = 0.3651(1 - \cos \psi) - \varepsilon_1 + \varepsilon_0 \quad . \tag{3}$$

In order to obtain the perturbations due to the motions of the
reference frame in the three angular elements it is necessary to take
partial derivatives with respect to time for (1) by assuming that only
θ and α are time-dependent and then to integrate them by assuming that

Ω is also time-dependent. As ψ and ε_1 move, usually, more slowly than Ω, the perturbations introduced are smaller than the motions of the reference frame. As the lunar longitude of the ascending node which enters into the argument of the principal nutation term, 18.6 year period term, moves more slowly even than that of Lageos satellite, the amplitude of the perturbation term produced by this nutation term is smaller than that of the nutation term itself. As the 18.6 year period nutation term's amplitude is known with the accuracy of $0\overset{.}{.}001$, the perturbations in the orbital elements can be computed with centimeter accuracy. The perturbations due to other nutation terms can be computed with the same accuracy by using the existing nutation theory unless any serious resonance is introduced.

When the mean figure axis is adopted as the z-axis, the pole coordinates should be known with centimeter accuracy to compute the perturbations due to them with the same accuracy. Even when the celestial pole axis is adopted, the pole positions should be known with centimeter accuracy to compute the values of time-dependent C_{21} and S_{21}.

Even when the equations of motion are formulated in a quasi-inertial system or in a non-uniformly moving system with additional terms, the perturbation behavior for the three angular elements is not essentially different from discussed here. The semimajor axis and the eccentricity are perturbed for the moving reference system. However, they can be derived with the centimeter accuracy when the expressions of the precession, the nutation and the pole motion are known with the same accuracy. If they are not known with this accuracy, the satellite motions can improve them from time to time.

References

Kozai, Y. and Kinoshita, H., 1973. Effects of Motions of the Equato-
    rial Plane on the Orbital Elements of an Earth Satellite,
    Celestial Mechanics, 7, 356-366.
Moritz, H., 1979. Concepts in Geodetic Reference Frames, Reports of
    Department of Geodetic Science, Ohio State University, No. 294.

ORIGIN AND SCALE OF COORDINATE SYSTEMS IN SATELLITE GEODESY

J. B. Zieliński
Department of Planetary Geodesy, Space Research Centre,
Polish Academy of Sciences, Warsaw, Poland

ABSTRACT

The center of mass of the Earth is commonly taken as origin for the coordinate systems used in satellite geodesy. In this paper the notion of the "geocenter" is discussed from the point of view of mechanics and geophysics. It is shown that processes in and above the crust have practically no impact on the position of the geocenter. It is possible however that motions of the inner core may cause variations of the geocenter of the order of 1 m. Nevertheless the geocenter is the best point for the origin of a coordinate system. Mather's method of monitoring geocenter motion is discussed, and some other possibilities are mentioned. Concerning the scale problem, the role of the constant GM and time measurements in satellite net determinations are briefly discussed.

INTRODUCTION

In a number of coordinate systems, applied in practice as well as proposed, the center of the Earth's mass, "geocenter", is designated as the origin. This is the case for systems realized by means of satellite observations as well as for the global geodetic system and astronomical systems connected with directions to very distant space objects. The notion of the "geocenter" was widely discussed during the Torun Colloquium by Bursa (1974), Moritz (1974), Groten (1974) and mentioned by many others. Some authors expressed the fear that in a non-rigid Earth the geocenter is not stable, but can change its position with respect to the Earth's surface. This undermines the significance of the geocenter as the best origin of the reference frame for geodynamics. Therefore it is worthwhile to reconsider the properties of this particular point.

DEFINITION, MECHANICAL PROPERTIES

Let the physical body be set up within some arbitrary but fixed coordinate system. Each material point is described by the position

239

E. M. Gaposchkin and B. Kołaczek (eds.), Reference Coordinate Systems for Earth Dynamics, 239–250.

vector $\underline{x}(x_1, i = 1, 2, 3)$ and a mass dM. The coordinates of the center of mass are

$$x_{0i} = \frac{\int_E x_i \, dM}{M} \quad , \qquad (i = 1, 2, 3) \tag{1}$$

where M is the total mass of the body, and E is the space of the body.

Integration is extended over the entire volume of the body and the boundary of integration must be specified. In the case of the Earth and the geocenter there exists a certain ambiguity because the limit of the atmosphere is not clearly determined. The author suggests that some conventional altitude be specified, below which the atmosphere will be considered as belonging to the Earth mass system. This altitude height coincide for instance with the juridical limit of the Earth space which probably be established as the 100 km elevation above sea level. However, as we shall soon see that the height of this limit is not very important.

In a freely-rotating rigid body the axis of rotation passes through the center of mass. In the case of the Earth, the observed instantaneous axis of rotation, which from now on we shall call the "spin axis", is connected with the solid Earth, while the definition of the geocenter comprises also the atmosphere. So, neither the spin axis nor the geographical axis, as defined by Munk-MacDonald (1960), need pass exactly through the geocenter. This fact should be taken into account when discussing the rigorous definition of the reference system.

Precise definition of the Earth's rotation axis is not easy because different fractions of the Earth have some differences in rotation. Therefore, a so-called "Tisserand axis" is introduced, on a minimum condition (Moritz 1979a):

$$\int_E (\underline{v} - \underline{\omega} \times \underline{x})^2 \, dM = \min \tag{2}$$

where $\underline{v}$ is the velocity vector of the particle dM, and $\underline{\omega}$ is the angular velocity vector of the total mass M. The integral is minimized by proper choice of the vector $\underline{\omega}$ which in turn defines the axis of rotation. This axis is not detectable in reality, but has the advantage that it is connected by definition with the geocenter.

Also connected with the geocenter are the axes and moments of inertia. The inertial tensor [I] has components

$$I_{ij} \int_E \left(\sum_{k=1}^{3} x_k^2 \delta_{ij} - x_i x_j \right) dM \qquad (3)$$

where i,j = 1, 2, 3 and δ_{ij} is the Kronecker delta. If the coordinate system is centered at the geocenter and the axes are directed in such a way that I_{ij} = 0 when $j \neq i$, then the diagonal components of [I] are called principal moments of inertia: I_{11} = A, I_{22} = B, I_{33} = C. They form a triaxial ellipsoid of inertia which is always centered at the geocenter.

If, on the contrary, the x_i-system is not centered at the geocenter, but the axes are parallel to the former ones, then

$$I_{ii} + M\Delta x_i^2 = A\delta_{i1} + B\delta_{i2} + C\delta_{i3} \qquad (4)$$

according to the Huyghens-Steiner theorem (Suslov, 1946). This property enables us to determine the influence of the center of mass shift on the rotation rate of the body using the angular momentum conservation law:

$$\underline{H} = [I] \cdot \underline{\omega} = \text{const.} \qquad (5)$$

The notation of "geocenter" plays an equally important role in the theory of the Earth's figure. If we expand the gravity potential of the Earth in terms of spherical harmonics, the first-degree terms are simple functions of the coordinates of the geocenter (Heiskanen and Moritz, 1967):

$$U_{11}S_{i1} + U_{12}\delta_{i2} + U_{10}\delta_{i3} = \frac{GM}{R^2} x_{01} \qquad (6)$$

where G is the constant of gravitation and R is the Earth's mean radius. Hence, these terms vanish if the origin of the system is located at the geocenter. Solving the boundary value problem by the Stokes formula we implicitly locate the origin of the reference frame at the geocenter because of the intrinsic property of the boundary value condition

$$\Delta g = - \frac{\partial T}{\partial r} - \frac{2T}{r} = \frac{1}{R} \sum_{n=0}^{\infty} (n - 1) T_n(\theta, \lambda) \qquad (7)$$

This expression does not contain the first-degree terms. So, if we do not make other assumptions, a gravimetrically determined geoid has the center of mass identical with the center of the reference ellipsoid (Heiskanen and Moritz, 1967).

The meaning of the center of mass is most evident in artificial satellite motion theory. The differential equation of motion of the material point in the force field is usually divided into three parts, e.g.

$$\ddot{r} = \underline{\nabla}V + \underline{\nabla}T + \underline{\nabla}F \tag{8}$$

The first term on the right hand side describes the influence of the so-called "central force" which is the gravitational attraction of a point mass. T is a perturbing potential reflecting the fact that the central body is not a point but has a certain structure and finite dimensions. Other forces have also some potential, either harmonic or not, denoted by F. Only the motion in the central force field is described by the closed analytical theory. Kepler's laws require the orbit to be a conic section and the central mass to lie in the plane of the conic at one of the foci.

In the real situation of an artificial Earth satellite, the orbit is never an ellipse, but if we know the perturbations we can reconstitute at any instant the so-called osculating orbit which is a Keplerian one whose plane passes through the geocenter. Having reconstructed this orbit we have a direct relation between the satellite position in space and the geocenter, in terms of direction as well as distance.

The problem is, how exact is our calculation of the perturbations? We have neither a complete model of all acting forces nor a perfect theory of perturbation. The process of improvement of these factors still continues, and the present accuracy is best reflected by the satellite position errors obtained by ephemeris computation, which is nowadays at the centimeter level.

STABILITY OF THE GEOCENTER, GEOPHYSICAL CONDITIONS

When talking about motion we must always define the frame with respect to which the motion takes place. If the discussion concerns the possible motion of the geocenter in relation to the non-rigid Earth the situation becomes really complicated. However, as we have access only to points situated on the surface of the Earth we are interested in relations between the geocenter and these points. Of course we have to accept that each of these points (stations) moves relative to the others, but at a given epoch each point has its fixed coordinates x_i in an arbitary system:

$$x_i^t \in E \quad , \qquad (i = 1, 2, 3)$$

where E is the Earth space, and t is the epoch.

Coordinates of the geocenter at epoch $t = t_0$ are found from (1)

$$x_{0i}(t) = \frac{\int_E x_i{}^t{}_0 dM}{M} \tag{9}$$

Now, suppose that at another epoch $t = t_1$ a majority of points have preserved their positions, so that the system x_i is preserved, but in the limited subspace E'_i there is a mass displacement

$$x_i{}^{t_1} \neq x_i{}^{t_0}$$

This displacement will be reflected by a corresponding shift in the position of the geocenter in the system x_i :

$$x_{0i}{}^{t_1} - x_{0i}{}^{t_0} = \frac{\int_{E'} (x_i{}^{t_1} - x_i{}^{t_0}) \, dM}{M} \tag{10}$$

This expression makes possible calculation of the change in the position of the geocenter resulting from mass deplacement in the limited volume, in relation to other points which are at rest. This is an intuitively supposed "motion" of the geocenter.

Let us make some very simple calculations to estimate the influence of some geodynamical phenomena.

The disappearance of the Antarctic polar ice cap would bring a change of about 30 m in the geocenter position. There is no need to consider such cataclysms in geodynamical investigations, but it gives some feeling about the sensitivity of the geocenter position to the mass changes occurring on the surface of the Earth.

In reality episodic changes occur in connection with earthquakes. Very large earthquake fields can extend over some 1000000 km^2 with displacements of the order of several meters. Suppose there is an uplift of 10 m over this area, and the depth of the displacement is 10 km. Such a tremendous earthquake would move

the geocenter by only 0.5 mm. These calculations show that
isolated episodic events on the surface of the Earth as well as
in the crust have practically no impact on the position of the
geocenter with respect to the rest of the globe.

The same is true of the seasonal changes in the atmosphere,
analyzed by Stolz (1976). Using data on seasonal variations of
the air pressure as well as on ground water storage be estimated
the possible range of the geocenter position oscillations as
2.8 mm within a six-month period. The model was certainly simplified, but the order of magnitude will be the same using more
sophisticated expressions. Anderson et al. (1975) found that the
center of mass of the solid Earth and oceans differs in position
from that of the geocenter by less than 5 cm. It seems that this
estimate is one order of magnitude too large and that the variations indicated by Stolz make up the most we can expect as the
influence of the atmosphere.

Special attention should be paid to the influence of Earth
tides. If the Earth were a uniformly elastic body, tidal forces
would produce only symmetrical deformations. It is possible,
however, that inhomogeneity of the mantle may produce unequal
response to the tidal forces causing small oscillations of the
geocenter. It seems unlikely that amplitude of these oscillations
could exceed a few mm, but we have not done any model calculations.

We do not know very much about mass displacements inside the
globe, i.e. in the mantle and/or in the core. Today mantle convection is a commonly accepted hypothesis. In this case the
quantity of mass involved in the motion is large in comparison to
the total Earth mass, but the rate of the motion is slow. The
phenomenon is of a global scale and secular character, deformations
extend over the entire globe and the system to which we could refer geocenter motion is lost. The same is true for tectonic plate
motion associated with convective streams in the mantle. Another
possible phenomenon — rotation of the mantle with respect to the
core — has an analogous pattern.

Yet another theoretical possibility of the geocenter motion
is mass displacement in the core. According to the currently
accepted Earth model, convection appears also in the outer liquid
core (Stacey 1977, p. 197-204). On the other hand, Barta (1974)
suggested asymmetry of the core structure, based on interpretation
of gravity anomalies. Teisseyre (1979) supposes that this
asymmetry can be supported by convection (Fig. 1).

Stability can easily be disturbed by processes occurring in
the Earth's outer layers — geochemical phase transformations.
These transformations occur constantly, causing changes in pressure, viscosity and temperature. They constitute a mechanism

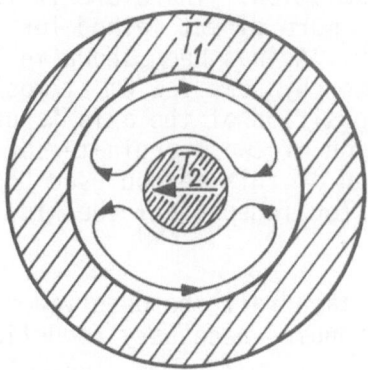

Figure 1.   Core assymmetry caused by core convection.

pushing the inner core in different directions (Teisseyre, private communication). In this situation some motions of the inner core cannot be excluded. Unfortunately, we do not have mathematical model permitting quantitative estimates. However, even this hypothetical phenomenon is not capable of changing the geocenter position very much. The density difference between the inner and outer core is of the order of $1g/cm^3$, hence the mass surplus influencing the geocenter is 7.5×10^{24} g. This is less than 10^{-3} of the total Earth mass and in consequences a 10 m displacement of the inner core (which may be supposed admissible) will give less than 1 cm change in the geocenter position. Such variations if detected would offer interesting information about the Earth's interior.

DETERMINATIONS OF THE GEOCENTER POSITION

By determining the positions of surface stations in the geocentric coordinate system we automatically determine the position of the geocenter in relation to these stations. This has been done in the Goddard Earth Model, the Smithsonian Standard Earth (SSE), the GRIM and others. It is not the purpose of this paper to compare different Earth models. Present estimates of the mean accuracy of the geocentric position in different models vary from approximately 5 m (Groten 1978). The author is inclined to believe that we are now close to the lower number. In recent years laser observations of passive satellites (Lageos and Starlette), as well as an immense quantity of Doppler observations of Transit satellites, contributed a great deal to the improvement of solutions. Comparison of different types of data made it possible to eliminate certain systematic errors or misinterpretation of results.

What is more dangerous is the correlation which appears between different unknowns in the huge systems of equations in-

dispensable for combination solution.   Therefore it is necessary
to envisage some independent, more direct method for finding
the position of the geocenter. As has been shown in §3, the spin
axis can pass off the geocenter by some few cm at most, hence the
determination of the offset position of the axis is practically
equivalent to the determination of two coordinates of the geocenter. The first application of this method used Doppler observations of deep space probes for independent checking of the SSE
(Gaposchkin 1973).

In the case of distant spacecraft the observed range rate
can be expressed, after introducing necessary reductions, as

$$s = r + \omega r, \cos \delta \sin t \qquad\qquad (11)$$

where r is the geocentric radius vector, δ, t are the declination
and hour angle of the spacecraft, and r_S is the spin-axis distance
of the observer. The last parameter can be calculated, assuming
that the ephemeris of the probe is known. The accuracy estimated
in SSE III was about ± 2m for r_S, but taking into account the
steady development in instrumentation and special planning of missions, an improvement by a factor of 10 can be expected. It appears
that this method can be efficiently used with the help of GPS
satellites.

Another method has been proposed by Domaradzki and Zielinski
(1979). It consists in measuring the angles and distances of the
object while the Earth is rotating at a certain angle (Fig. 2).

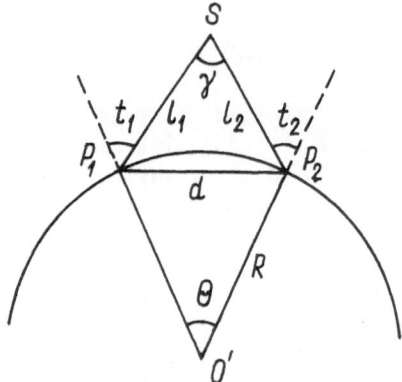

Figure 2.   Determination of geocenter using angle and range
            measurements.

This method was first applied at four SAO stations and in three
cases the solution was found. The degree of accuracy of the
calculated spin-axis distance was about 15 m, too much to investigate the geocenter motion, but this error resulted from lack of

precision of the photo camera direction data. The same method
can be applied using GPS satellites or the Moon, measuring distance by laser beams and angles by interferometry, with an accuracy
expected to rise to the decimeter level.

A totally different method based on the properties of the
gravity field has been proposed by Mather et al. (1977). The
idea is to use the fact that the first-degree harmonic terms in
development of the Earth gravitational potential vanish only if
the origin of the reference system coincides with the center of
mass. By differentiating (6) and substituting spherical coordinates one gets an expression

$$\delta g = c(\Delta x_1 \cos \psi \cos \lambda + \Delta x_2 \cos \psi \sin \lambda + \Delta x_3 \sin \psi) \quad (12)$$

where $c = -3.08$ µgal/cm, δg is the measured absolute gravity difference between two epochs, and Δx_i is the displacement of the
geocenter. In principle three equations will suffice for determination of the three unknowns Δx_i, but Mather et al. suggested
a minimum of at least 25 stations. This suggestion cannot be
accepted without certain reservations. As we have seen, any
changes of the geocenter position are caused by the redistribution
of mass, but this redistribution also produces changes in gravity
anomalies i.e. in terms of every degree in the development of the
geopotential, not only in the first. When measuring gravity,
absolute or relative, we are always faced with having to solve
the well-known boundary value problem. However, by accepting the
hypothesis that only the inner core is responsible for geocentric
motion, Mather's method can work. Only a limited number of terms
of low degree will be significantly affected by motion of the
inner core, so even with a limited number of stations a solution
to the problem can be found.

SCALE

One can consider the scale as a metric property of the space
stretched upon the triad of unit vectors defining the three-
dimensional coordinate systems (Grafarend et al. 1979). Nevertheless one can also approach the problem more pragmatically, considering the scale as one of seven parameters of transformation
between two reference frames. In this sense we understand a
scale difference to be the ratio of numbers describing the same
length in both frames.

In the past we have observed some scale difference between
particular solutions. They were especially apparent in comparisons
between combination solutions and pure Doppler solutions. However, a thorough analysis of the software revealed the sources
of the scale discrepancies (Langley et al. 1979, Hothem 1979).
They proved to be connected with some of the constants used,

and with the reduction method, rather than with the physics of
observations.

In this connection let us note one of the peculiarities of
the coordinate system realized by satellite observations. The
size of a satellite orbit is essentially determined by the third
of Kepler's laws:

$$n^2 a^3 = GM \qquad (13)$$

where n is the mean motion of the satellite, and a is the semi-
major axis of the orbit. This is a trivial statement for anyone
working on orbit determination, but it is a very important constraint ensuring a uniform scale for a global net. Thanks to
(13) the distance measurement errors can be effectively adjusted.
The final accuracy depends on the precision of time measurement
(Zielinski 1968). At present there are two types of approach:
one is to accept some conventional value for GM which establishes
the scale for the solution, the other is to improve this value
in the course of the solution. In the latter case the primary
scale source is the length measurement standard defined by the
velocity of light, but after adjustment the corrected GM value
is compatible with an obtained scale. In any case GM plays a
crucial role in the determination of the linear scale of the
coordinate system.

In physical geodesy we have again the possibility of determ-
ing the size of the reference ellipsoid and of the geoid if GM
is known (Heiskanen and Moritz 1967):

$$\Delta a = N_0 = \frac{GM}{R_0 \gamma} - \frac{\Delta W}{\gamma} \qquad (14)$$

where $\Delta a = N_0$ is the correction to the semimajor axis of the
reference ellipsoid, γ is the mean value of gravity, and Δw is
the correction to the potential on the geoid surface. Knowing
the exact value of GM we can find Δa and improve the parameters
of the geodetic datum.

I would like to stress, however, that both procedures are
valid only if our calculations are based on the system connected
with the center of mass of the Earth. Satellite orbit theory,
as well as the theory of the geoid, can be used in a convenient
way if the system origin is equivalent to the geocenter. Only
then can we obtain uniform scaling for our reference system.

Numerical values of the geocentric gravitational constant
and the Earth equatorial radius were discussed at the General
Assembly of IUGG at Canberra by the Special Study Group on

Fundamental Geodetic Constants. The values adopted were (Moritz 1979b):

$$GM = 3986005 \pm 0.5 \times 10^8 \; m^3 s^{-2}$$

$$a = 6378137 \pm 2 \; m$$

These are certainly the best current estimates based on the latest results. Nevertheless it would seem that the compatibility of these two figures should be made a subject of further study.

CONCLUSIONS

From the above deliberations we can derive the following conclusions:

1. The motion of the geocenter can be discussed only if a number of points conserving this frame exist on the Earth's surface. We must confine our considerations to cases in which mass displacements are limited in space. If deformations affect the entire Earth, the frame is destroyed and the idea of "motion of the geocenter" becomes meaningless.

2. This implies that we can consider motion of the geocenter only on a short-term scale. But, as was shown above, any short-term phenomena occurring near the surface, e.g. earthquakes, ocean tides, atmospheric changes etc., exert a negligible influence on the position of the geocenter.

3. Processes within the core, and possible motion of the inner core, are the only admissible sources of detectable motion of the geocenter, of short-period and attaining a possible level of a few centimeters. The detection of such motion would be extremely interesting from the point of view of geophysics.

4. The geocenter is a well defined and extremely stable point, accessible to measurement. The location of the origin of coordinate systems at this point is thus fully justified.

5. To ensure a uniform length scale, the scale of the global coordinate system must be connected with the geocenter.

ACKNOWLEDGEMENT

The author wishes to acknowledge the help of Mr. A. Brzeziński in calculations and discussions of mechanical problems, as well as valuable comments by Dr. B. Kolaczek.

REFERENCES

Anderson,E.G., Rizos,C., and Mather,R.S.:1975. Unisurv G, 23, 23.
Barta,G.:1974. Satellite Geodesy and the International Structure of the Earth. Space Research, XIV. Akad. Verlag, Berlin.
Bursa,M:1974. Proc. IAU Colloquium No. 26 on Reference Coordinate Systems in Earth Dynamics, Torun, Poland, 133.
Domaradzki,S., and Zielinski,J.B.:1979, Artificial Satellite, 14,19.
Gaposchkin,E.M.:1973. Smithsonian Standard Earth III. Smith-sonian Astrophysical Observatory, Cambridge, Massachusetts, Special Report No. 353.
Grafarend,E.W., Mueller,I.I., Papo,H.B., and Richter,B:1979. Investigations on the Hierarchy of Reference Frames in Geodesy and Geodynamics. Ohio State University, Dept. of Geodetic Sciences, Columbus, Ohio, Report No. 289.
Groten,E.:1974. Proc. IAU Colloquium No. 26 on Reference Coordinate Systems in Earth Dynamics, Torun, Poland, 247.
Groten,E.:1978. The present (1977) State of the Art in Gravimetry Proc. European Workshop on Space Oceanography, Navigation and Geodynamics, Schloss Elmau, DBR.
Heiskanen, W.A., and Moritz,H.:1967. Physical Geodesy, W.H. Freeman, San Francisco, California.
Hothem,L.D.:1979, Proc. 2nd Int. Geodetic Symp. on Satellite Doppler Positioning, Austin, Texas.
Langley,R.B., Cannon,W.H., Petrachenko,W.T., and Kouba,J.:1979. Proc. 2nd Int. Geodetic Symp. on Satellite Doppler Positioning, Austin, Texas.
Mather,R.S., Masters,E.G., and Coleman,R.:1977. Unisurv G, 26, 1.
Moritz,H.:1974. Proc. IAU Colloquium No. 26 on Reference Coordinate Systems in Earth Dynamics, Torun, Poland, 161.
Moritz,H.:1979a. Ohio State University, Dept. of Geodetic Sciences, Columbus, Ohio, Report No. 294.
Moritz,H.:1979b. Report of Special Study Group No. 5.39, Int. Assoc. Geodesy, submitted to XVII General Assembly of Int. Union of Geodesy and Geophysics, Canberra, Australia.
Munk, W.H., and MacDonald,G.J.F:1960, The Rotation of the Earth, Cambridge University Press.
Stacey,F.D.:1977. Physics of the Earth, 2nd ed., J. Wiley.
Stolz,A.:1976. Geophys. J. R. Astr. Soc., 44, 19.
Suslov,G.K.:1946. Theoreticheskaia Mechanika, Gostechizdat, Moscow, U.S.S.R.
Teisseyre,R.:1979. Acza Geophysica Polonica, 27, 369.
Zieliński,J.B.:1968. Application of the Radius Vector of Artificial Satellites as Length Measure for Geodetic Purposes. Politechnika Warszawska, Prace Naukowe, Geodezja, 1, 7.

CRITICAL REMARKS ON THE POSSIBILITY OF DETERMINING VARIATIONS OF THE GEOCENTER POSITION USING GEOSTATIONARY SATELLITE OBSERVATIONS

J. Klokočník
Astronomical Institute of the Czechoslovak
Academy of Sciences, Ondřejov Observatory
J. Kostelecký
Research Institute of Geodesy
Topography and Cartography, Observatory Pecný Ondřejov

ABSTRACT[*]

   The possibility of determining the variations of the Earth's center (expressed in terms of the geopotential harmonic coefficients $\bar{C}_{11}$, $\bar{S}_{11}$, or of some functions of them), using geostationary satellites, was discussed by Burša and Šíma (1978) after discerning the difference between observed and theoretical positions of libration points. Although their effect on geostationary orbits cannot be expected to be of high magnitude, the question of determination of these terms arose. It was suggested that the "resonant phenomenon" observed here might be analogous to the resonant effects in close satellite orbits, because the satellite is at the 1/1-resonance (1 nodal revolution per 1 sidereal day).

   The rate of change of a satellite's semimajor axis $\underline{a}$ (Lagrange planetary equation) at the 1/1-resonance can be written as follows (Klokočník and Kostelecký, 1979):

$$\left(\frac{da}{ds}\right)_{1/1} = 2n_s a_e \sum_{\gamma=1}^{\infty} \left[C_\gamma \sin(\gamma\phi_{1/1}) - S_\gamma \cos(\gamma\phi_{1/1}) \right] + 0(e) \quad,$$

(1)

where $\phi_{1/1} = \omega + \Omega + M - S$ is the resonant angle, E (a, e, I, ω, Ω, M) are the usual orbital elements, n_s is the satellite's mean (Keplerian) motion, a_e is the semimajor axis of the best-fitting Earth's ellipsoid, and C_γ, S_γ are the lumped geopotential coefficients. The coefficients are defined as follows:

$$\begin{Bmatrix} C \\ S \end{Bmatrix}_\gamma^E = \sum_{i=0}^{\infty} (-1)^i \, \tilde{F}_{\gamma+2i,\gamma,i} \left(\frac{a_e}{a}\right)^{\gamma+2i-1} \begin{Bmatrix} \bar{C} \\ \bar{S} \end{Bmatrix}_{\gamma+2i,\gamma}$$

[*]Previously published in Bull. Astron. Inst. Czechosl. 1980: <u>31</u>, 123.

E. M. Gaposchkin and B. Kołaczek (eds.), Reference Coordinate Systems for Earth Dynamics, 251–253.
Copyright © 1981 by D. Reidel Publishing Company.

with the harmonic geopotential coefficients (Stokes constants) $\overline{C}_{\gamma+2i,\gamma}$, $\overline{S}_{\gamma+2i,\gamma}$ of order γ and degree $\gamma + 2i$, $\gamma = 1, 2, \ldots$; $i = 0, 1, 2, \ldots$; F are certain functions of the inclination functions.

Formal setting $\gamma = 1$, $i = 0$ into Eq. 1 shows the dependence of a on $\overline{C}_{11}$, $\overline{S}_{11}$; these terms may be interpreted as a shift (Δ) of true geo-center in some reference frame in which geostationary satellites are observed. If we substitute sufficiently large values of $\overline{C}_{11}$, $\overline{S}_{11}$ and and integrate Eq. 1, then their influence in the total da/ds begins to compete with the equatorial flattening $\overline{C}_{22}$, $\overline{S}_{22}$ - effect. It is then natural to ask whether such a non-geocentricity of the reference frame may show itself a dominant resonant effect in a geostationary orbit? The answer is negative.

The reason for the negative answer is just the above mentioned formal substitution, which is illegitimate. Our Lagrange planetary equation (LPE) for the geostationary orbits was derived from general LPE, where all tesseral harmonics are kept. Both these equations are valid only in an actual inertial reference frame. The satellite's motion (and all observations) are considered in a shifted non-inertial frame, there $\overline{C}_{11}$, $\overline{S}_{11} \neq 0$. Therefore, the original LPE must be trans-formed into another form compatible with that quasigeocentric system.

According to (Klokočník and Kostelecký, 1980, Eq. 14):

$$\frac{d\overline{a}}{ds} = \frac{2}{n_s\overline{a}}\frac{\partial R}{\partial M} + \Delta(\dot{S} - n_s)\,\sin\,(\overline{\Phi}_{1/1} - \lambda_{11}) + 0 \quad , \tag{2}$$

where Δ, λ_{11} stay instead of $\overline{C}_{11}$, $\overline{S}_{11}$ (since they are more convenient in the derivations); the values with the bars are related to the quasigeocentric reference frame and R is the disturbing potential. It is important that the terms $\overline{C}_{11}$, $\overline{S}_{11}$ (or Δ, λ_{11}) are not included in this potential.

When integrating (2) by means of successive approximations, and supposing the resonant case (i.e. $\dot{\overline{\Phi}}_{1/1} \to 0$, $\dot{S} - n_s \to 0$), we arrive at the situation, where the limit of the second term on the right side of Eq. 2 approaches 1. This results in the conclusions that there is no resonant effect on the geostationary orbit owing to $\overline{C}_{11}$, $\overline{S}_{11}$ and that the resulting small effect would be of the same amplitude on any satel-lite's orbit. Our "non-inertiality" is just a geometrical feature, which could be measured only if the "measuring accuracy" (orbit determi-nation) would be comparable with the magnitude of the effect (i.e., with the amplitude of Δ). It should be noticed that Burša and Šíma (1979a,b) obtained the same result by another approach.

Orbital variations due to $\overline{C}_{11}$, $\overline{S}_{11}$ can be found from very precisely determined orbits of close satellites, e.g., as a part of the experiment with two counter-orbiting drag-free satellites suggested by Van Patten et al. 1976 and Breakwell et al. 1978 for testing Lense-Thiring effect for improving the individual harmonic coefficients of low order m and degree n (originally suggested for $n,m > 1$). (Relative geocentric co-ordinates of observing stations also need to be known with the accuracy comparable with Δ).

REFERENCES

Breakwell, J. V., Everitt, C. W. F., Schaechter, D. B., and Van Patten, R. A., 1978. Journal Brit. Interpl. Soc. 31, 465.
Burša, M., and Šíma, Z., 1978. Bull. Astron. Inst. Czechosl. 29, 232.
Burša, M., and Šíma, Z., 1979a. Studia Geophys. Geodet. 23, 1.
Burša, M., and Šíma, Z., 1979b. Bull. Astron. Inst. Czechosl. 30, 193.
Klokočník, J., and Kostelecký, J., 1979. Earth gravitational field effect on a near-geostationary satellite's orbit, Zprávy a pozorování geodetické observatoře Pecný, Ed. VUGTK, 2.
Klokočník, J., and Kostelecký, J., 1980. Bull. Astron. Inst. Czechosl. 31, 123.
Van Patten, R. A., and Everitt, C. W. F., 1976. Celest. Mech. 13, 429.

THE REFERENCE FRAMES AND A TRANSFORMATION OF THE SPHERICAL FUNCTIONS

V. G. Shkodrov
Section of Astronomy
Bulgarian Academy of Sciences
Sofia, Bulgaria

ABSTRACT

The use of spherical functions in dynamical problems is very common. As a rule, they arise in perturbing functions. It is well known that passing from one reference frame to another is accompanied by a double transformation of the perturbation function. That is why problems lose their simplicity and elegance. The problem of two solid bodies is a typical example in this respect.

In the present paper the questions connected with the transformation of the spherical functions when passing from one reference frame to another frame are considered. Traditional functions are generally unsuitable as they introduce a series of difficulties in the problems. That is why complex spherical functions are used. The transformation of spherical functions due to rotation of the coordinate frame is made by means of the Wigner's functions. When translating the frame the Clebsch-Gordon's coefficients are used.

1. INTRODUCTION

In dynamical problems different modifications of the right-handed Cartesian-coordinate system $\{0,x,y,z\}$ and connected with it, spherical reference frame $\{r,\theta,\lambda\}$, are used:

$$x = r \sin\theta \cos\lambda, \quad y = r \sin\theta \cos\lambda, \quad z = r \cos\theta, \tag{1}$$

in which the traditional spherical functions are defined:

$$Y_{nm}(\theta,\lambda) = P_n^m(\cos\theta)\{{\cos m\lambda \atop \sin m\lambda}\}. \tag{2}$$

E. M. Gaposchkin and B. Kołaczek (eds.), Reference Coordinate Systems for Earth Dynamics, 255–259.
Copyright © 1981 by D. Reidel Publishing Company.

In (2) $P_n^m(\cos\theta)$ are Associated Legendre functions.

The expression of the perturbation function by means of (2) leads to some difficulties and complicated transformations when passing from one reference frame to another. The Kaula's transformation (1961) is the best example of that. There the tesseral harmonics and Kepler orbital elements are connected within the two body problem. When refining the analysis of the planet's gravitational potential one can meet the same difficulty (Shkodrov, 1975).

Above all these difficulties relate the transformation of (2) connected with the translation or rotation of the reference system $\{r,\theta,\lambda\}$, as well as the choice of proper Euler angles. In the mathematical apparatus of Quantum mechanics these difficulties are eliminated almost completely. At a large extent one can realise it by replacing (1) with the new complex frame, defined as follows:

$$x_{+1} = \frac{1}{\sqrt{2}} r\sin\theta e^{i\lambda}, \quad x_0 = r\cos\theta, \quad x_{-1} = \frac{1}{\sqrt{2}} r\sin\theta e^{-i\lambda}, \tag{3}$$

which is known as a cyclical frame. The spherical functions (2) in (3) have the following form:

$$\mathscr{D}_{nm}(\theta,\lambda) = \bar{P}_n^m(\cos\theta)e^{im\lambda}, \quad \mathscr{D}_{nm}^*(\theta,\lambda) = \bar{P}_n^m(\cos\theta)e^{-im\lambda}, \tag{4}$$

where (*) means the complex-conjugated spherical function, and $P_n^{-m}(\cos\theta)$ are normalised Associated Legendre functions.

The formalism, based on (3) and (4), does not hide the qualitative picture of the dynamical problems, and leads to greater elegance. The only difficulty in this case is probably the novelty of the formalism proposed, but it seems to be not so bad, as complex quantities are usual things in Celestial mechanics and have been used in it's apparatus since it was formed as a scientific discipline.

Here we have to note that in this case our efforts to use this formalism is just contrary to that of B. Jeffreys (1965). Our wish is to use it in dynamical problems in the form close to that in Quantum mechanics.

2. TRANSFORMATION OF THE COMPLEX SPHERICAL FUNCTIONS UNDER TRANSLATION OF THE REFERENCE SYSTEM

Let the reference system $S_1\{G_1,x_1,y_1,z_1\}$ pass into the system $S_2\{G_2,x_2,y_2,z_2\}$, due to the translation of the vector $\vec{R}$. As it is known that under this transformation, the coordinates of the point $P(\vec{r}_1)$, given in S_1, are transformed according to

$$\vec{r}_2 = \vec{r}_1 - \vec{R}, \tag{5}$$

where $\vec{r}_2$ is the radius-vector of P in S_2. In order to obtain the rule for transformation of the spherical functions

$$[\mathscr{D}_{n_1 m_1}(S_1) \Rightarrow \mathscr{D}_{n_2 m_2}(S_2)]$$

as a result of the translation (5) we shall use the relation

$$\cos\theta_2 = \frac{r_1}{r_2} \cos\theta_1 - \frac{R}{r_2} \cos\theta, \tag{6}$$

bearing in mind that r_1, θ_1, λ_1 and r_2, θ_2, λ_2 are spherical coordinates of the point P connected with the reference system S_1 and S_2, respectively, and R, θ, Λ are spherical coordinates of G_2 in S_1, defined by $\vec{R}$. From (6) we obtain:

$$\mathscr{D}_{10}(\theta_2,\lambda_2) = \sum_{k=0}^{1}(-1)^{k+1}\left(\frac{R}{r_2}\right)\left(\frac{r_1}{R}\right)^k \mathscr{D}_{k0}(\theta_1,\lambda_1)\,\mathscr{D}_{1-k,0}(\theta,\Lambda). \tag{7}$$

The general rule for transformation of the spherical functions $\mathscr{D}_{nm}(\theta,\lambda)$ in this case is obtained from (5), (6) and (7) by the method of induction. It is:

$$\mathscr{D}_{n_2 m_2}(\theta_2,\lambda_2) = \sum_{k=0}^{n_2} \sum_{(p,q)}(-1)^{n_2+k} C_{kpn_2-kq}^{n_2 m_2}\left(\frac{R}{r_2}\right)^{n_2}\left(\frac{r_1}{R}\right)^k$$

$$\times\ \mathscr{D}_{kp}(\theta_1,\lambda_1)\,\mathscr{D}_{n_2-kq}(\theta,\Lambda), \tag{8}$$

where $C_{kpn_2-kq}^{n_2 m_2}$ are the Clebsch-Gordon coefficients. The summation in (8) is made for all integer values of p and q, for which $C_{kpn_2-kq}^{n_2 m_2} \neq 0$.

3. TRANSFORMATION OF $\mathscr{D}_{nm}(\theta,\lambda)$ UNDER ROTATION OF THE REFERENCE SYSTEM

In order to obtain the rule for transformation of $\mathscr{D}_{nm}(\theta,\lambda)$ due to rotation of reference frame let us define the Euler angles α, β and γ. After (Edmonds, 1957) we call α a rotation about the z-axis, bringing the frame of axes from the initial position S into the position S', and β a rotation about the y-axis of the frame S' called the line of nodes. The resulting position of the frame of axes is denoted by S", and γ a rotation about the z-axis of the frame S". That is the final position of the frame. For that selection of Euler angles, the spherical functions $\mathscr{D}_{nm}(\theta,\lambda)$ are transformated as follows:

$$\mathscr{D}_{nm'}(\theta_1,\lambda_1) = \sum_{m=-n}^{n} \mathscr{D}_{nm}(\theta,\lambda)\, D_{mm'}^{n}(\alpha,\beta,\gamma), \tag{9}$$

where θ,λ are the spherical coordinates in initial reference frame S, and θ_1,λ_1 are the spherical coordinates in the new rotating system S_1. The angles θ,λ and θ_1,λ_1 are related by:

$$\cos\theta_1 = \cos\theta\cos\beta + \sin\theta\sin\beta\cos(\lambda-\alpha),$$
$$\text{ctg}(\lambda'+\gamma) = \text{ctg}(\lambda-\alpha)\cos\beta - \text{ctg}\theta\sin\beta\sin^{-1}(\lambda-\alpha). \tag{10}$$

The inverse transformation $(S \Leftarrow S_1)$ is:

$$\mathscr{D}_{nm}(\theta,\lambda) = \sum_{m'=-n}^{n} \mathscr{D}_{nm'}(\theta',\lambda')D_{mm'}^{n*}(\alpha,\beta,\gamma). \tag{11}$$

Wigner's D-functions, or generalised spherical functions, are complex functions of three real arguments. They have the following structure:

$$D_{mm'}^{n}(\alpha,\beta,\gamma) = e^{-im\alpha}d_{mm'}^{n}(\beta)e^{-im'\gamma}, \tag{12}$$

where the real function $d_{mm'}^{n}(\beta)$ may be defined:

$$d_{mm'}^{n}(\beta) = \left[\frac{(n+m')!(n-m')!}{(n+m)!(n-m)!}\right]^{\frac{1}{2}} \sum_{\sigma}\binom{n+m}{n-m-\sigma}\binom{n-m}{\sigma}(-1)^{n-m-\sigma}$$
$$\times \cos^{2\sigma-m'+m}(\tfrac{\beta}{2})\sin^{2n-2\sigma-m'-m}(\tfrac{\beta}{2}) \tag{13}$$

In a particular case when $m=0$, or $m'=0$ we have:

$$D_{0m'}^{n}(\alpha,\beta,\gamma) = (-1)^{m'}(2n+1)^{-\frac{1}{2}}\mathscr{D}_{nm'}^{*}(\beta,\gamma),$$
$$D_{m0}^{n}(\alpha,\beta,\gamma) = (2n+1)^{-\frac{1}{2}}\mathscr{D}_{nm}^{*}(\beta,\alpha). \tag{14}$$

If $m=m'=0$, then

$$D_{00}^{n}(\alpha,\beta,\gamma) = P_{n}(\cos\beta). \tag{15}$$

4. CLEBSCH-GORDON COEFFICIENTS

The Clebsch-Gordon coefficients are of great importance for our coordinate transformations. In the theory of groups the representation of the elements of a group by linear functions of an irreducible representation is realized by means of these coefficients. They are real and satisfy the conditions:

$$c_{j_1m_1j_2m_2}^{j_1+j_2m_1+m_2} > 0, \quad c_{j_1m_1j_2-j_1}^{jm} > 0, \quad c_{j_1j_1j_2j_2}^{j_1+j_2j_1+j_2} = 1. \tag{16}$$

and differ from zero if $m_1+m_2 = n$, $j+j_1+j_2$ is even.

The following Clebsch-Gordon series:

$$\mathscr{D}_{n_1 m_1}(\theta,\lambda)\, \mathscr{D}_{n_2 m_2}(\theta,\lambda) = \sum_{\ell=\max\{n_1+n_2,m_1+m_2\}}^{n_1+n_2}$$

$$\times\; C_{n_1 m_1 n_2 m_2}^{\ell m_1+m_2}\, C_{n_1 0 n_2 0}^{\ell 0}\, \mathscr{D}_{\ell m_1+m_2}(\theta,\lambda) \qquad (17)$$

may be of great importance for the dynamical problems. The Clebsch-Gordon series may be obtained using the recurrence relations of spherical functions (Shkodrov, 1975).

5. CONCLUSION

The formalism briefly stated here considerably facilitates the transformation of the perturbation function when passing from one frame to another. The fact, that the functions of Wigner and the Clebsch-Gordon coefficients are perfectly mastered in Quantum mechanics, is an additional advantage. The attempt in (Shkodrov, Gechev, 1979, 1980; Shkodrov, 1980, 1980a) to use this mathematical formalism in different problems of Celestial mechanics showed, that it brings elegance and makes the solutions simpler. Its effectiveness is specially evident in the problem of two solid bodies. Due to this formalism a simple representation of the force function is obtained. In the two planet problem and in the two body problem the use of (7) leads to all known expressions for the transformations of the perturbation function.

REFERENCES

Edmonds, A. R.: 1957, Angular Momentum in Quantum Mechanics, New Jersey.
Jeffreys, B.: 1965, Geophys. J., 10, 141.
Kaula, W.: 1961, Geophys. J. Roy. Astron. Soc., 5.
Shkodrov, V. G., Gechev, T. G.: 1979, Compt. Rend. Bulg. Acad. Sci., 32, 1014.
Shkodrov, V. G., Gechev, T. G.: 1980, Ibid., 33 (in press).
Shkodrov, V. G.: 1980, Ibid., 33, 8 (in press).
Shkodrov, V. G.: 1980, Ibid., 33, 9 (in press).
Shkodrov, V. G.: 1975, Doctoral Thesis.

CONCEPTS OF REFERENCE FRAMES FOR A DEFORMABLE EARTH

Burghard Richter
Geodätisches Institut der Universität Stuttgart
Kepler Strasse 11
D-7000 Stuttgart
GERMANY

ABSTRACT

In order to describe geometrical facts in geodesy, geodynamics, or astronomy, one must have suitable reference frames. This paper deals with some basic concepts for the definition of both terrestrial and celestial systems which make it possible to describe position and motion of mass elements of the earth or stars, respectively. It is suggested to use such frames with respect to which coordinate changes of points on the earth's surface or of stars, respectively, become minimal. The rotation vector should be defined by applying the rotation operator on the velocity field.

1. GENERAL REQUIREMENTS OF REFERENCE FRAMES, CLASSIFICATION IN LOCAL, GLOBAL, AND UNIVERSAL FRAMES

The use of the expression "motion" requires a statement as to what is regarded as fixed. Such a definition is necessary to identify points or directions at the present epoch with respect to points or directions at an earlier epoch. There are two extreme choices to define what is fixed: 1) Local: The mass on which one is standing or on which one has a geodetic point and the observational instrument is regarded as fixed. Then all other mass elements of the deformable earth and all other celestial bodies will move with respect to this fixed point. 2) Universal: The axes of an inertial system can be considered as universally fixed. However, a strict practical realization of an inertial system is hardly possible. Later on, a certain approximation will be introduced.

261

E. M. Gaposchkin and B. Kołaczek (eds.), Reference Coordinate Systems for Earth Dynamics, 261–265.
Copyright © 1981 by D. Reidel Publishing Company.

Neither of these two choices is satisfying the needs of geodesy or geodynamics. For the deformable earth, one needs something that is earth-fixed in some "average" sense over the whole globe. Let us call this "globally fixed". After one has defined what is to be regarded as fixed, one can establish a reference frame, by defining for one epoch the origin and the direction of the axes. This is in principle quite arbitrary; but once one has done this, the frame will be defined for all subsequent epochs. The special frame one may choose will, however, be less important in this contribution, since all frames which do not move with respect to each other are equivalent.

2. PRINCIPLES OF DEFINING WHAT IS FIXED, IN GEODESY AND ASTRONOMY

2.1 Global Concepts for the Earth

If we exclude such principles which privilege a single point or few points of the earth surface, there remain essentially two concepts: a dynamical one and a geometrical one. For a dynamical definition, one can take the mass centre and the principal axes of inertia. Any system with respect to which these elements do not move is globally fixed in the dynamical sense. The practical disadvantages of this definition are well known, namely that the mass centre is not directly accessible, that the equatorial axes are ill defined, and that both have motions with respect to the earth surface that may be larger than the relative motions of different parts of the surface.

These circumstances favour the geometrical concept: Every geodesist would be happy if the crust of the earth were rigid because coordinates of terrestrial points would not change in this case. Therefore, it is a good choice to fix one's reference frame in such a way that the square sum of coordinate changes or, equivalently, the square sum of the displacements of all points of the earth surface becomes minimal from one epoch to the immediately following. This means that the square sum of all velocities $\underset{\sim}{v}$ is minimal.

$$\iint_{\Sigma} |\underset{\sim}{v}|^2 \, d\sigma = \min \tag{1a}$$

where Σ = surface of the earth and $d\sigma$ = element of Σ. Taking into consideration that along the edges of crustal plates mass flows out or sinks into the interior, one could extend the integration area:

$$\iint_{W} |\underset{\sim}{v}|^2 \, dw = \min \tag{1b}$$

where W = earth crust and a certain domain below the crust,
dw = element of W. This is, of course, in practice not strictly
realizable, because observations would have to continuous both
in space and in time. This leads to the approximation:

$$\sum_i p_i \; |\underset{\sim}{x}_i(t_j) - \underset{\sim}{x}_i(t_{j-1})|^2 = min \qquad\qquad (1c)$$

where $\underset{\sim}{x}$ is the position vector, i a discrete point, j the discrete epoch, p_i a positive weight factor.

Thus, if at an initial epoch t_0 a reference frame has been defined and coordinates $\underline{x}(t_0)$ been computed, the frame can be updated from one epoch to the next. Therefore, it is necessary to
determine for each epoch t_j preliminary coordinates $\underline{y}(t_j)$ in an
arbitrary free frame from geodetic observations at epoch t_j.
Then the transformation parameters $\theta_1, \theta_2, \theta_3$ (for rotation) and
s_1, s_2, s_3 (for translation) from the free frame are determined
from the adjustment condition

$$\sum_i p_i \; \| \underline{R}(\theta_1, \theta_2, \theta_3) \; \underline{y}_i(t_j) + \begin{bmatrix} s_1 \\ s_2 \\ s_3 \end{bmatrix} - \underline{x}_i(t_{j-1}) \|^2 = min \qquad\qquad (2)$$

using Cartesian coordinates.

Let the solution, for which the minimum is achieved, be $\theta_k = \bar{\theta}_k$,
$s_1 = \bar{s}_1$, (k,l = 1,2,3); then

$$\underline{x}_i(t_j) = \underline{R}(\bar{\theta}_1, \bar{\theta}_2, \bar{\theta}_3) \; \underline{y}_i(t_j) + \begin{bmatrix} s_1 \\ s_2 \\ s_3 \end{bmatrix}$$

will be the corrdinates in the updated globally fixed frame at
epoch t_j. It should be noted that this frame at epoch t_j and
that at epoch t_{j-1} are by definition identical. Only the coordinates have changed a little due to deformation. The establishment of the frame at the initial epoch t_0 should be done in such
a way that it coincides then with a traditional frame, as, for
example, the one defined by the mass centre, the CIO, and the
meridian of Greenwich. We shall call such a frame an earth-
fixed one.

2.2 Generalization and Specialization of the Minimum Principle

The same problem of deformation as on the earth arises "in the
sky"; for the relative positions of the stars, too, change in
time (2- or 3-dimensional proper motion). Therefore, eq. (2) can
be applied for defining a "fixed" celestial frame (as an approximation to an inertial frame) in the same way as for an earth-
fixed frame. The index i denotes then the star (including
quasars) instead of the point. The fact that the proper motion

of quasars is much smaller (at least the directional component)
than the proper motion of visible stars can be taken into account
by assigning a lower or zero weight to visible stars. In
astrometry one is rather interested only in the direction than in
the absolute position of a star. Therefore, it makes sense not
to minimize the square norm of the absolute displacement vector
($\Delta x^2 + \Delta y^2 + \Delta z^2$, in Cartesian coordinates), but the projection
of the displacement vector on the unit sphere ($\cos^2\delta\ \Delta\alpha^2 + \Delta\delta^2$,
in spherical coordinates). Further, one cannot determine rota-
tion and translation parameters, but rotation parameters only.

Also in the terrestrial case, we can split the absolute displacement vector in its radial and its directional component, if we
use spherical coordinates r,λ,ϕ. As the tides cause mainly short
period radial deformations, one can reduce eq. (2) on the radial
displacement for a short period updating. On the other hand,
plate tectonics cause mainly secular directional deformations (as
seen from the origin near the geocentre). After reducing the
observations for short period tidal deformations, one can, therefore, define a long-period earth-fixed system by specifying
eq. (2) on the directional displacement.

3. THE ROTATION VECTOR

The rotation vector, which usually plays an important role for
defining terrestrial or celestial reference frames, has not yet
been mentioned. It is, however, necessary for the definition of
a time system which is consistent with the natural run of day.
There are two ways for defining the rotation of a body. Rotation
in the first sense is, loosely speaking, when the direction of
the position vector from the mass centre to some mass element is
changing. Rotation in the second sense is a change of orientation
of a mass particle. It can be realized without any reference to
the centre of mass and is defined by $\omega = 1/2$ rot $\underset{\sim}{v}$, where $\underset{\sim}{v}$ is
the velocity and ω the rotation vector, both referring to the
same "fixation". A suitable separation of deformation from rotation, according to the principle of polar decomposition, is already included in this definition. As it is independent of the
mass centre it is better fitting the earth-fixed frame introduced
above in eq. (2,3) than rotation of the first kind and should,
therefore, be preferred. In the special case of a rigid earth,
both definitions, of course, coincide.

In the general case of a deformable earth, however, the rotation
vector changes not only in time, but also from point to point.
This is quite clear because the relative rotation of plates with
respect to the earth-fixed frame is superposed on the rotation of
this frame with respect to a universally fixed frame; the resulting rotation will, therefore, be different for each plate. For
defining a global rotation vector there are two possibilities:

one can either take the rotation of the earth-fixed frame or the average of all local rotation vectors. These two choices will in general not coincide.

REFERENCES

GRAFAREND, E. (1977): Space-Time Differential Geodesy: The Changing World of Geodetic Science (ed. U. Uotila), Vol. I, Rep. 250, Dep. Geod. Sci., The Ohio State Univ., Columbus.
GRAFAREND, E.; MUELLER, I.; PAPO, H ; RICHTER, B. (1979): Investigations on the Hierarchy of Reference Frames in Geodesy and Geodynamics, Rep. 289, Dep. Geod. Sci., The Ohio State Univ., Columbus.
HEITZ, S. (1975): Bezugssysteme geodätischer Erdmodelle, Zeitschrift für Vermessungswesen loo, pp. 251-260.
HEITZ, S. (1980): Mechanik fester Körper, Dümmler Verlag, Bonn.
MORITZ, H. (1979): Concepts in Geodetic Reference Frames, Rep. 294, Dep. Geod. Sci., The Ohio State Univ., Columbus.
TRUESDELL, C. (1954): The Kinematics of Vorticity, Science Series No. 19, Indiana University Publications, Bloomington.

COMBINED SPACE GEODETIC AND GEOPHYSICAL MEASUREMENTS FOR STUDIES OF CRUSTAL MOVEMENT IN SCANDINAVIA

ALLEN JOEL ANDERSON
INSTITUTE OF GEOPHYSICS
DEPARTMENT OF GEODESY
THE UNIVERSITY OF UPPSALA
S-755 90 UPPSALA
SWEDEN

INTRODUCTION

Scandinavia rests upon one of the most stable crustal platforms on the Earth, the Baltic shield. None the less the study of <u>crustal movement</u> is of great importance. Inparticular studies of <u>vertical movement</u> induced by <u>post glacial</u> <u>rebound</u> have been carried out for over 100 years. These have been accomplished using water level variations, and more recently in this century using precise leveling methods. The uplift measurements of Fennoscandia has led to a clearer understanding about the rheology of the crust and upper mantle of the Earth.

ACTIVE TECTONICS OF THE BALTIC SHIELD

Although the region is free from major earthquake activity, <u>strain release</u> at shallow depth occures along the Gulf of Bothnia in the center of the <u>Baltic shield</u>. This is almost certainly due to the accumulated strain caused by the different manner inwhich land uplift effects land and water covered areas (Anderson, 1980 and references therein).

Larger earthquakes of deeper origin occur at the edge of the of the shield, in the Oslo fjord region, and in South Western Sweden.

SPACE GEODETIC VLBI ACTIVITY

Present <u>VLBI</u> measurements in Scandinavia are carried out at <u>Onsala Space Observatory</u> under the framework of the NASA Geodynamics Program (see Annual Report for 1979). Further measurements have been made at Metsähovi in Finland and in Norway using a small dish (Nes, et al, 1979). Plans have been discussed to use the large <u>EISCAT</u> antennas in the far north as well as a cooperative venture between the Scandinavian countries for a portable system. Table 1 indicates the VLBI baseline rate changes to be expected on the grounds of present known strain release mechanisms operating on these areas. Both horizontal and vertical effects have been compared.

E. M. Gaposchkin and B. Kołaczek (eds.), Reference Coordinate Systems for Earth Dynamics, 267–270.

Table 1
Rates of uplift, baseline rates, estimated horizontal strain release vrs uplift
baseline rates for Scandinavian VLBI baselines.

VLBI BASELINE	ΔZ DIFFERENCE (CM/YR) VERTICAL UPLIFT RATE	ΔZ·SIN(θ/2) BASELINE RATE	HORIZONTAL STRAIN RELEASE (CM/YR) VRS UPLIFT BASELINE RATE (%)		
1. Onsala-Kiruna	+0.6 (cm/yr)	+0.05 (cm/yr)	+0.1 (cm/yr)		+50%
2. Onsala-Sodankylä	0.7	0.07	0.1	Fennoscandia plate	+70%
3. Onsala-Tromsö	0.0	0.00	0.1		00%
4. Dwingeloo-Onsala	0.2	0.01	0.2	Uplift bulge	+05%
5. Dwingeloo-Sodankylä	0.9	0.13	0.2		+65%
6. Bonn-Onsala	0.1	0.01	0.5	Oslo-Rhein graben	+02%
7. Bonn-Sodankyla	0.8	0.13	0.5		+26%
8. Haystack-Onsala	0.2	0.09	1.8	Mid-Atlantic ridge	+05%
9. Haystack-Sodankylä	0.9	0.43	1.8		+24%
10. Goldstone-Onsala	0.1	0.06	1.8		+03%
11. Goldstone-Sodankylä	0.8	0.47	1.8		+26%

Inorder to control the geodetic parameters for the VLBI solution extensive measurements using Doppler Satellite point positioning have been carried out for Onsala (Anderson, 1979, Ekmann, 1980 and others).

A tidal gravimeter has been operated at Onsala and one is to be installed on a permanent basis in late 1980. Figure 1 indicates the anomalous M_2 tidal component at Onsala found in the earlier measurement. It is planned to use Onsala as a tie for several types of terrestrial and space geodetic experiments during the 1980's.

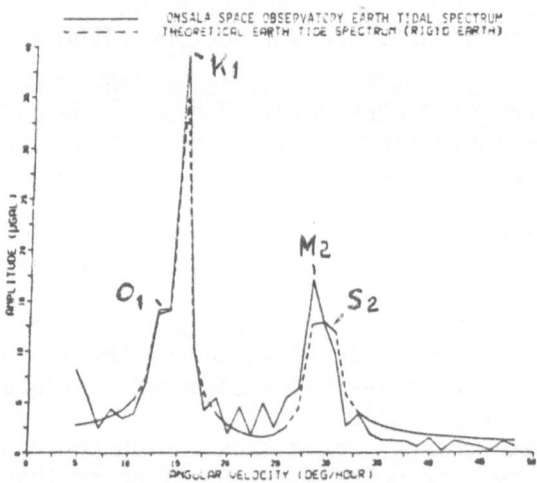

FIGURE 1. SPECTRAL ANALYSIS OF TIDAL GRAVIMETER RECORD TOGETHER WITH THEORETICAL TIDAL SPECTRA FOR ONSALA SPACE OBSERVATORY.

TABLE 9. REPEATIBILITY OF INTERSTATION DISTANCES AND HEIGHT DIFFERENCES FROM EDOC-II TO SCANDOC-79.

Line	Distance (m)	Δ Dist. (m)	(ppm)	Δh (m)	Δ(Δh) (m)
TRS...TRD	790873.55	-0.49	-0.62	5.10	-0.15·
TRS...EGB	1358712.91	-1.05	-0.77	-4.07	0.48
TRS...JNA	1047592.41	0.53	0.51	11.31	-0.08
TRD...EGB	570717.84	-0.52	-0.91	-9.17	0.63
TRD...JNA	975322.57	0.79	0.81	6.20	0.06
EGB...JNA	1301868.44	0.10	0.07	15.37	-0.57
Mean		-0.11	-0.15		0.06
Rms		0.70	0.72		0.44

TABLE 8. REPEATIBILITY OF DOPPLER POSITIONS IN SCANDINAVIA

Station	EDOC-SCANDOC Δx (m)	Δy (m)	Δz (m)	NORDOC-SCANDOC Δx (m)	Δy (m)	Δz (m)
TRS	-0.63	-1.03	-0.18			
TRD	-0.90	-0.45	0.11	-0.59	0.49	-0.54
EGB	-1.47	0.12	-0.26			
MSB				-0.21	0.69	-1.00
JNA	-1.23	0.31	0.06			
Mean	-1.06	-0.26	-0.07	-0.40	0.59	-0.77
Rms	0.37	0.61	0.18			

Figure 2. The SCANDOC 79 Results compared with the earlier EDOC and
NORDOC results. (after Ekmann,1980)

4. DOPPLER SATELLITE ACTIVITY

Since 1973 Doppler Satellite measurements have been carried out
throughout Scandinavia. Presently there are 9 Satellite receivers in
Sweden, and over 20 considering Scandinavia as a whole. This situation
allows for the establishment of regularly measured networks within
the Scandinavian Nations which can be used together with other methods
to monitor the stability of the geodetic solutions as well as to be
used as a reference field to measure crustal movements.

In 1979 the first of the SCANDOC measurements were made. Figure 2
indicates the network, and several tables have been given indicating
the baseline stability which has been acheived in these Nordic networks
(Ekmann, 1980).

5. CONCLUSIONS

The Scandinavian Area has several Space Geodetic programs which
when coordinated with Terrestrial projects lend themself very well to
the study of recent crustal movements. Recent results (Clark,et al,1980)
indicating repeatability to 4 CM on the Onsala-Haystack baseline,
together with an extensive Doppler Translocation Network in Scandinavia
showing baseline repeatibility to better than 0.7 ppm clearly show that
these methods are acheiving the level of precision needed for studies
of crustal movements in Scandinavia.

REFERENCES

ANDERSON, A.J.(1979) Evaluation of Doppler Point Positioning and tidal Gravimeter data at Onsala Space Observatory obtained during 1975-1979. Report from Institute of Geophysics, Uppsala.

ANDERSON, A.J.(1980) Scandinavian Studies of Recent Crustal Movements and the Space Geodetic Baseline Network. in Radio Interferometry Techniques in Geodesy, NASA Conf. Publ. 2115, pp 73-82, GSFC.

CLARK, T.A., LINDQUIST, G., RÖNNÄNG, B., and I.I. SHAPIRO (1980) TransAtlantic VLBI measurements between Haystack and Onsala with 4 CM repeatability. Journal of Geophysical Research (in press)

EKMANN, G. (1980) The Scandinavian Doppler Observation Campaign 1979 (SCANDOC-79) Results and Comparisons. Report from Geographical Survey of Norway, Oslo.

NASA Geodynamics Program: Annual Report for 1979. NASA tech. mem. 81978 Washington, D.C.

NES, H., HAGFORS, T., and G. SETTE (1979) A Very Long Baseline Interferometry Experiment with mobile equipment. Zeitschrift für Vermessungswesen 104, 224-235.

PÂQUET, P.(1980) European Doppler Observational Campaigns EDOC. in International Coordination of Space Techniques for Geodesy and Geodynamics Bulletin, pp 12-18, Dept. Geod. Sci., Columbus, Ohio.

DYNAMICS OF AN ARTIFICIAL SATELLITE IN AN EARTH-FIXED REFERENCE FRAME : EFFECTS OF POLAR MOTIONS

P. Farinella
Observatorio Astronomico di Brera, Merate (Como), Italy
A. Milani
Istituto di Matematica "L. Tonelli", Università di Pisa, Italy
A.M. Nobili
Istituto di Scienze dell'Informazione, Università di Pisa, Italy
F. Sacerdote
Istituto di Matematiche Applicate "U. Dini", Università di
Pisa, Italy

ABSTRACT

In an Earth-fixed reference frame, polar motions (precession, lunisolar nutation, free nutation) introduce small apparent forces in the equations of motion of an Earth satellite. We discuss the possibilities (a) of integrating the orbit in an Earth-fixed frame when tracking data are used for geophysical applications, and (b) of determining from orbital data a set of unknown parameters describing the long-period wandering of the pole.

1. INTRODUCTION

In an inertial reference frame the effects of polar motions on the range data which allow orbit determination for artificial satellites are mainly kinematical, since the observing stations are bound to the Earth.
On the other hand, for geophysical applications it is easier to study the satellite motion with respect to an Earth-fixed reference frame because in this case (a) the geopotential does not depend on kinematics of Earth rotation, (b) the observational model is very simple because the stations have a fixed position (except for lunisolar tides), (c) if the satellite orbit is geosynchronous, the motion in a body-fixed frame is very slow and this fact improves the stability of the numerical integration of the orbit. However, in an Earth-fixed frame apparent forces arise due to the fact that the Earth rotates with the angular velocity

$$\vec{\Omega}(t) = \vec{\Omega}_o + \Delta\vec{\Omega}(t) \tag{1}$$

271

E. M. Gaposchkin and B. Kołaczek (eds.), Reference Coordinate Systems for Earth Dynamics, 271–274.
Copyright © 1981 by D. Reidel Publishing Company.

where $\vec{\Omega}_o$ is a constant vector directed along the z-axis fixed within the body ($\Omega_o = 7.29 \times 10^{-5}$ rad s^{-1}), and $\Delta\vec{\Omega}$ is a small time-dependent vector which contains the contributions of precession and lunisolar nutation, free nutation and periodic variations in the length of the day.
In the following we discuss the difficulties and the advantages of the modelling of all these effects by means of apparent forces.

2. APPARENT FORCES

The motion of the Earth's rotation axis in an inertial frame can be described by a Fourier series giving the components of the angular velocity vector

$$\dot{\theta} = -\Omega_o \sum_i \varepsilon_i \sin \nu_i$$
$$\dot{\phi} \sin \theta = \Omega_o \sum_i \varepsilon_i \cos \nu_i \tag{2}$$

where θ, ϕ and ψ are the Euler angles, $\nu_i = n_i t + \nu_{io}$ is an angular variable with period $T_i = 2\pi/n_i$. In a body-fixed frame the same motion of the angular velocity vector can be described by a different Fourier series

$$\Omega_x = \Delta\Omega_x = -\Omega_o \sum \varepsilon_i \sin(\nu_i - \psi)$$
$$\Omega_y = \Delta\Omega_y = \Omega_o \sum \varepsilon_i \cos(\nu_i - \psi) \tag{3}$$

The different frequencies correspond to different physical effects, $n_i = 0$ and $\nu_{io} = 0$ for lunisolar precession, $T_i \simeq 40000$ yr for planetary precession, T_i between 18.6 yr and a few days for lunisolar nutation. The free nutation of the Earth appears in the body-fixed reference frame with periods $\tau_i = T_i/(1-T_i)$ days, of the order of 1 yr, hence in the inertial frame with T_i close to 1 day.
The acceleration due to apparent forces is

$$\vec{A} = \dot{\vec{P}} \wedge \vec{\Omega} + 2\dot{\vec{P}} \wedge \vec{\Omega} + \vec{\Omega} \wedge (\vec{P} \wedge \vec{\Omega}) \tag{4}$$

where $\vec{P}$ is the satellite position vector. By alling $\vec{A}_o$ the acceleration due to $\vec{\Omega}_o$ and neglecting the terms proportional to $\Delta\Omega^2$, we get

$$\vec{A} = \vec{A}_o + \dot{\vec{P}} \wedge \Delta\vec{\Omega} + 2\dot{\vec{P}} \wedge \Delta\vec{\Omega} + 2(\vec{\Omega}_o \cdot \Delta\vec{\Omega})\vec{P} - (\Delta\vec{\Omega} \cdot \vec{P})\vec{\Omega}_o - (\vec{\Omega}_o \cdot \vec{P})\Delta\vec{\Omega} \tag{5}$$

Therefore the frequencies of the apparent forces are linear combinations of the $n_i - \dot{\psi}$ with the mean motion n_o of the satellite.
Long-period effects arise from terms with long-period arguments $\nu_i - \psi$ (like those corresponding to Chandler wobble) and from terms with short-periodic arguments $\nu_i - \psi$ giving a beat with the orbital mean motion. This latter case can be easily illustrated by a geosynchronous satellite. We assume an equatorial circular orbit :

$$\vec{P}(0) = (x_o,0,0) \quad , \quad \dot{\vec{P}}(0) = (0,0,0) \quad , \quad n_o^2 = \Omega_o^2 = \frac{GM_o}{x_o^3} \tag{6}$$

By linearizing in the perturbative parameters ε_i, if the z-component of $P(t)$ is written as

$$z(t) = \sum \varepsilon_i \zeta_i(t) \tag{7}$$

we have

$$\ddot{\zeta}_i + \Omega_o^2 \zeta_i = x_o \Omega_o (\Omega_o + \dot{\psi} - n_i) \sin(\nu_i - \psi) \tag{8}$$

The solutions of equation (8) show a beat with frequencies $(n_i - \dot{\psi} + \Omega_o)/2$ and $(n_i - \dot{\psi} - \Omega_o)/2$. For instance, for the lunisolar precession term $n_i = 0$ and $\Omega_o - \dot{\psi} = \dot{\phi} \cos\theta$, so that the z-component of $\vec{P}$ oscillates with nearly diurnal period and amplitude modulated with a period of about 26000 yr. The lunisolar nutation harmonics with periods longer than 2 days appear in a body-fixed frame as beats due to a forcing term in the second member of equation (8), with a period shorter than 2 days (because $\tau_i = T_i/(1-T_i)$ days).

In conclusion the orbit of a satellite can be integrated in a body-fixed frame by using the apparent forces given by equation (5), where $\Delta\dot{\vec{\Omega}}$ and $\Delta\vec{\Omega}$ must include nearly diurnal variations coming from precession and lunisolar nutation, in addition to the effects of the long-period wandering of the pole. In order to obtain $\Delta\dot{\vec{\Omega}}$ and $\Delta\ddot{\vec{\Omega}}$ the angular astronomical data must be differentiated twice. This can be easily done if the angular data are represented by a Fourier series (as in analytical theories) or by a polynomial fit. The differentiation does not necessarily provide a good fit to the actual behaviour of $\Delta\dot{\vec{\Omega}}$ and $\Delta\ddot{\vec{\Omega}}$. However, we assume that in the integration of the equations of motion an algorithm is used which handles in a stable way perturbations with a timescale of about one day (this is anyway needed in most cases). Then precession and lunisolar nutation are reproduced in the integrated orbit with about the same accuracy as that of the available data.

A similar treatment applies to length-of-day variations by deducing $\Delta\dot{\Omega}_z$ and $\Delta\Omega_z$ from UT measurements, then calculating the apparent forces. Integration of the equations of motion with this term added will give - on the same assumptions as before - the correct longitude drift and acceleration.

3. DETERMINATION OF POLAR MOTION

The method currently used to determine polar motion by satellite tracking is kinematical, i.e., the orbit is integrated in an inertial frame with polar motion affecting only the station positions.

In a body-fixed frame, polar motion can be determined by a different method. It can be modelled by a suitable fitting depending on a set of

unknown parameters, while precession and nutation are modelled from
observational data. Then the unknown parameters will appear in the equat
of motion via the apparent forces. Therefore they can be determined by a
differential corrections iterative process, with the same method used to
determine any other set of solve-for parameters appearing in the force
model (e.g.,geopotential coefficients). We remark that for polar motion
determinations a high satellite orbit is better, while in general for
geophysical studies low orbits are more useful.

We are studying the possibility of using this approach to analyse laser
range data from the LASSO system which will be carried by the geosynchro
nous satellite SIRIO 2, to be launched by ESA in 1981 (Serene and Albert
noli, 1980; Bertotti et al., 1980). In this case the acceleration due to
polar motion is of the order of 2×10^{-5} cm s^{-2}, mainly in the z directi
To estimate the attainable accuracy, this value must be compared with
the "true" dynamical perturbations. Among these, some are well known
(e.g.,Earth's oblateness, lunisolar gravitational forces) and others
produce effects with a different signature (e.g. resonant harmonics of
the geopotential, which cause a semimajor axis libration - Kamel et al.,
1973). The most critical perturbation is due to solar radiation pressure
because   (a) it cannot be accurately modelled neither in direction, nor
in absolute value (b) its z-component has an annual period, hence it
masks the Fourier components of polar motion with similar periods.
For a satellite of an area-to-mass ratio of 0.05 cm^2 g^{-1}, the solar
radiation pressure produces an acceleration of the order of about
3×10^{-6} cm s^{-2}. The z-component will have a maximum of about
10^{-6} cm s^{-2}. If we assume for this component a 20% uncertainty, apparent
forces due to polar motion cannot be determined better than 1%.
As a matter of fact, the accuracy limits depend mainly on the precision
and geometry of the available range data. At present the laser method
(as applied in the LASSO mission) seems capable of about the same accura
in polar motion determinations as that of the traditional methods Howev
more advanced laser systems are planned and the attainable accuracy
will be improved.

REFERENCES

Bertotti,B., Bevilacqua,R., Farinella,P., Gianni,P., Milani,A. and
    Nobili,A.M.: 1980, "Geophysical LASSO - proposal to ESA", Int.Rep.
    Osservatorio Astronomico di Brera.
Kamel,A. , Ekman,D. and Tibbitts,R.: 1973, Celest.Mech. 8, p. 129.
Serene,B. and Albertinoli,P.: 1980, ESA Journal 4, p. 59.

A NOTE ON THE ORIGIN, OBJECTIVES AND PROGRAMME OF PROJECT MERIT

George A Wilkins
Chairman: Joint IAU/IUGG Working Group on the Rotation of
           the Earth

    Project MERIT is a special programme of international collabora-
tion to Monitor Earth-Rotation and Intercompare the Techniques of
observation and analysis. It was conceived in 1978 at IAU Symposium
No 82 on Time and the Earth's Rotation and a draft proposal was
prepared by a working group set up by the Presidents of IAU Commissions
19 and 31. The proposal was endorsed at the IAU General Assembly at
Montreal in 1979 August and at the IUGG General Assembly at Canberra
in 1979 December, when the organisation and membership of the Working
Group were modified accordingly. The Group is affiliated to the
Commission on the International Coordination of Space Techniques for
Geodesy and Geodynamics (CSTG), which is sponsored by the International
Association of Geodesy (IAG) and by COSPAR. Project MERIT has received
the support of the International Council of Scientific Unions and of
many national organisations and observatories throughout the world.

    Project MERIT has three principal objectives: (1) To foster the
development of new techniques for the measurement of the variations
in the rate and axis of rotation of the Earth. These variations give
rise to non-uniformities in the scale of universal time (UT) and to a
quasi-cyclic revolution of the pole of rotation around the pole of
figure of the Earth. (2) To obtain precise data on earth-rotation
in order to improve our knowledge and understanding of the causes and
effects of these variations in the rotation of the Earth. Analyses
of earth-rotation data provide information about the properties of
the interior of the Earth and about dynamical processes in the oceans
and atmosphere as well as in the solid Earth. (3) To make recommendations on the observational basis and organisational arrangements for
future international services in earth-rotation. These services are
of great importance in navigation and surveying, and especially in
geodesy and space navigation when the highest-possible precision is
required.

    The programme of activities includes special periods for intensive
observation and opportunities for the participating scientists to
meet to discuss the techniques of observation, data reduction and

E. M. Gaposchkin and B. Kołaczek (eds.), Reference Coordinate Systems for Earth Dynamics, 275–276.
Copyright © 1981 by D. Reidel Publishing Company.

analysis and to present the scientific results as they are obtained.
There will be a short campaign of observations during 1980 August to
October to test the techniques and the arrangements for international
cooperation; for the new techniques of laser ranging and radio interferometry this will be the first time that any attempt has been made
to obtain such results regularly and quickly. This short campaign
will be reviewed at a MERIT Workshop to be held at Grasse on 1981
May 19-21; this will be followed by an IAU Colloquium at which recent
scientific results in earth-rotation and Earth-Moon dynamics will be
discussed. Meetings to plan the main campaign will be held during
the IAG and IAU General Assemblies in 1982 and at other suitable
opportunities. There will be a year-long period of regular obsera-
tions by all techniques during 1983/84, followed by detailed analyses
and by a careful assessment of the contributions that the different
techniques should make to an improved international service for the
provision of data on universal time and polar motion. These data
provide the practical link between the terrestrial and celestial
reference frames and so are of fundamental importance to both geodesy
and astronomy.

THE APPLICATION OF FAINT MINOR PLANET DYNAMICS TO THE PROBLEMS OF IMPROVING THE FUNDAMENTAL REFERENCE SYSTEM

P. D. Hemenway and R. L. Duncombe
The University of Texas
Austin, Texas

ABSTRACT

The underlying method of forming the adopted Fundamental Reference System is the compilation of observations referred to some (independent) well defined coordinate systems. If the observations themselves are used to define the independent coordinate system, they may be used to help establish the Fundamental <u>System</u>. If the observations are differential, they may be used to refer star positions to the Fundamental System and/or smooth systematic irregularities within the Fundamental System.

We are presently studying the possible use of faint (16th mag.) minor planets as "test particles" to establish a dynamically based reference frame. Several methods of increased observational accuracy lend themselves to the determination of such a system. The present studies are directed toward determining a) the attainable accuracies for various observation types and b) the effects that varying the distribution of observation types would have on the internal consistency of such a system.

The study also includes analyzing the application of the very high projected accuracy of the Space Telescope to the problem of relating faint reference frames (e.g. the above system, the Hipparcos system, and the VLBI system) to each other and to the extant Fundamental System.

Close cooperation between this project and other ongoing programs will optimize the usefulness of the results.

Finally, some results to date are presented and some overall advantages of such a dynamically based reference frame are discussed.

The basic goal of fundamental astrometric observations is the formation of a reference frame by which the kinematic properties of celestial bodies may be determined. The ability of dynamical theories to account for the observations may then be tested by comparing predictions

E. M. Gaposchkin and B. Kołaczek (eds.), Reference Coordinate Systems for Earth Dynamics, 277–282.
Copyright © 1981 by D. Reidel Publishing Company.

with observations. The quality of the comparison depends on three factors:

1. The accuracy of the observations.

2. The sufficiency of the theory.

3. The correctness of the reference frame to which both
    theory and observation are referred.

Since most dynamical theories are based on Newtonian mechanics, as modified by Einstein, the coordinate system one attempts to use should be as close to inertial as possible. Therefore, the dynamics of moving bodies tend to be used in the definition of the coordinate system itself.

Because astrometric observations have been tied to the Earth as an observing platform, the dynamics of the Earth have played an important role in the formation of the adopted Fundamental Reference System. The most accurate astrometric observations (radio interferometer observations, for example) must either solve for, or account for, the variation of station coordinates as a function of time with respect to a celestial coordinate system. The most accurate geodetic observations (observations of satellite motion, for example) have divorced themselves from measurements with respect to a stellar reference frame altogether; but their need to understand the variation of station coordinates with respect to a dynamically determined reference frame becomes paramount. The small variations now observed in the motion of the Earth and in the relative motion of individual stations makes the determination of even a consistent reference frame a very difficult task. Applying adopted geophysical definitions to forming a celestial coordinate system has become the full-time occupation of many astronomers throughout the world.

Aside from the vagaries of the Earth's motion, high precision optical observations of star images with respect to the reference frame are difficult. Observations of the Sun and planets define the astronomical coordinate system, and they are even harder to relate to the system because of extended amorphous images, and the Sun distorts any instrument used to observe it.

Minor planets afford the possibility of improving the coordinate system through more accurate observations. We are studying the feasibility and scientific validity of using accurate ground-based and space observations of a selected set of small minor planets to obtain corrections to the fundamental celestial coordinate system and provide a zero circle (the ecliptic) for a system of positions and proper motions. Many attempts to use minor planets for fundamental purposes have been undertaken in the past. The most recent discussions are Branham (1979, 1980), which illustrate accuracies of, and problems with, classical techniques applied to bright minor planets.

Two observing techniques hold the promise of increased accuracy in a coordinate system incorporating faint minor planet observations. The position of an image on a photographic plate with respect to a set of faint background star images can be measured with very high relative accuracy, e.g. 0.4 microns, plate to plate for high-information content emulsions such as IIIaJ, (Chiu, 1977). Long focus telescopes (reflectors as well as refractors) have the potential to provide accurate relative positions of minor planets with respect to background stars with unknown absolute positions.

The usual equations of condition for reducing observations of the minor planets for corrections to their orbital elements are:

$$(\alpha_0 - \alpha_c) = \sum_{j=1}^{6} \frac{\partial \alpha}{\partial E_j} \Delta E_j$$

$$(\delta_0 - \delta_c) = \sum_{J=1}^{6} \frac{\partial \delta}{\partial E_j} \Delta E_j$$

(1)

where (α, δ) are the coordinates of the minor planet at the time of observation. The subscripts refer to the observed (0) and computed (c) positions, and E_j the jth orbital element.

In forming a reference frame, zone corrections to an initial coordinate system as well as corrections to the Earth's orbital parameters are included on the right hand side. The least accurately known entries in equations (1) are α_0 and δ_0, the observed position. All the other terms are numbers computed using the initial conditions and an assumed model of the solar system. The differential corrections are obtained relative to the initial model. The accuracy of a "true" position is limited by the accuracy with which the transformation to the initial reference frame may be made. This accuracy is presently on the order of ±0.2 arcseconds rms (±0.15 arcseconds rms for transit circle observations, ±0.2 to ±0.3 arcseconds for photographic transfers to the FK4).

By combining observations of more than one minor planet with respect to the same set of faint background stars, the absolute positions of the background stars may be eliminated. If two minor planets pass through the same star field at different times t_1 and t_2 (see figure 1) the equations of condition become:

$$\alpha_0^{(1)}(t_1) - \alpha_c^{(1)}(t_1) = \sum \frac{\partial \alpha^{(1)}(t_1)}{\partial E_j^{(1)}} \Delta E_j^{(1)}$$

(2)

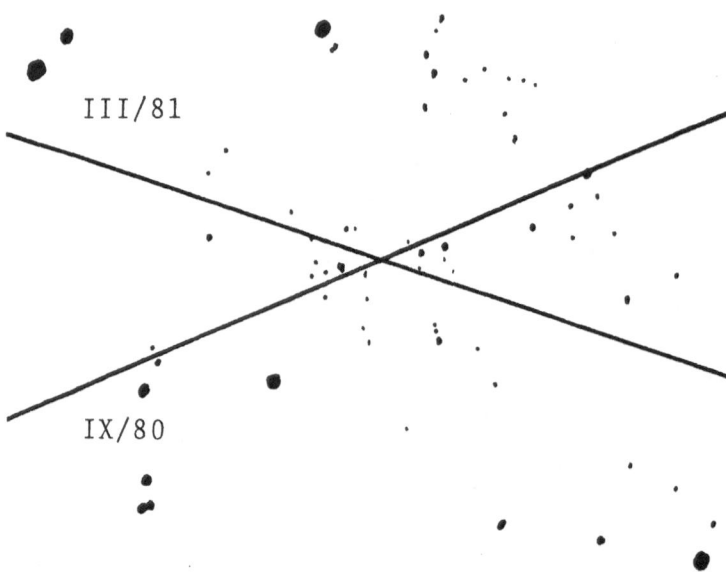

III/81

IX/80

Figure 1. Paths of two minor planets through the same star field.

$$\alpha_0^{(2)}(t_2) - \alpha_c^{(2)}(t_2) = \sum \frac{\partial \alpha^{(2)}(t_2)}{\partial E_j^{(2)}} \Delta E_j^{(2)} \qquad (2)$$

The difference becomes:

$$\alpha_0^{(2)}(t_2) - \alpha_0^{(1)}(t_1) = \alpha_c^{(2)}(t_2) - \alpha_c^{(1)}(t_1) + \left(\Sigma_2 - \Sigma_1 \right) \qquad (3)$$

We may represent the right ascension of the field center of a plate of a long focus telescope by α_{fc} and the offset of the minor planet image from that position by $\Delta\alpha_0$. Then:

$$\alpha_0^{(2)}(t_2) = \alpha_{fc}^{(2)} + \Delta\alpha_0^{(2)}(t_2) \qquad (4)$$

From a central overlap plate solution,

$$\alpha^{(1)}{}_{fc} = \alpha^{(2)}{}_{fc} \quad ,$$

so we have

$$\alpha_0^{(2)} - \alpha_0^{(1)} = \Delta\alpha_0^{(2)} - \Delta\alpha_0^{(1)}$$

which depends only on the long focus plate reductions. An equation of
condition among the orbital elements of two minor planets (or one minor
planet at different times) results:

$$\Delta\alpha_0^{(2)}(t_2) - \Delta\alpha_0^{(1)}(t_1) = \alpha_c^{(2)}(t_2) - \alpha_c^{(1)}(t_1) + \sum_{j=1}^{6} \frac{\partial\alpha^{(2)}(t_2)}{\partial E_j^{(2)}} \Delta E_j^{(2)}$$

$$- \sum_{j=1}^{6} \frac{\partial\alpha^{(1)}(t_1)}{\partial E_j^{(1)}} \Delta E_j^{(1)} \tag{5}$$

This equation of condition has two remarkable properties: 1) It is independent of the systematic errors of the assumed reference frame; and
2) It provides a direct mathematical relation between the orbital elements of two minor planets. Furthermore, it is an equation identical
in form to the classical equation of condition, so that the methods of
solution apply.

In forming a reference frame, zone corrections to the assumed coordinate system as well as corrections to the Earth's orbital elements
must be determined explicitly. Combining suitably modified observational
equations of the classical type and of the crossing point type can
result in the determination of such corrections. The present goal is
to estimate the accuracies of various observing and reduction techniques,
and to determine the accuracies of correction to an adopted reference
system using such techniques.

The plan is to compare the speed and accuracy of hand measurements,
simple centroiding, as well as one and two dimensional Fourier decon-
volution of trailed star and minor planet images scanned automatically
by a microdensitometer. The measuring (setting) error will be determined by fitting several measurements of one exposure to each other.
The relative positional accuracy will be determined by fitting the
short-arc motion of a minor planet to several exposures over a short
time interval. Minor planet Hypatia (238) was observed at 5 hour angles
throughout one night, with the 0.76 meter telescope (f/13) at McDonald
Observatory. The plates have been raster-scanned and measured by hand.

Nemausa (51) has a long history of observation for coordinate
system work, (Møller, 1978). We are undertaking a program with Leif
Kristensen to observe Nemausa at a crossing point on September 17, 1980
and March 18, 1981. (See IAU circular #3480). Astrometric plates will
be obtained at several observatories with long focus instruments. The
observations will be reduced to a common frame of faint 13th to 16th
magnitude, background stars. The consistency of the times of "crossing"
will be compared with the ephemeris of Nemausa to determine the actual
accuracy of the technique relative to Nemausa's motion. (Nemausa has
one of the best known motions of all minor planets). While Nemausa is
larger (150 km) than the minor planets which are being contemplated for

this project (30 km), it will provide a good test of the method.

From the meager data, one can see that sub-1/100th arcsecond rms accuracy is a real possibility. With telescopes of 20-meter focal-length or longer, (e.g. the 2.1 meter f/13 relfector at McDonald Observatory) a few milli-arcseconds accuracy for the relative positions of two minor planet images may be possible. The estimated accuracy of a relative position with the Space Telescope is ±0.002 arcseconds/observation. The accuracy of a Hipparcos position relative to the "rigid" Hipparcos system also has that level of expectation. Thus, a combination of Hipparcos, Space Telescope, and ground based observation of minor planets could provide a stellar reference system which is inherently based on the dynamics of the solar system, and is systematically error free at a very low level.

The regular motion of minor planets across large arcs of the celestial sphere are analogous to the regular motion of the optical axis of a transit circle over large arcs due to the rotation of the Earth. For any given accuracy (say ±0.001 arcsecond) the classical methods (e.g. transit circles) would require a knowledge of the geocentric position of the instrument to a corresponding linear accuracy (3 cm at 6000 km). For the same positional accuracy, the minor planet methods would require a knowledge of the position of the instrument with respect to the minor planet to the corresponding linear accuracy (0.75 km at 1 A.U.). One kilometer accuracy is trivial for the knowledge of the station coordinates on the Earth, but may be difficult to realize from observations and dynamics of minor planets. Hence, the burden of observational and theoretical precision is shifted from the realm of geodesy to the realm of celestial mechanics. We are attempting to determine whether the new techniques will be sufficient to carry the burden. If so, we will propose to undertake a program using the techniques outlined here. The coupling of such a system to a geodetic system on the scale of centimeters is a problem for your consideration.

We are indebted to Leif Kristensen for the crossing point data on 51 Nemausa. We particularly thank Paul Herget for his active participation in this study. We appreciate the support of the National Science Foundation under Grant AST-79-16870 and the Organizing Committee of this Conference.

REFERENCES

Branham, R. L., 1979, Astron. Jour., 84, pp. 1399.
Branham, R. L., 1980, Astronomical Papers on the AENA, XXI, Part 3.
Chui, L. G., 1977, Astron. Jour. 82, pp. 842-848.
Møller, O., "The Work on a Possible Contribution to the Fundamental System of Reference by Observation of Minor Planet 51 Nemausa", in Reiz and Andersen (1978).
Reiz, A. and Andersen, T., 1978, Astronomical Papers Dedicated to Bengt Stromgren, Copenhagen University Observatory, J. J. Trykteknik a/s, Copenhagen.

RELATIVISTIC REDUCTION OF ASTRONOMICAL MEASUREMENTS AND REFERENCE FRAMES

V. A. Brumberg
Institute of Theoretical Astronomy
191187 Leningrad
U.S.S.R.

ABSTRACT

With the accuracy of modern observations the relativistic treatment of the basic astronomical reference frames only requires the consideration of comparatively simple types of metrics such as heliocentric Schwarzschild metric, geocentric Schwarzschild metric and metric of the Earth-Sun system. Dynamical (related to the motion of the bodies) and kinematical (related to the light propagation) characteristics of these metrics enable one to perform the accurate relativistic reduction of astronomical measurements. In this reduction, the choice of specific quasi-Galilean coordinates may remain arbitrary. This paper presents expressions for the main relativistic terms in coordinates of the principal planets and Moon using the PPN formalism parameters β, γ and coordinate parameter α. General formulae for the reduction of radar, radio-interferometric and astrometric observations of planets and for the interpretation of lunar laser ranging are given. For estimating the actual magnitude of relativistic effects, the ephemeris data should be expressed in terms of physically measurable quantities.

INTRODUCTION

Relativistic treatment of the problem of astronomical reference frames involves some distinctive features compared to the Newtonian treatment.

First of all, much attention is given now to the physical approach to this problem. Just as the problem of time measurement changed from astronomy to physics, the definition of three orthogonal directions in space will possibly be made in the future by laboratory means (such as gyroscopes) rather than with the use of astronomical objects. Such a laboratory measurement of time and determination of space directions provide a reference frame suitable for astronomy as well. Mathematically this system is represented by four vectors subjected to Fermi-Walker propagation along the world-line of the laboratory. They

E. M. Gaposchkin and B. Kołaczek (eds.), Reference Coordinate Systems for Earth Dynamics, 283–294.

consist of the time-like tangent to the world-line and three space-like vectors normal to it. This approach is developed in detail by Synge (1960). In spite of extensive recent investigations of the mathematical aspects of this problem, the presentation by Synge still remains the most adequate for astronomical practice.

In addition, a relativistic reduction of astronomical observations becomes increasingly important. In essence, astronomical observations reduce to measuring angles between light rays or time intervals between events marked by light signals. Relativistic treatment of these quantities is due mainly to the effect of the solar gravitational field. It is necessary to take into account both the direct influence of this field on light propagation and the effect of performing measurements in a curved space. Observations reduced in this way may be further used for determining astronomical reference frames by classical methods.

Another feature of the relativistic analysis is of a purely mathematical nature and is related with the possibility to use arbitrary quasi-Galilean coordinates for describing events in the solar system. Even though the mathematical character of the motion of bodies and the propagation of light are different in distinct quasi-Galilean coordinates, the physically measurable quantities do not depend on the choice of coordinates. It is only important to calculate dynamical characteristics and to perform relativistic reduction of kinematic data in a single coordinate system.

This paper deals with questions pertaining to the last two facts. Sections 2 and 3 are devoted to the reduction of radar, radio interferometric and astrometric observations. In extending the results of Brumberg (1979, 1981) and Brumberg and Finkelstein (1979), the geocentric Schwarzschild metric is considered as enabling us to combine, in a unified manner, both relativistic gravitation and aberration effects. In Section 4, the lunar laser ranging is considered. It is shown that, in opposition to the views of Baierlein (1967), the influence of the relativistic effects does not result in a mere multiplication of Newtonian quantities by a constant factor. The whole treatment is performed in the Parametrized Post-Newtonian approximation (PPN) formalism (Will, 1974) but in arbitrary (to the sufficient degree) quasi-Galilean coordinates.

HELIOCENTRIC SCHWARZSCHILD METRIC

Metric

Heliocentric Schwarzschild metric is of the form where $m = GM_{\odot}/c^2$

$$ds^2 = \left\{1 - 2(m/r) + 2[\beta - \alpha(r)]\ (m/r)^2\right\} c^2 dt^2$$
$$-\left\{1 + 2[\gamma - \alpha(r)]\ (m/r)\right\} (d\underline{r})^2 - 2[\alpha(r) - r\alpha'(r)]\ (m/r^3)(\underline{r}d\underline{r})^2 \quad , \tag{1}$$

β, and γ are principal constants of the PPN formalism (for GRT, β=γ=1), $\alpha(r)$ is an arbitrary function of r satisfying the conditions of a quasi-Galilean metric: $\alpha'(r) \to 0$, $\alpha(r)/r \to 0$ with $r \to \infty$ (the dash denotes differentiation with respect to r). Transformation

$$\overset{\sim}{\underline{r}} = \underline{r} - m\alpha(r)\underline{r}/r \tag{2}$$

converts (1) to the Eddington-Robertson metric (metric (1) with α=0). The most widely used coordinate systems correspond to α=0 (harmonic coordinates) or α=1 ("standard" coordinates). Introduction of $\alpha(r)$ in a literal form has two objectives. First of all, the appearance of α in the expression of any particular quantity demonstrates its coordinate dependence, in other words its unobservability. Besides this, in final relations in terms of the ephemeris of physically measurable quantities, α should disappear, which would confirm the correctness of the calculations.

Motion of the Major Planets

For approximate analytical estimations, it is convenient to have the following expansions of polar orbital coordinates (radius-vector r, argument of latitude u) in powers of eccentricity e:

$$\frac{r}{a} = 1 + \frac{1}{2}\left[1 + \mu k_0(a)\right]e^2 - \left[1 + \mu k_1(a)\right] e \cos(\lambda - \pi)$$

$$- \frac{1}{2}\left[1 + \mu k_2(a)\right] e^2 \cos 2(\lambda - \pi) + \ldots \quad , \tag{3}$$

$$u = \lambda - \Omega + 2e \sin(\lambda - \pi) + \left[\frac{5}{4} + \mu(-3-3\gamma + \frac{5}{4}\beta)\right] e^2 \sin 2(\lambda - \pi) + \ldots \tag{4}$$

These expansions result from the closed expressions of the Schwarzschild problem (Brumberg, 1972). Here

$$k_0(a) = -\frac{4}{3} - \frac{4}{3}\gamma + \frac{2}{3}\beta - \alpha(a) + a\alpha'(a) + \frac{1}{2} a^2 \alpha''(a) \quad ,$$

$$k_1(a) = -1 - \gamma + \beta - \alpha(a) + a\alpha'(a) \quad ,$$

$$k_2(a) = -4 - 4\gamma + 2\beta - \alpha(a) + a\alpha'(a) - \frac{1}{2} a^2 \alpha''(a) \quad ,$$

$\mu = m/a$, λ, π, and Ω are angular arguments (mean longitude, longitude of the perihelion, longitude of the node) with Ω being constant, λ and π being linear functions of time with mean motions:

$$\dot{\lambda} = n, \qquad \dot{\pi} = \mu(2 + 2\gamma - \beta) n/(1 - e^2) \quad , \tag{5}$$

$$n = \left(GM_\odot/a^3\right)^{1/2} \left\{1 + \mu\left[-\frac{1}{2}\gamma - \beta + \frac{3}{2}\alpha(a)\right]\right\} \tag{5}$$

Light Propagation

The law of light propagation under conditions

$$\underline{r}(t_0) = \underline{r}_0 \quad , \quad \underline{\dot{r}}(-\infty) = c\underline{\sigma} \quad , \quad \underline{\sigma}^2 = 1$$

is given by:

$$\underline{r}(t) = \underline{r}_0 + c(t - t_0)\,\underline{\sigma} + m\left\{(\gamma + 1)\left[\frac{\underline{\sigma} \times (\underline{r}_0 \times \underline{\sigma})}{r_0 - \underline{\sigma}\underline{r}_0} - \frac{\underline{\sigma} \times (r \times \sigma)}{r - \underline{\sigma} r}\right.\right.$$

$$\left.\left. - \underline{\sigma}\ln\frac{r + \underline{\sigma}\,r}{r_0 + \underline{\sigma}\,\underline{r}_0}\right] + \frac{\alpha(r)}{r}\,\underline{r} - \frac{\alpha(r_0)}{r_0}\,\underline{r}_0\right\} \quad . \tag{6}$$

The velocity of light at the time t is obtained from:

$$\frac{1}{c}\,\underline{\dot{r}}(t) = \underline{\sigma} + \frac{m}{r}\left\{[\alpha(r) - \gamma - 1]\,\underline{\sigma} - [\alpha(r) - r\alpha'(r)]\frac{(\underline{\sigma}\,r)}{r^2}\,\underline{r}\right.$$

$$\left. - (\gamma + 1)\frac{\underline{\sigma} \times (\underline{r} \times \underline{\sigma})}{r - \underline{\sigma}\,\underline{r}}\right\} \quad . \tag{7}$$

For a solar ray,

$$\underline{\sigma} = (\underline{r}_0/r_0) = (\underline{r}/r) \quad .$$

The solution of the boundary value problem

$$\underline{r}(t_0) = \underline{r}_0 \quad , \quad \underline{r}(t_1) = \underline{r}_1$$

is determined by (6) and (7) with

$$\underline{\sigma} = \frac{\underline{D}_{01}}{D_{01}} + \frac{m}{D_{01}}\left\{(\gamma + 1)\frac{r_1 - r_0 + D_{01}}{(\underline{r}_0 \times \underline{r}_1)^2} + \frac{1}{D_{01}^2}\left[\frac{\alpha(r_0)}{r_0} - \frac{\alpha(r_1)}{r_1}\right]\right\}$$

$$\times \left[\underline{D}_{01} \times (\underline{r}_0 \times \underline{r}_1)\right] \tag{8}$$

where

$$\underline{D}_{01} = \underline{r}_1 - \underline{r}_0$$

That time of flight will be

$$
t_1 - t_0 = \frac{D_{01}}{c} + \frac{m}{c} \left\{ (\gamma + 1) \ln \frac{r_0 + r_1 + D_{01}}{r_0 + r_1 - D_{01}} + \alpha(r_0) \frac{r_1^2 - r_0^2 - D_{01}^2}{2r_0 D_{01}} \right.
$$

$$
\left. + \alpha(r_1) \frac{r_0^2 - r_1^2 - D_{01}^2}{2r_1 D_{01}} \right\} \quad . \tag{9}
$$

For a solar ray (with $r_1 > r_0$) ,

$$
t_1 - t_0 = \frac{1}{c} \left\{ r_1 - r_0 + m \left[(\gamma + 1) \ln \frac{r_1}{r_0} - \alpha(r_1) + \alpha(r_0) \right] \right\} \quad . \tag{10}
$$

Doppler Shift

Let ν_0 be the frequency of a light emitter at a given point $\underline{r_0}(t_0)$ and ν_1 be the frequency of the light at a receiver situated at a point $\underline{r_1}(t_1)$. Then

$$
\frac{\nu_0}{\nu_1} = \frac{1 - m/r_1 - \underline{\dot{r}}_1^2/(2c^2)}{1 - m/r_0 - \underline{\dot{r}}_0^2/(2c^2)} \frac{dt_1}{dt_0} \quad . \tag{11}
$$

GEOCENTRIC SCHWARZSCHILD METRIC

Metric

The transformation

$$
\underline{r} = \underline{R}(t) + \underline{\rho} \tag{12}
$$

where $\underline{R}(t)$ is the heliocentric vector of the Earth transforming (1) to the geocentric Schwarzschild metric

$$
ds^2 = \left\{ 1 - 2(m/r) - (\underline{\dot{R}}/c)^2 + 2 [\beta - \alpha(r)] (m/r)^2 - 2(m/r) [\gamma - \alpha(r)] \right.
$$

$$
\times (\underline{\dot{R}}/c)^2
$$

$$
\left. - 2(m/r^3) [\alpha(r) - r\alpha'(r)] (\underline{r}\underline{\dot{R}}/c)^2 \right\} c^2 dt^2 - 2 \left\{ \underline{\dot{R}} + 2(m/r) [\gamma - \alpha(r)] \underline{\dot{R}} \right.
$$

$$
\left. + 2(m/r^3) [\alpha(r) - r\alpha'(r)] (\underline{r}\underline{\dot{R}})\underline{r} \right\} d\underline{\rho} \, dt - \left\{ 1 + 2(m/r) [\gamma - \alpha(r)] \right\} d\underline{\rho}^2
$$

$$
- 2(m/r^3) [\alpha(r) - r\alpha'(r)] (\underline{r} d\underline{\rho})^2 \quad . \tag{13}
$$

This metric describes the solar gravitational field but from the geocentric point of view. The line-element $d\ell$ of space distance of this metric is determined by

$$d\ell^2 = \left\{1 + 2(m/r) \left[\gamma - \alpha(r)\right]\right\} d\underline{\rho}^2 + 2(m/r^3) \left[\alpha(r) - r\alpha'(r)\right] (\underline{r}d\underline{\rho})^2$$

$$+ (\underline{\dot{R}}d\underline{\rho}/c)^2 \quad . \tag{14}$$

For the case of the synchronization of clocks it results that instants t_0 at a point $\underline{\rho}$ and t_0' at a point $\underline{\rho} + d\underline{\rho}$ are simultaneous provided that

$$t_0' = t_0 + \underline{\dot{R}}d\underline{\rho}/c^2 \quad . \tag{15}$$

Lengths and Angles

If $\underline{P}, \underline{Q}$ are three-dimensional vectors applied at a point $\underline{r}(t)$, their scalar product and length in space (14) is

$$(\underline{PQ})_{rel} = \underline{PQ} + 2(m/r) \left[\gamma - \alpha(r)\right] (\underline{PQ}) + 2(m/r^3) \left[\alpha(r) - r\alpha'(r)\right]$$

$$\times (\underline{Pr})(\underline{Qr}) + (\underline{\dot{R}P})(\underline{\dot{R}Q})/c^2 \quad , \tag{16}$$

$$P_{rel} = P\left\{1 + (m/r) \left[\gamma - \alpha(r)\right] + (m/r^3) \left[\alpha(r) - r\alpha'(r)\right] (\underline{Pr})^2/P^2 \right.$$

$$\left. + (\underline{\dot{R}P})^2/(2c^2P^2)\right\} \quad , \tag{17}$$

where $\underline{PQ}$ and $P = |\underline{P}|$ denote the scalar product and the length in Euclidean sense. For the angle ψ between vectors $\underline{P}, \underline{Q}$ if we define $(\underline{PQ})_{rel} = P_{rel} Q_{rel} \cos \psi$, one has

$$\cos \psi = (\underline{PQ})/(PQ) + \left\{(m/r^3) \left[\alpha(r) - r\alpha'(r)\right] \left[(\underline{Pr})(\underline{P} \times \underline{r})/P^2 - (\underline{Qr})\right.\right.$$

$$\times (\underline{Q} \times \underline{r})/Q^2] + [(\underline{\dot{P}R})(\underline{P} \times \underline{\dot{R}})/P^2 - (\underline{Q\dot{R}})(\underline{Q} \times \underline{\dot{R}})/Q^2]/(2c^2)\right\}$$

$$\times (\underline{P} \times \underline{Q})/(PQ) \quad . \tag{18}$$

Geocentric Angle Between Light Rays

The direction of a light ray (7) crossing the earth in a position $\underline{R}$ at the time t may be determined by a vector $\underline{P} = \underline{\dot{\rho}}/c$ with $\underline{\dot{\rho}}$ to be computed from (12) and (7) where we set $\underline{r} = \underline{R}$. Calculating the Euclidean length P of this vector, one finds

$$\underline{P}/P = \underline{\sigma} + [\underline{\sigma} \times (\underline{\sigma} \times \underline{\dot{R}})]/c + (\underline{\sigma}\underline{\dot{R}}) \, [\sigma \times (\sigma \times \underline{\dot{R}})]/c^2 - \sigma(\underline{\sigma} \times \underline{\dot{R}})^2/(2c^2)$$

$$+ \frac{m}{R} \left\{ [\alpha(R) - R\alpha'(R)] \frac{(\underline{\sigma}R)}{R^2} + \frac{\gamma + 1}{R - \underline{\sigma}\underline{R}} \right\} [\underline{\sigma} \times (\underline{\sigma} \times \underline{R})] \quad . \tag{19}$$

Hence, from (18) there results an expression for geocentric angle ψ between two light rays having at t = -∞ the directions $\underline{\sigma}_1$, $\underline{\sigma}_2$ and crossing the earth, in the position $\underline{R}$ at time t:

$$\cos \psi = \underline{\sigma}_1\underline{\sigma}_2 + \frac{1}{c} (\underline{\sigma}_1\underline{\sigma}_2 - 1)(\underline{\sigma}_1\underline{\dot{R}} + \underline{\sigma}_2\underline{\dot{R}}) + \frac{1}{c^2} (\underline{\sigma}_1\underline{\sigma}_2 - 1) \left[(\underline{\sigma}_1\underline{\dot{R}})^2 + (\underline{G}_2\underline{\dot{R}})^2 \right.$$

$$\left. + (\underline{\sigma}_1\underline{\dot{R}})(\underline{\sigma}_2\underline{\dot{R}}) - \underline{\dot{R}}^2 \right] + (\gamma + 1) \frac{m}{R} \left(\frac{R \times \underline{\sigma}_1}{R - \underline{R}\underline{\sigma}_1} - \frac{R \times \underline{\sigma}_2}{R - \underline{R}\underline{\sigma}_2} \right) (\underline{\sigma}_1 \times \underline{\sigma}_2) \quad . \tag{20}$$

Relativistic Reduction of Astronomical Measurements

For relativistic reduction, of astronomical measurements one may use relations (9) (radar ranging, radio-interferometry), (11) (Doppler's observations) and (20) (astrometry). The detailed analysis of different cases of this reduction is given in Brumberg, (1981) for the Earth at rest ($\underline{\dot{R}}=0$). Let us restrict ourselves to some basic statements.

Radar ranging. A round-trip transit reduces in essence to a double expression (9). If the coordinate dependent distance D_{01} is expressed in terms of initial measured quantities, then coordinate dependence on α in (9) disappears. The numerical value of the corresponding relativistic effect and its functional dependence on β, γ may be determined by the initial measurements (Brumberg, Finkelstein, 1979).

VLBI. Let a radio wave coming from an infinitely distant source reach station 2 on the Earth at a time t_2 and station 1 at a time t_1. At station 1, one measures a time delay $\tau = t_1 - t_2'$, t_2' being the time at station 1 simultaneous with t_2 at station 2. The expression for τ may be found from $\tau = (t_1 - t_0) - (t_2 - t_0) - (t_2' - t_2)$ (t_0 is the time of the wave emission) by using (15) for $t_2' - t_2$ and (9) for $t_i - t_0$ (i = 1,2) with the limit $r_0 \to \infty$. Further use of (17)-(19) permits to express τ in terms of physically measurable quantities such as the proper length of the base vector and the angle between the base vector and the direction to the radio souce (Brumberg, 1981).

Angular measurements. Relation (20) covers a great variety of types of astrometric observations. Let, for instance, $\underline{\sigma}_1$ be the direction of a light ray at t = -∞ from an infinitely distant source. Then if we let $\underline{\sigma}_2 = \underline{R}/|\underline{R}|$ we obtain the special relativistic formula for the aberration, if we let $\underline{\sigma}_2 = \underline{R}/R$ we obtain the formula for the light deflection, and if $\underline{\sigma}_2$ is determined by (8) then we get an expression of the angle between an observed planet and a distant source. Choosing the last case, $\underline{\sigma}_1 = \underline{R}/R$, one gets the expression for the angular distance of a planet from the Sun. Presenting coordinates of the planet and the Sun in terms of initial measurements, one obtains coordinate independent relativistic effects determined only by measurements (Brumberg and Finkelstein, 1979).

Regarding $\underline{R}(t)$ in (12) and (13) as heliocentric vector of the point of observation on the Earth, we obtain the topocentric Schwarzschild metric enabling to take into account both the annual and diurnal motions of the Earth. In this matter we may combine the relativistic gravitational effects with those due to the special theory of relativity. In practical calculations these reductions may be separated as it was assumed in the above-mentioned papers.

LUNAR LASER RANGING

The Field of n Point Masses

Using the parameters β, γ of the PPN formalism (Will, Nordtvedt, 1972) and the coordinate parameters α, ν (Brumberg, 1972), the field of n non-rotating point masses M_i is described by the following metric

$$ds^2 = \left\{ 1 - 2\sum_i \frac{m_i}{\rho_i} + 2(\beta - \alpha)\left(\sum_i \frac{m_i}{\rho_i}\right)^2 + (4\beta - 2)\sum_i \frac{m_i}{\rho_i}\sum_{j\neq i}\frac{m_j}{r_{ij}} \right.$$

$$+ 2\alpha\sum_i \frac{m_i}{\rho_i^3}\sum_{j\neq i} m_j\left(\frac{1}{r_{ij}} - \frac{1}{\rho_j}\right)(\underline{\rho}_i\underline{r}_{ij}) - \frac{1}{c^2}(2\gamma + 1)\sum_i \frac{m_i}{\rho_i}\dot{\underline{r}}_i^2$$

$$+ \frac{1}{c^2}(\nu - 1)\frac{\partial^2}{\partial t^2}\sum_i m_i\rho_i \left.\right\} c^2 dt^2 + \frac{2}{c}\sum_i \frac{m_i}{\rho_i}\left[(2\gamma + 2 - \alpha - \frac{\nu}{2})\dot{\underline{r}}_i \right.$$

$$+ (\alpha + \frac{\nu}{2})\frac{1}{\rho_i^2}(\underline{\rho}_i\dot{\underline{r}}_i)\underline{\rho}_i \right] d\underline{r}\, cdt - 2\alpha\sum_i \frac{m_i}{\rho_i^3}(\underline{\rho}_i\, d\underline{r})^2 - \left[1 + 2(\gamma - \alpha) \right.$$

$$\left. \times \sum_i \frac{m_i}{\rho_i} \right] (d\underline{r})^2 \tag{21}$$

where $\underline{\rho}_i = \underline{r} - \underline{r}_i$, $\underline{r}_{ij} = \underline{r}_i - \underline{r}_j$, $m_i = GM_i/c^2$. In contrast to (1), α is constant. In papers on PPN formalism, the coordinate system is fixed by the choice $\alpha = 0$, $\nu = 1$. In papers on GRT, harmonic system, with $\alpha = \nu = 0$ are used. Denoting harmonic coordinates by $\sim$ we have

$$\tilde{t} = t + \frac{\nu}{2c^2} \frac{\partial}{\partial t} \bigg|_i \quad m_i \rho_i \quad , \tag{22}$$

$$\tilde{\underline{r}} = \underline{r} - \alpha \sum_i \frac{m_i}{\rho_i} \underline{\rho}_i \quad , \qquad \tilde{\underline{r}}_i = \underline{r}_i - \alpha \sum_{j \neq i} \frac{m_j}{r_{ij}} \underline{r}_{ij} \quad . \tag{23}$$

In the post-Newtonian approximation the parameter ν does not affect the motion of bodies and light propagation.

Main Relativistic Terms in Lunar Theory

Let indices 1 and 2 in (21) be referred to the Earth and Sun respectively. Taking into account that $m_1 \ll m_2$, put $\underline{r}_1 = R$, $r_2 = 0$, $\underline{R}$ being the heliocentric vector of the Earth determined by (3), (4) converted to the geocentric system $\underline{r} = \underline{R} + \underline{\rho}_1$. The Moon will be considered as a probe particle moving in the field (21) of the Earth and the Sun. Then the equations of lunar motion follow from the geodesic principle and are described by the Lagrangian $L = c^2 - c(ds/dt)$, if we consider here only a point-mass the results obtained eliminate the Nordtvedt effect (breaking the principle of equivalence for massive bodies). Thus the variational inequalities turn out to be of the most importance. Neglecting the sun's eccentricity, the eccentricity and inclination of the lunar orbit and parallactic terms, one may find for the radius-vector ρ_1 and longitude v of the Moon the following expressions

$$\frac{\rho_1}{\underline{a}_0} = 1 - m^2 \cos 2D + \ldots + \mu \left\{ -\frac{2}{3}\gamma - \frac{4}{3}\beta - \frac{1}{4} + \frac{1}{2}\alpha + \frac{1}{3}(2\gamma + 1)m \right.$$

$$+ \left(\frac{1}{3}\beta - \frac{2}{3}\gamma - \frac{109}{96} + \frac{15}{16}\alpha \right)m^2 + \left[\frac{1}{4} - \frac{1}{2}\alpha + \left(\frac{1}{4} + 2\gamma + 2\beta - \frac{1}{2}\alpha \right)m^2 \right] \cos 2D$$

$$+ \left. \left(\frac{7}{32} - \frac{7}{16}\alpha \right)m^2 \cos 4D \right\} \quad , \tag{24}$$

$$v = nt + \epsilon + \frac{11}{8} m^2 \sin 2D + \ldots + \mu \left\{ \left[\frac{1}{2}\alpha - \frac{1}{4} - \frac{11}{12}(\beta + 2\gamma)m^2 \right] \sin 2D \right.$$

$$+ \left. \left(\frac{11}{16}\alpha - \frac{11}{32} \right)m^2 \sin 4D \right\} \quad . \tag{25}$$

Here n is the lunar mean motion, N is the mean motion of the Sun related to the semimajor axis A of the Earth orbit by (5), $\mu = N^2A^2/c^2$ is consistent with values for μ as in (3), (4), $m = N/(n - N)$, D is the difference of the mean longitudes of the Moon and the Sun, $\underline{a}_0$ is the Hill's parallactic factor

$$a_0 = a_0(1 - \frac{1}{6} m^2 + \ldots), \qquad n^2 a_0^3 = GM_1 \quad .$$

These terms are obtained up to m^2 inclusively. For $\alpha = 0$ they agree with the expressions of Finkelstein and Kreinovich (1976).

Lunar Laser Ranging

In accordance with (21), the time $t - t_0$ of light propagation between points $\underline{r}_0$ and $\underline{r}$, neglecting the motion of the bodies, is determined by

$$c(t - t_0) = |\underline{r} - \underline{r}_0| + \sum_i m_i \left\{ (\gamma + 1) \ln \frac{\rho_i + \rho_i^{(0)} + |\underline{\rho}_i - \underline{\rho}_i^{(0)}|}{\rho_i + \rho_i^{(0)} - |\underline{\rho}_i - \underline{\rho}_i^{(0)}|} \right.$$

$$\left. + \frac{\alpha}{2\rho_i \rho_i^{(0)}} \frac{(\rho_i + \rho_i^{(0)})}{|\underline{\rho}_i - \underline{\rho}_i^{(0)}|} \left[\left(\underline{\rho}_i - \underline{\rho}_i^{(0)} \right)^2 - \left(\underline{\rho}_i - \underline{\rho}_i^{(0)} \right)^2 \right] \right\} \; . \quad (26)$$

Introduce the distance S between the Earth station and a reflector on the Moon, the distance d between the centers of mass of the Earth and Moon, the heliocentric distance R of the Earth, the geocentric angle H between the directions to the Sun and Moon and the radii d_e, d_m of the Earth and Moon respectively. Considering the lunar laser ranging is usually performed near the meridian of the laser station one may put in relativistic terms $S = d - d_e - d_m$ and $\rho_1^{(0)} = d_e$, $\rho_1 = d - d_m$, $\rho_2^{(0)} = R$, $\rho_2 = R - d \cos H$. Therefore, the round-trip coordinate time interval will be

$$T = \frac{2S}{c} \left[1 + \frac{m_2}{R} (\gamma + 1 - \alpha \sin^2 H) \right] + \frac{m_1}{c} (2\gamma + 2) \ln \frac{d - d_m}{d_e} \quad . \quad (27)$$

For harmonic coordinates $\alpha = 0$ this formula was derived first by Baierlein (1967). Noticing the proportionality of the right-hand member of (27) to S/c with practically a constant factor (the last term in (27) may be multiplied within the adopted accuracy by S/d), Baierlein came to conclusion that it is impossible to reveal relativistic effects in measuring T. But the right-hand member of Baierlein's formula represents a coordinate-dependent expression valid only in harmonic coordinates. In the Newtonian, part the distance S is not a physically measurable quantity. The expression of S in terms of measurable quantities (mean motions of the Moon and Sun and gravitational parameters of the Earth and Sun) contains time-dependent relativistic corrections. In accordance with (24), Newtonian value S_N may be presented as follows

$$S_N = \underline{a}_0 (1 - m^2 \cos 2D + \ldots) - d_e - d_m \quad . \quad (28)$$

Putting S = d in relativistic terms and using (24), there results

$$
T = \frac{2S_N}{c} + \frac{m_1}{c}(2\gamma + 2) \ln \frac{d - d_m}{d_e} + \frac{2\mu a_0}{c} \left\{ \frac{1}{3}\gamma - \frac{4}{3}\beta + \frac{3}{4} + \frac{1}{3}(2\gamma + 1)m \right.
$$

$$
\left. + \left(\frac{1}{3}\beta - \frac{2}{3}\gamma - \frac{109}{96}\right)m^2 + \left[\frac{1}{4} + (\gamma + 2\beta - \frac{3}{4})m^2\right] \cos 2D + \frac{7}{32} m^2 \cos 4D \right\} \quad .
$$

$$(29)$$

This relation enables us to obtain a real estimate of the relativistic effects in lunar laser ranging. The most significant periodic relativistic effect with the argument, 2D (with period of 14.76 days) depends rather faintly on parameters β, γ and its amplitude is determined in fact by $\mu a_0/2 = 2$ meters ($\mu = 10^{-8}$, m = 0.08, $a_0 = 4 \cdot 10^5$ km). It may be added that according to Finkelstein and Kreinovich (1976), parameters β, γ make a contribution to the parallactic term with the argument D and a coefficient proportional to $\mu(a_0/A)/m$. But the magnitude of this term is smaller by at least one order than the magnitude of the main relativistic variational term.

The transformation to the proper time τ of the laser station is performed by

$$
d\tau = \left(1 - \frac{m_1}{d_e} - \frac{1}{c^2} \dot{R}V - \frac{3}{2} \mu\right) dt \tag{30}
$$

where $\dot{R}$ is heliocentric velocity of the center of mass of the Earth, V is geocentric velocity of the laser station. Integrating the right-hand member of (30) yields the known formula relating τ and t (Mulholland, 1972). Such a relation is necessary for an accurate computation of solar and lunar ephemerides. But in order to estimate the time delay in lunar laser ranging, one may neglect the variations of the functions appearing in (30) and consider this formula as a direct relation between proper τ and coordinate T time delays.

CONCLUSION

The problem of determining astronomical reference frames meets with difficulties even at the Newtonian level (Kovalevsky, 1975). The theory of relativity increases these difficulties. Much remains to be done to perform all part of the Newtonian theory in a relativistic basis. In this paper some questions of the relativistic reduction of astronomical measurements have been considered.

But the difficulties of the relativistic treatment of reference frames should not be exaggerated. The practical problem is to correlate results of measurements performed at different times in distinct observatories. Knowing with some accuracy the metric of the gravitational field, one may calculate the motions of observatories and reduce all measurements to one actual or fictitious laboratory at some moment of time (placed, for instance, in the center of mass of the Earth, Sun or Solar system). A discussion of such reduced measurements leads in turn to an improvement of the metric.

REFERENCES

Baierlein, R., 1967. Phys. Rev. 162, 1275.

Brumberg, V. A., 1972. Relativistic Celestial Mechanics. Nauka, Moscow (in Russian).

Brumberg, V. A., 1979. Celes. Mech. 20, 329.

Brumberg, V. A. and Finkelstein, A. M., 1979. J. Exper. Theor. Phys. (U.S.S.R.) 76, 1474 (Sov. Phys. JETP 49, 749, 1979).

Brumberg, V. A., 1981. Astron. J. (U.S.S.R.) 58, 1.

Finkelstein, A. M. and Kreinovich, V. Ja., 1976. Celes. Mech. 13, 151.

Kovalevsky, J., 1975. In: On Reference Coordinate Systems for Earth Dynamics (eds. B. Kolaczek and G. Weiffenbach), p. 123, Poland.

Mulholland, J. D., 1972. Publ. Astron. Soc. Pacific 84, 357.

Synge, J. L., 1960. Relativity: The General Theory. North-Holland Publ. Com.

Will, C. M. and Nordtvedt, K., 1972. Astrophys. J. 177, 757.

Will, C. M., 1974. In: Experimental Gravitation (ed. B. Bertotti), p. 1, Academic Press.

PLANETARY EPHEMERIDES

Jay H. Lieske
Astronomisches Rechen-Institut
Heidelberg, Federal Republic of Germany

and

E. Myles Standish
Jet Propulsion Laboratory
Pasadena, California, USA

ABSTRACT

   In the past twenty years there has been a great amount of growth in
radiometric observing methods, as well as in classical optical observations. Through radar ranging and Doppler observations of the planets
and spacecraft, we have been able to improve our knowledge of the location and motion of the planets by several orders of magnitude and have
succeeded in planning and executing space missions which would have been
difficult if not impossible to plan and to perform utilizing the classical ephemerides. We will outline the goals and methods employed by
the Jet Propulsion Laboratory in its effort to develop improved ephemerides which accurately reflect the motions of planets in an inertial
system.

   We will demonstrate that in the ideal situation a radar observation is largely independent of the problems usually associated with the
precessional motion of the earth's axis and is, in fact, a reliable
method for obtaining inertial mean motions. Based upon our hypothesis
that the modern JPL ephemerides are valid in a fixed (i.e., non-rotating)
coordinate system, we will explore the implications concerning optical
observations of the Sun, precession, equinox drift and the relationship
between dynamical and universal time scales, as well as comparisons with
Newcomb (1898) ephemerides.

I.   RANGE OBSERVATIONS

   It may be shown that interplanetary ranging data by itself is sufficient to determine mean motions with respect to an inertial system.
Starting with a very simplified example, we consider two planets in co-
planar, circular orbits. We observe the round-trip time delay τ and by
using the relationship $\rho = c\tau/2$ which relates the round-trip time delay
to the one-way distance ρ we observe that at conjunction $\rho_c = a + a_e$ and at

E. M. Gaposchkin and B. Kołaczek (eds.), Reference Coordinate Systems for Earth Dynamics, 295–304.
Copyright © 1981 by D. Reidel Publishing Company.

opposition $\rho_0 = a - a_e$, where a and a_e represent the semi-major axes of the planet and the Earth. We can then obtain ratios of semi-major axes to remove dependence upon laboratory units and we can employ Kepler's law to obtain

$$n_e/n = (a/a_e)^{3/2} \ . \tag{1}$$

Furthermore, from the observed synodic period, T_{syn}, we can determine both mean motions from

$$n_e = 2\pi \ T_{syn}^{-1} \ (1 - n/n_e)^{-1} \ , \tag{2}$$

with a similar equation for the other planet – which shows that one is indeed able to determine the inertial mean motions of the planets from the observations at conjunction and opposition and from the synodic period.

In practice, of course, the situation is more complicated. A more complete expression for the actual one-way distance would be

$$\rho = | \ \vec{r} - \vec{R} - \vec{R}_e + \vec{R}_p \ | \tag{3}$$

where $\vec{r}$ and $\vec{R}$ are the heliocentric vectors to the planet and the Earth, where $\vec{R}_e$ is the geocentric position of the observer, and where $\vec{R}_p$ is the planetocentric point from where the signal is returned. In Eq. (3) there is only slight dependence upon precession, since $\vec{r}$ and $\vec{R}$ are referred to the inertial coordinate system (B1950.0) in which the planetary ephemerides are numerically integrated. Hence precession only enters into the calculation of $\vec{R}_e$ when one relates the geocentric body-fixed position of the observer to a space-fixed system. An additional source of error enters into Eq. (3) through the calculation of sidereal time since the sidereal time as calculated with respect to the FK4 equinox will differ from that calculated with respect to the dynamical equinox by E, the equinox error. In this paper we will only be interested in the rate of motion of an equinox and we will ignore any constant angular offset between the fixed dynamical equinox and one contained in the JPL ephemerides that may differ from it by a constant angle. Hence, the total error committed in the reduction of the station coordinates to a fixed equinox is given by

$$T \ \Delta k = (\Delta p \ \cos\varepsilon - \dot{E}) \ T \tag{4}$$

where $\Delta k = -0\rlap{.}''3$. This effect is extremely small for radar observations, amounting to only 0.1 m/y, the effect being proportional to $R_e \ \Delta k$. Thus, for practical purposes we can say that radar data are not affected by errors in precession and equinox motion, the effects being absorbed in station locations since they are of a diurnal nature.

Of greater consequence are problems in the determination of $\vec{R}_p$, which contains errors from various sources such as (a) topographical variations in the case when a radar signal is reflected from a surface feature of the planet, (b) orbit determination uncertainties when the transponder is located on board an orbiting spacecraft (e.g., Mariner IX), or (c) uncertainties in the physical ephemeris of the planet and the location of the landed spacecraft when the transponder is on the surface of a planet (e.g., Viking Lander). Formal covariance analysis predicts that we can determine inertial mean motions of the Earth and Mars to better than $0.''01$ per century, although a more realistic value for this number might be $0.''03$ per century. Simple error estimates from Eq. (2) indicate that the Earth's mean motion can be determined to

$$\sigma(n_e) = 0.''001 \ T \ q(m) \tag{5}$$

where $q(m)$ is the one-way range uncertainty in meters. Thus, if $q = 30m$ — a very conservative value – we can determine the Earth's mean motion to $0.''03$ per century. On the other hand, an error of $1''$ per century in Δn could not optically be observed in the semi-major axis since the error is only $1000\,m$ or $0.''001$.

II. TIME SCALES

For radar observations atomic clocks are always employed, and since Universal Time only enters the problem through the diurnal rotation of the Earth, we will assume that our observations are effectively on a dynamical time system. In actual practice, if we let t_d represent dynamical time and let t_u represent Universal Time, then we have

$$t_d = t_u + \Delta T = t_u + (t_e - t_u) + (t_d - t_e) = t_u + \Delta T_e + \delta T \tag{6}$$

where we have divided ΔT into two parts:

$$\Delta T = \Delta T_e + \delta T \tag{7}$$

with ΔT_e signifying "Ephemeris Time" minus Universal Time and δT representing dynamical time minus ephemeris time. The time scale t_e is that appropriate to some ephemeris (Newcomb's for example) and was generally tabulated in the astronomical ephemerides before atomic time systems were developed. Values of ΔT are ultimately obtained from the Bureau International de l'Heure (BIH) and, assuming that the stars are on the FK4 system, the corresponding Universal Times are correct and will not change with the adoption of the FK5.

In our processing of older optical observations we employ the values of ΔT_e given by Brouwer (1952), rather than the generally unknown ΔT of Eq. (7). The Brouwer values of ΔT_e are consistent with Brown's lunar theory containing a quadratic term in the Moon's mean longitude of $\dot{n}_{\mathbb{C}}/2 = -11.''22$. The term δT in Eq. (7) is present in order to allow one to employ another, perhaps more realistic, motion for the Moon's secular acceleration. If one adopts a different secular acceleration of the Moon, the approximate value of δT is

$$\delta T = -(\dot{n}_{\mathbb{C}} + 22\overset{..}{.}44)\ T^2\ (sec).\tag{8}$$

If Morrison's (1979) value of $\dot{n}_{\mathbb{C}} = -26''$ is correct, then we would obtain $\delta T = 3\overset{s}{.}6\ T^2$ as one estimate of δT. On the other hand, use of the value $\dot{n}_{\mathbb{C}} = -38\overset{.}{.}3$ given by Duncombe et al. (1975) would produce a value $\delta T = 16^s T^2$. A recent determination by Williams (1980), based upon lunar laser ranging data, yields $\dot{n}_{\mathbb{C}} = -23\overset{..}{.}1 \pm 2''$, so he would obtain $\delta T = 0\overset{s}{.}7\ T^2$. For very old observations (which are not processed by JPL) the difference can be appreciable. Whatever the true value of δT may be, we should investigate its influence on the meridian circle data.

III. OPTICAL OBSERVATIONS

For optical observations, the Universal Time of transit is calculated as follows. By definition, at transit the right ascension of the planet referred to the equinox of date is equal to the sidereal time. We therefore determine the Universal Time t_u at which transit occurs, as predicted by the ephemeris. If the sidereal time definition is $ST(t_u)$ on the FK4 system or $ST'(t_u)$ on the FK5 (dynamical equinox) system with (in an of-date system)

$$\alpha_{dyn} = \alpha_{cat} + E ,\tag{9}$$

then by international agreement $ST'(t_u) = ST(t_u) + E$, so that one obtains the identical Universal Time on the FK4 system as well as on the FK5 system.

Let a $\sim$(tilde) refer to a right ascension with respect to the equinox of date and let a subscript F refer to a right ascension with respect to a fixed equinox. Let an apostrophe (') denote a correct value, so that $\tilde{\alpha}'$ is the correct right ascension with respect to the dynamical equinox of date and α'_F is the correct right ascension with respect to the fixed equinox of B1950. If $\tilde{\alpha}'(t')$ represents the correct right ascension referred to the real equinox of date at the real dynamical time t', then $\tilde{\alpha}'(t') = \alpha'_F(t') + p_\alpha T$, where p_α is the correct (not Newcomb) precession in right ascension. Then the real transit occurs at t'_u when

$$\tilde{\alpha}'(t'_u + \Delta T) = ST'(t'_u) = ST(t'_u) + E\tag{10}$$

if one has an ephemeris which is valid with respect to the real equinox. At JPL, however, we always use the Newcomb precession p^0 which differs from the correct value due to Fricke (1971) by $p = p^0 + 1\overset{.}{.}1T$. Hence, at JPL, we compute the time of transit from

$$\tilde{\alpha}(t_u + \Delta T_e) = ST(t_u)\tag{11}$$

where $\tilde{\alpha}$ is the right ascension referred to the equinox of date using Newcomb's precession p^0: $\tilde{\alpha}(t) = \tilde{\alpha}'(t) - \Delta p_\alpha T$. Thus, if C' represents the true RA of transit and if C represents the RA calculated via Eq. (11),

then we make the error $C' - C = \Delta p_\alpha T + 0\overset{\prime\prime}{.}04\ \delta T$. In addition to the error
made by JPL in computing the meridian transit, the observed right as-
cension (0) is in error (if it is on the FK4 system) by the amount given
in Eq. (9). Hence,

$$0 - C = (0' - C') + (\Delta p_\alpha T - \dot{E} T + 0\overset{\prime\prime}{.}04\ \delta T), \qquad (12)$$

and it is the quantity $\Delta k = \Delta m - \dot{E} + 0\overset{\prime\prime}{.}04\ \delta T/T$ which JPL estimates from the
optical data (Standish 1976).

The preceding, then, is a short summary of the manner in which we
process both radar and optical observations. In the case of optical
transits it is seen that we make the error $C' - C$, while the data contain
the equinox error of Eq. (9). Since the parameter Δk is estimated
(along with a similar equation Δn for declination) in the analysis, no
harm is done and one obtains from the optical observations the values
of Δk and Δn.

V.   EPHEMERIS DATA

Before comparing the ephemerides with Newcomb, we would like to
outline the types of data which have been employed. The so-called
"Export Ephemerides" such as DE-69 (O'Handley et al. 1969) and DE-96
(Standaish et al. 1976) are generally available to all interested scientists. They are better documented than our "interim" ephemerides which
are either (a) export ephemerides which have not yet been thoroughly documented or (b) ephemerides which represent temporary milestones (e.g.,
for a specific space mission). In recent years the interim ephemerides
(DE-102, DE-108, DE-111) have seen rather widespread use by scientists
and consequently it is appropriate for us also to discuss them here.
The data upon which the ephemerides are based are summarized in Table 1.

The U.S. Naval Observatory optical observations were transformed
to the FK4 system using USNO tables for Ephemerides DE-96 to DE-108.
For DE-111 the transformations of Schwan (1977) were employed. Limb
corrections for Mercury and Venus were applied through DE-96, while
phase modeling was employed later.

Radar bounce data (radar signal being reflected from a planet)
reached its greatest influence in DE-96 and DE-102, subsequently being
superseded by Viking observations. Radar observations prior to 1967
are weighted at approximately 15km while subsequent data are weighted
at 1.5km. Spacecraft ranging and normal points were employed in DE-102
and in subsequent ephemerides. Mariner IX data were weighted at 40m
away from conjunction and 400m near conjunction. Viking orbiters and
landers were weighted at 50m and 15m in DE-108 and DE-111. Mars radar
closure observations (whereby one employs two observations of Mars
spaced one synodic period apart to remove effects of topographic variations), employed in DE-96 through 108, were weighted at 150m, while
the Mars radar-occultation (employment of spacecraft-determined

occultation radii to eliminate the radius dependence of the radar range)
comparisons used in DE-108 were weighted at 500m. Finally, in DE-111,
a combined planetary-lunar ephemeris solution was obtained in which luna
laser ranging data (weighted at 34cm) were employed for the first time
to obtain a simultaneous solution for the Moon and planets.

In Table 2 we summarize some of the constants employed in the ephem
erides. Prior to DE-96 a different speed of light was employed, which
accounts for the major changes in the astronomical unit. The values in
Columns 4 (Δk) and 5 (Δn) are the parameters which result from the optical data only in estimating optical (and precession) drifts. The interpretation of Δk and Δn should be viewed with caution. They are quantities which relate the presumed inertial ephemeris to the apparent observations made with a meridian circle. Only if the solar observations
are indeed on the FK4 can the corrections be interpreted in a manner
equivalent to the results obtained by Fricke (1980). It should be re-
membered that the term $\dot{E}$ results from the observational errors while the
Δp terms are present because of inconsistencies in calculating the
events. The values of $\dot{n}_{(}$ given in Column 6 are primarily for outside
users of JPL ephemerides. They should only be considered by users of
the ephemerides who analyze lunar observations and who derive their own
values of δT.

V.   COMPARISON OF EPHEMERIDES

We now come to a discussion of the modern JPL ephemerides in comparison with the classical ones of Newcomb. Several scientists such as
Stumpff (1977, 1979, 1980), Kristensen (1980), van Flandern (1980),
Schubart (1980) and others have compared the JPL ephemerides with subroutines or tapes which are generally described as being "Newcomb" ephem·
erides. Schubart, in comparing DE-102 with Herget's (1953) evaluation
of Newcomb's tables in a B1950 coordinate system, finds "no significant
secular deviation" of the JPL longitude from that of Newcomb. Kristensen
however, finds that DE-108 minus Newcomb theory in an of-date system is
about +0."83T and van Flandern obtains 1."01T in an of-date comparison of
DE-102 minus Newcomb, even though the mean motion differences between
DE-102 and DE-108 are very slight. In both cases the authors employ
different computer subroutines to evaluate Newcomb's theory with respect
to the equinox of date and then use the new IAU precession ($\Delta p = 1."1T$) to
precess the JPL ephemerides from B1950 to the equinox of date, while
Schubart did not employ any precession since both DE-102 and Herget's
evaluation of the Newcomb Tables are in a fixed B1950 system. Stumpff
obtains a value based on a DE-96 comparison (which has a poorer determination of the sun's mean motion), but which when reduced to the DE-108
system is in general agreement with that of Kristensen if Stumpff employs
the new IAU precession.

If, in fact, the JPL ephemerides are correct and are relative to a
fixed equinox in the B1950 system (it may be rotated slightly from the
dynamical equinox), then one must employ the new precession in

calculating correct positions with respect to the dynamical equinox of date. Hence, the use of the new precession by Kristensen, van Flandern and Stumpff is proper. If the Newcomb ephemerides are valid with respect to the equinox of date, then the Kristensen - van Flandern - Stumpff analyses should show no significant secular deviations for DE-102 and subsequent JPL ephemerides (evaluated in an of-date system with the new precession). On the other hand, one might expect the Schubart comparison of DE-102 with Herget's evaluation of Newcomb in a B1950 coordinate system to show a residual trend of 1"T since we believe the Newcomb precession (which Herget employed to precess the of-date Newcomb Tables positions to B1950) to be in error.

These apparently contradictory (to our hypothesis that the JPL ephemerides are inertial) results may still have an explanation, however. That possible answer lies in questioning the time-honored assumption that Newcomb's solar ephemeris is accurate in representing the real solar motion with respect to the dynamical equinox of date. We would, in fact, suggest that Newcomb's theory is not accurate with respect to the dynamical equinox of date.

In developing the FK4 catalogue, Fricke and Kopff (1963) employed the same equinox N1 (apart from a constant offset) as Newcomb employed (1872, 1882) in Newcomb's analysis of inner planet observations which were used to produce the solar ephemeris (Newcomb 1895). Hence, let us make the following three assumptions: (1) the FK4 and N1 catalogues have the same equinox motion (Fricke and Kopff, 1963); (2) Newcomb's solar ephemeris (Newcomb 1898) is with respect to "Newcomb's dynamical equinox" α_{DN} where

$$\tilde{\alpha}_{DN} = \tilde{\alpha}_{N1} + 0\overset{''}{.}3 \ T \tag{13}$$

(Newcomb 1895, pp. 88, 126); (3) the real dynamical equinox $\tilde{\alpha}'$ is related to the FK4 equinox by Fricke's recent (1980) determination

$$\tilde{\alpha}' = \tilde{\alpha}_{FK4} + 1\overset{''}{.}27 \ T. \tag{14}$$

From these three assumptions, then, we can derive

$$\dot{\tilde{\alpha}}' - \dot{\tilde{\alpha}}_{DN} = + 0\overset{''}{.}97 \tag{15}$$

where a dot signifies a time derivative (per century). Hence, since Newcomb's solar ephemeris was with respect to $\dot{\tilde{\alpha}}_{DN}$ it follows that the sun's real mean motion $\tilde{n}'$ with respect to the equinox of date differs from that calculated by Newcomb ($\tilde{n}_N$) in the amount

$$\tilde{n}' - \tilde{n}_N = 0\overset{''}{.}97 \ \cos\varepsilon = 0\overset{''}{.}89 \ \text{per century.} \tag{16}$$

Thus, we should expect that an accurate solar ephemeris would have a mean motion 0"89 per century larger than Newcomb's mean motion, when measured in an of-date system. This result is in accord with the findings of Stumpff, Kristensen and of van Flandern given earlier. A comparison of an inertial mean motion n' with a Newcomb ephemeris mean motion n_N in a fixed system should then show

$$n' - n_N = 0\overset{''}{.}89 - \Delta p = -0\overset{''}{.}21 \text{ per century} \qquad (17)$$

in agreement with the finding of Schubart.

Hence, it seems that the hypothesis that the solar data of Newcomb are affected by an error of approximately $0\overset{''}{.}9\,T$ in an of-date system may be the key to explaining the results noted earlier.

VI. THE LATEST DATA

So far, then, we have put forth some possible, if not plausible, reasons for asserting that the modern ephemerides are indeed valid representations in an inertial system. One final example will serve to support that belief. In its efforts to develop a Very Long Baseline Interferometry (VLBI) system for use in navigating future space missions, JPL has conducted some radio-interferometric experiments measuring the angular separation between radio sources and the Mars Viking Orbiter. Employing orbit determination methods to reduce the orbiter data, we can effectively "observe" the radio source relative to Mars.

In these experiments a radio source catalogue is employed and the angular separations are calculated in an of-date system. The residuals, calculated by Newhall (1980) on DE-108 exhibit fairly large means, being $\overset{''}{.}176$ in right ascension and $\overset{''}{.}122$ in declination. However, we may assume that the radio source catalogue is valid at a current epoch and equinox (quite similar to a star catalogue), while the Mars ephemeris is valid in a fixed B1950 frame. If one then applies the precession corrections to reduce the Mars ephemeris to a current equinox and if one assumes that the Newcomb precession is in error by $1\overset{''}{.}1\,T$, then the declination residuals become smaller by a factor of 3. The right ascension residuals, originally calculated on DE-108, become smaller by a factor of 5 if DE-111 is used with the new precession. Thus, we again have some evidence that the latest ephemerides, produced from radiometric and optical data, are indeed valid representations of the planetary motions in an inertial system.

Acknowledgments

We are indebted to Dr. J. Schubart of the Astronomisches Rechen-Institut for many thorough discussions on this matter. This work was supported in part by NASA through its JPL contract. One of us (Lieske) was supported by the Alexander von Humboldt Foundation during his stay in Heidelberg as a Humboldt Prize awardee.

Table 1
JPL Ephemerides – Data Sets

Observations	Dates	No. of Obs.	Ephemeris
USNO Transits			
Wash. system	1911–1967	34 304	69
FK4 system	1911–1971	37 583	96,102
	1911–1976	38 942	108
	1911–1976	39 396	111
Radar (Bounce)			
Mercury, Venus, Mars	1964–1968	704	69
	1964–1973	5 052	96,102
Mercury, Venus	1964–1977	1 199	108
	1964–1977	1 307	111
Spacecraft Ranging			
Mariner V (Venus)	1967	214	69
Mariner IX (Mars)	1971–1972	804	96
	1971–1972	803	102,108,111
Viking Orbiter	1976–1977	4 463	102
	1976–1977	2 892	108
Viking Lander	1976–1977	147	108
	1976–1978	665	111
Pioneer X, XI	1973–1974	2	96,102,108,111
Mars Radar, Closure			
	1971–1973	291	96
	1971–1976	306	102
	1971–1978	321	108
Mars Radar, Occult.	1967–1978	2 890	108
Saturn Sat. Astrometry	1973–1979	4 790	111
Lunar Laser Ranging	1969–1978	2 531	111

Table 2
Planetary Ephemerides – Miscellaneous Constants

Ephemeris	Astr. Unit 149 597 800+	E/M 81.+	Δk	Δn	$\dot{n}_{(\!(}$	c(km/s) 299792.+
69	93.0 km	.301	---	---	$-38\overset{''}{.}3$	.5
96	71.411...	.3007	$-1\overset{''}{.}19$	$+0\overset{''}{.}15$	−38.3	.458
102	70.684...	.3007	−0.76	+0.29	−27	.458
108	70.705...	.300492	−0.57	+0.42	−38	.458
111	70.653...	.300587	−0.78	+0.44	−23	.458

REFERENCES

Brouwer, D.: 1952, Astron. J. 57, 133.
Duncombe, R. L., Seidelmann, T. C., and Van Flandern, T. C.:
    1975, On Reference Coord. Systems, IAU Colloq. 26, Ed. B.
    Kołaczek, G. Weiffenbach, Warsaw.
Fricke, W.: 1967, Astron. J. 72, 1368.
Fricke, W.: 1971, Astron. Astrophys. 13, 298.
Fricke, W.: 1975, On Reference Coord. Systems, IAU Colloq. 26,
    Ed. B. Kołaczek, G. Weiffenbach, Warsaw.
Fricke, W.: 1980, Mitteil. Astron. Gesellschaft 48, 29.
Fricke, W., Kopff, A.: 1963, Fourth Fundamental Catalogue (FK4),
    Veröff. Astr. Rechen-Inst. Heidelberg Nr. 10, 1445.
Herget, P.: 1953, Astr. Pap. Wash. 14.
Kristensen, L.: 1980, Private Communication.
Morrison, L. V.: 1979, Geophys. J. Roy. Astron. Soc. 58, 349.
Newcomb, S.: 1872, Wash. Obs. for 1870, App. III.
Newcomb, S.: 1882, Astr. Pap. Wash. 1, Part 4.
Newcomb, S.: 1895, Fundamental Const. of Astr., Washington.
Newcomb, S.: 1898, Astr. Pap. Wash. 6.
Newhall, X. X.: 1980, Private Communication.
O'Handley, D. A., Holdridge, D. B., Melbourne, W. G., and
    Mulholland, J. D., 1969, JPL Development Ephemeris Number 69,
    JPL TR 32-1465.
Schubart, J.: 1980, Private Communication.
Schwan, H.: 1977, Veröff. Astron. Rechen-Inst. Nr. 27.
Standish, E.M., Keesey, M. S. W., and Newhall, X. X.: 1976, JPL
    Development Ephemeris Number 96, JPL TR 32-1603.
Stumpff, P.: 1977, Astron. Astrophys. 56, 13.
Stumpff, P.: 1979, Astron. Astrophys. 78, 229.
Stumpff, P.: 1980, Private Communication.
van Flandern, T.: 1980, Private Communication.
Williams, J. G.: 1980, Private Communication.

NEW CELESTIAL REFERENCE SYSTEM

P. K. Seidelmann, G. H. Kaplan, T. C. Van Flandern
U.S. Naval Observatory
Washington, D.C. 20390

Abstract

The IAU (1976) System of Astronomical Constants, the FK5 and
new lunar and planetary theories are being introduced in 1984.
The investigation and planning for the transition has revealed
the complex interdependencies between observational techniques
and the reference systems, and their strong link to the rotating and orbiting Earth. The inaccuracies in our knowledge of
the star positions, astronomical constants and the rotation
and motion of the Earth are embedded in subtle ways in the
observations and the reference coordinate systems. For example
the FK4 reference system in 1950.0 coordinates rotates with
respect to an inertial system. Details are given for the
conversion to the new system.

The concepts for a future reference system are developed,
based on separating the real motions involved such that
observations from various moving platforms can be related
to the appropriate coordinate system, without involving
motions which are not intrinsically involved in the observations. Therefore, reference systems determined or utilized
in space, while affected by aberration and parallax, would
logically be defined with no dependence on precession, nutation
polar motion, or Universal Time, which are all concerned with
motions of the Earth's surface.

I. INTRODUCTION

The recommendations adopted by the 1976 IAU General Assembly
included the following changes to the fundamental reference
system: (1) a revised value for the constant of precession;
(2) a new standard epoch, designated J2000.0; (3) the use of
the Julian century as the unit of time for precession and
proper motion; (4) a new stellar reference frame, the FK5, to
correspond as closely as possible to the dynamical reference

E. M. Gaposchkin and B. Kołaczek (eds.), Reference Coordinate Systems for Earth Dynamics, 305–316.
Copyright © 1981 by D. Reidel Publishing Company.

frame as determined from relevant modern observations; (5) an
amended expression for Greenwich mean sidereal time, based on
the equinox correction adopted for the FK5; (6) a new procedure for the computation of stellar aberration, together
with the removal of the E-terms from the mean places of
stars; (7) a new theory and reference pole for nutation; and
(8) new definitions of time scales to be used with dynamical
theories and ephemerides.

At the IAU General Assembly of 1979 in Montreal, the IAU
(1979) Theory of Nutation was adopted. While the advisability
of the details of this resolution are still being discussed,
it is certain that a change will be introduced both for the
series coefficients and for the reference pole. Also at the
1979 IAU General Assembly, the principle that the FK5 reference frame correspond as closely as possible to the dynamical
reference frame was reiterated and the form of the correction
was adopted. Subsequently, Fricke (1980) has recommended the
following correction to the right ascensions of the FK4 stars:

$$E(T) = 0\overset{s}{.}035 \pm 0\overset{s}{.}003 + (0\overset{s}{.}085 \pm 0\overset{s}{.}010)(T-19.50).$$

These corrections to the present reference coordinate system
are important improvements necessary to satisfy the observational data (Fricke (1975)). However, as the implementation of
these changes is considered, the problems of the past, which
have become embedded in the present reference system, become
apparent. Therefore, some of the complexities involved in
these changes will be discussed in detail and some suggestions
for consideration for future reference systems will be
presented.

II. TRANSITION: FK4 TO FK5, 1950.0 TO 2000.0

    A. Equinox Correction

A critical element of the new reference frame is the location
of the equator and equinox. In order to specify the location
of the equinox, it is first necessary to define the equinox.
The origin of right ascension of a star catalog reference
system is designated as the catalog equinox. The dynamical
equinox is usually defined as the intersection of the mean
ecliptic plane and celestial equator, but this definition is
not adequately precise. Specifically, the dynamical equinox
is defined as the average location of the ascending node of
the Earth's moving mean orbit on the equator. This is different by about 0$\overset{.}{.}$1 from the average location of the assending
node of the Earth's instantaneous orbit on the equator,
because in this case the Earth's velocity also has a
component due to the motion of the mean orbital plane.

The correction to the FK4 equinox (E(T)) and its time
derivative (called the "motion of the equinox") represent an offset of the origin of right ascensions of
the FK4 catalog from the dynamical equinox, and the offset is increasing with time (see Figure 3). The recommended value of E(T) is valid for stars of magnitude
4.9, approximately the mean magnitude of the FK4. Since
there is a detectable magnitude dependent systematic
error in the FK4 right ascensions, the brightest and
faintest star in the FK4 will receive corrections which
average about $\pm$ 0$\overset{s}{.}$005 from the recommended value of E.

Since the correction E will be added to the right ascensions
of all the stars in the FK4, the FK5 right ascensions of the
local meridian for any observer for any instant (i.e., the
local sidereal time (LST)) will also be increased by the
value of E. Since the Greenwich Sidereal Time (GST) is equal
to the LST plus the longitude λ (west positive), either the
GST or λ must change. Since it is undesirable to have a
time-dependent correction to the observer's longitude, the
choice is to introduce the change in the GST.

When the nutation in right ascension is removed from GST, we
have the Greenwich Mean Sidereal Time which is directly
related to the Universal Time scale (UT1) by a formula.
Since GMST is to increase by the amount E, either UT1 or the
formula must change. Rather than introduce a discontinuity
of the size of E in UT1, the equation is changed to:

$$\text{GMST of } 0^h\text{UT1} = 6^h41^m50\overset{s}{.}5484 + 8640184\overset{s}{.}8129T_u + 0\overset{s}{.}0931T_u^{2}$$

where T_u is measured in Julian centuries from J2000.0
(JD 2451545.0).

There will be small discontinuities and periodic variations
in the measured value of UT1 due to systematic differences
between FK4 and FK5 star positions and due to the new theory
of nutation. The small periodic changes in UT1 can be calculated retroactively, so new tables of Δ UT1 = UT1-UTC can
be prepared based on the FK5.

Since the revised formula defines the "fictitious mean Sun,"
another possible consequence of the equinox correction is an
increase in the difference between the fictitious mean Sun
and the true Sun (projected onto the equator). The ephemeris
of the Sun is being corrected based on observational data,
so the formula for the mean longitude of the true Sun will
change somewhat. In any case, the change due to the equinox
correction is small compared to the other differences between
the fictitious mean Sun and the true Sun.

B.   Other differences between FK5 and FK4 reference
     systems

1.   Precession Constant

The new value of the general precession in longitude
(a correction at B1900.0 of +1".10 per century to Newcomb's
value of luni-solar precession and -0".029 per century to
planetary precession (Lieske, et al (1977)) will be used
with the FK5, while Newcomb's value was used with the FK4.
Considering now only the change in the constant of precession and not the equinox correction, in both right
ascension and declination the mean and apparent places of the
stars and their centennial variations are not affected by
the change in the constant of precession, because the precession correction is compensated for by corrections to the
proper motions. For this reason also, the new value of the
precession constant should be used with the FK5 catalog, but
not with the FK4 and prior star catalogs.

The primary impact of the change in the constant of precession
will be for the motions of solar system objects, where the
equivalent of a correction to proper motions of the stars is
a correction to the sidereal mean motions and to the rates
of the mean longitudes of nodes and pericenters.

2.   E-Terms of Aberration

The E-terms of aberration, or elliptical terms, have a
maximum amplitude of 0".34 and were embedded in the mean
places of stars to simplify the computation of aberration
corrections. With the use of computers and the interest in
improving accuracies, the separation of E-terms is a source
of confusion, complication, and inaccuracy, so they are
being removed from the mean places of epoch J2000.0. The
calculation of apparent places of stars should now be based
on stellar aberration computed rigorously from the Earth's
velocity vector with respect to the solar system barycenter.
There should be no impact from this change as long as all
mean places or astrometric positions in the FK4 system are
converted to FK5 mean places or astrometric positions without
E-terms, and algorithms for the computation of stellar aberration do not remove the E-terms. The E-terms at epoch
B1950.0 in the sense of FK5-FK4 are of the form:

$$\Delta\alpha = + 0".341 \sin (\alpha + 11^{h}15^{m}) \sec \delta$$

$$\Delta\delta = + 0".341 \cos (\alpha + 11^{h}15^{m}) \sin \delta + 0".029 \cos \delta$$

3. New Standard Epoch and Julian Century

The conversion from tropical to Julian units of time in the
FK5 means that precession, proper motion, and centennial variations will be given per Julian century of 36525 days
instead of per tropical century of 36524.2198781 days.   The
new standard epoch is J2000.0 which is 2000 January
1.5 (TDB) = JD 2451545.0. On this system, J1950.0 is 1950
January 1.0 = JD 2433282.5, whereas the old standard epoch
designated 1950.0, (now denoted by B1950.0) was 1950 January
0.923 = JD 2433282.42345905.

The computational convenience of the new system is obvious,
but algorithms for the computation of local apparent places
must be adjusted to the new epoch and unit of time. If done
correctly, this change will have no significant effect on the
apparent hour angles or declinations of stars.

4. FK5-FK4 Systematic Differences

The systematic differences between the FK5 and FK4 represent
distortions of the FK4 system in position and proper motion,
to the extent that they are known. These are expected to
be less than 0".2, except in parts of the southern sky, where
observational data for both the FK4 and FK5 are sparse.
Because the differences are systematic, they will not cancel
out in comparisons between theory and observations and they
will result in changes of astronomical measurements, for
example UT1, that previously used the FK4 system.

5. Nutation

The new theory of nutation will cause periodic variations
in the computation of apparent places, the largest of which
have amplitudes of 0".045 in longitude and 0".007 in obliquity
with a period of 18.6 years, and 0".042 in longitude and 0".020
in obliquity with a period of 183 days. These changes will
affect the comparison of theory with observations and the
determinations of UT1, so new tables of UT1-UTC will be
required. In a similar way, determination of the Earth's
polar motion will be affected.

The new theory of nutation incorporates a new reference pole
called the Celestial Ephemeris Pole (CEP). This reference
pole has no diurnal motion with respect to a space-fixed
coordinate system or an Earth-fixed coordinate system and it
is the center of the quasi-circular diurnal paths of the stars
in the sky. This means that the new nutation theory implicitly
includes the dynamical variation of latitude. Therefore,
separate corrections for this effect should not be used with
the new theory. This change will not affect complete reduc-

tions of observations, but will remove a diurnal variation
that was left when observations were reduced using nutation
only.

6. Relativistic Light-Bending

The relativistic bending of light rays in the Sun's gravitational field can reach 1".8 at the solar limb and decreases
almost linearly with angular distance from the Sun. Anywhere
within 120° of the Sun the deflection is greater than 0".002.
The magnitude of the deflection (θ) which is radially away
from the Sun's center is:

$$\theta = k(\frac{1+\cos D}{\sin D} + \frac{1}{4} \sin 2D)$$

where D is the angular distance from the Sun and $k = 1.9742$
x 10^{-8} radians = 0".0040720 is the deflection at $D = 90°$
(Brandt (1974), Wade (1976)). This correction is valid for
the stars and outer planets. For the inner planets a more
complex expression is required to rigorously represent the
various possibilities.

C. Effective Coordinate Systems

The various corrections mean that there are different
effective coordinate systems currently in use and different
relations between these systems and the FK5 reference system.
There are four distinct effective coordinate systems, two are
identified with the FK4. The changes from the FK4 to the FK5
are designed to eliminate, as far as current knowledge permits,
the difference between these systems. However, time and
improved observational accuracy will eventually again reveal
the four separate systems.

        1. The effective apparent FK4 coordinate system
(figure 1) is developed from observations of apparent places
of stars and planets. Observed positions, other than photographic, are usually apparent places. To convert these to
the FK5, add the systematic differences FK5-FK4, the correction
to the zero point of right ascension (E), and the light-bending
effect. No correction for E-terms of aberration, precession,
nutation, or the epoch change should be made. In Figure 1,
the moving true origin of right ascension fails to maintain a
constant distance from the moving dynamical equinox by the
amount of the "motion of the Equinox," which is actually the
motion of the origin of right ascension. In principle, the
motions of all three defining planes are independent.

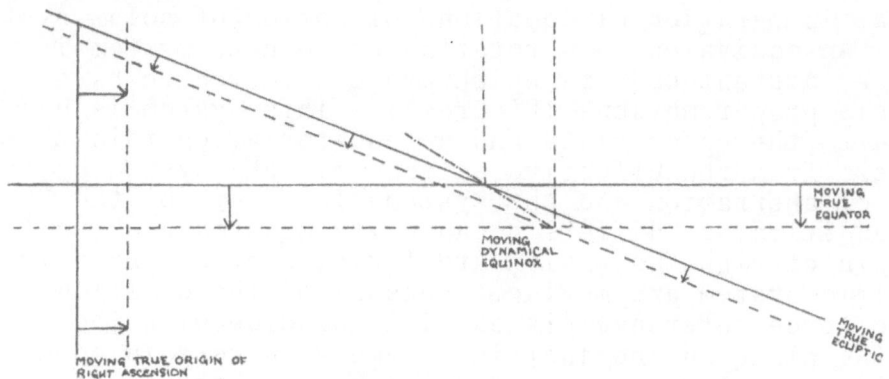

FIG. 1- EFFECTIVE APPARENT SYSTEM

2.   The effective astrometric FK4 coordinate system
(Figure 2) is ordinarily referred to the mean equator and
equinox of B1950, and is the system defined by star catalogs,
such as the FK4 itself. Astrometric positions of solar system
objects are computed to be immediately comparable with star
positions. To convert these to the FK5, add the systematic
differences and zero point corrections, remove the E-terms,
change the precession constant (which changes the proper
motions of the stars), and change to epoch J2000. No correction for the change in nutation or light-bending is
ordinarily required. In Figure 2, the correction of -0".03
per century to planetary precession implies that the B1950
mean ecliptic has a slight residual motion; the correction
of +1".10/cy to luni-solar precession implies a residual
motion of the B1950 mean equator; and the resultant residual
motion of the B1950 mean dynamical equinox is almost compensated for in right ascension by the residual motion of the
"equinox" (B1950 mean origin of right ascension).

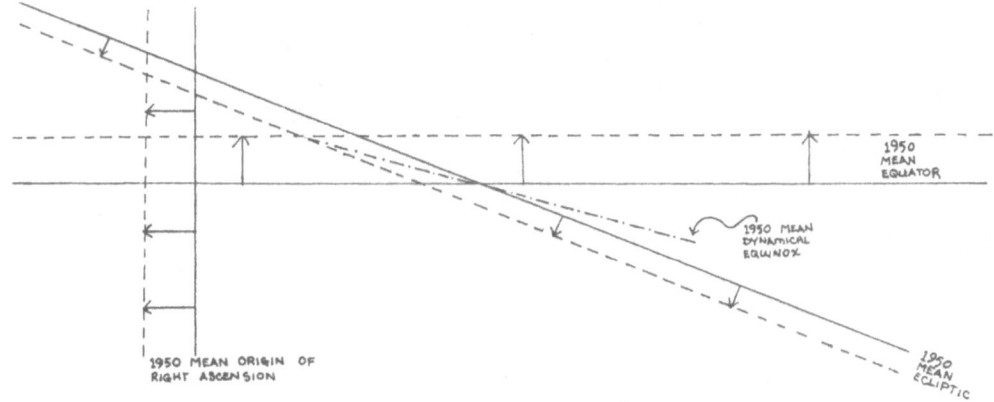

FIG. 2- EFFECTIVE ASTROMETRIC SYSTEM

3. A non-rotating coordinate system is defined by the
numerical integration of equations of motion of solar system
bodies. An equivalent non-rotating coordinate system is
defined by distant objects which are considered to have
negligible proper motions (Figure 3). This system is usually
referred to the epoch B1950 and an equatorial coordinate syst
It differs from the effective astrometric FK4 system by the
E-terms of aberration and the systematic errors of the FK4;
but in addition, it differs because its mean equator, eclipti
and origin of right ascension are fixed, whereas those in the
astrometric system are moving. Because of these motions, eac
of these three reference planes will coincide with the corresponding plane in the inertial frame at a certain epoch. I
principle, therefore, one must specify the three epochs of
coincidence of the reference planes to completely define the
relationships between the inertial and astrometric systems.

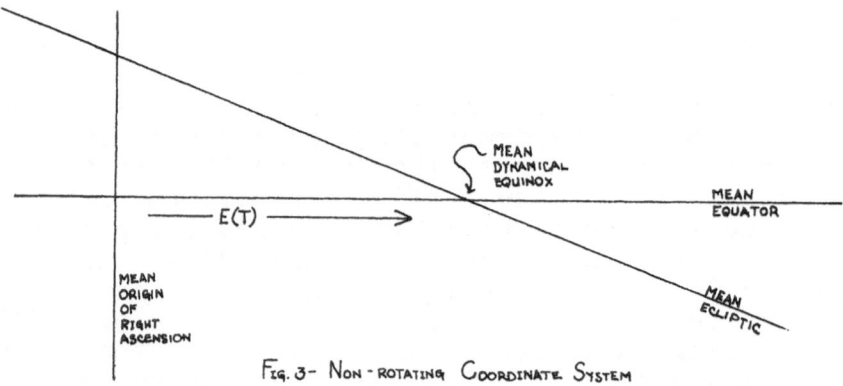

Fig. 3- Non-rotating Coordinate System

4. The effective geometric system (Figure 4) results fro
the conversion of computed lunar or planetary ecliptic longitude and latitude coordinates to B1950 equatorial coordinates
using Newcomb's precession constant. The geometric frame is
intended to be an inertial frame, but fails to achieve this
because of the error in the precession constant. The origin
right ascension is automatically set to the intersection poin
of the ecliptic and the mean equator. Because of the inad-
vertent motion of the 1950 mean equator, the mean origin of
right ascension also moves, but at a rate rigorously related
to the error in precession, having no connection with the
motion of the origin of right ascension in the effective
astrometric system. The rectangular coordinates of the Sun
tabulated in the American Ephemeris are in this system.

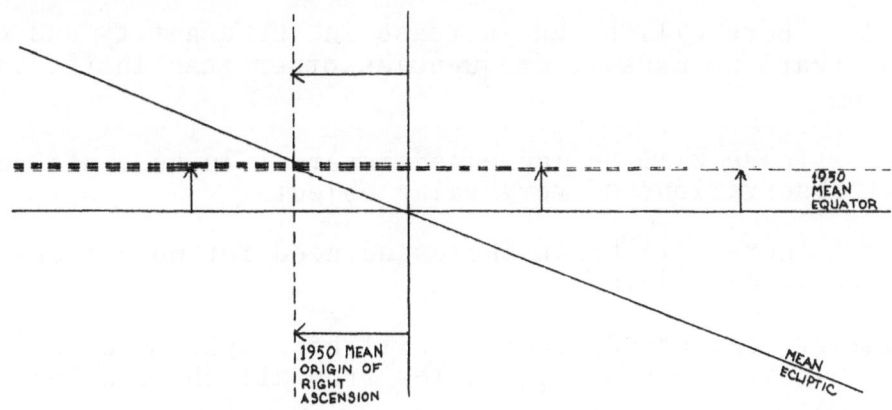

Fig. 4 - Effective Geometric System

III. FUTURE REFERENCE SYSTEM

The FK5 reference system and the IAU (1976) System of Astronomical Constants have been designed to provide a workable system free from known errors. Where changes have been introduced, it is felt that the remaining uncertainties are an order of magnitude smaller than they were before. It is hoped that this system will have a long and useful lifetime and will serve the astronomical requirements well.

There have been several thought-provoking papers concerning future reference systems with various concepts and reasons for the recommendations being proposed (Guinot (1979), Eichhorn (1979), Murray (1979), Kovalevsky (1979), Grafarend et al. (1979)). Perhaps now, while the problems of the new system and the interrelationships involved in conversions are fresh in our minds, it is appropriate to consider the general concepts, not the particular details, of an ideal reference system for the future. The existence of an ideal or goal might also provide guidance as observing techniques are revised and decisions are made concerning what observational data should be published.

In considering the ideal reference system, a fresh start should be initiated, unburdened by the past. The character of future astronomical observations should be considered. It appears that the following characteristics of future astronomical observations might have a significant effect on the specifications of a future reference system.

1. There will be an increase in the quantity, quality and variety of observations which are made from instruments that are not located on the surface of the Earth.

2.   There will be an increase in the quantity and quality
of observations made at frequencies other than in the visible
spectrum.

3.   There will be increased interest in accurate positional observations of very faint objects.

4.   There will be an increased need for more accuracy
in observational data.

In summary, increased accuracy will be sought in observations
at all magnitudes and frequencies from all observational
platforms.

If the above characteristics are accepted, the increased
observations from instruments independent of the Earth's
surface will have the most effect on the reference system.
Therefore, it appears desirable to establish the reference
system independent of a particular observing platform.
Then a stellar reference system would be independent of the
Earth's orbital or rotational motions.

The independence from the Earth, in addition to the observations of faint objects and observations at nonvisual frequencies, indicates a space-fixed nonrotating reference system.
The motions of solar system bodies could be referred to the
same system, except that for each planet, including the Earth,
there would have to be a planetocentric coordinate system.

The principal difference between this concept and the present
system is the separation of the origin from the dynamical
equinox.   This idea has been suggested before (Guinot (1979)),
but apparently the reasons given have not been convincing. The
distinction has already been introduced by the terminology
catalog equinox which refers to the origin of right ascensions
in a star catalog.

In practice, the dynamical equinox is not a desirable origin
either theoretically or observationally. A theoretical difficulty of definition has been discussed earlier in this
paper, but at some level of accuracy the location of the dynamical equinox is dependent on the orbital theory of the
Earth which is being used. The observational difficulty is,
simply, that the dynamical equinox cannot be observed. Rather,
many observations must be used to determine retroactively the
location of the equinox. The most direct object to be observed,
the Sun, is subject to the greatest systematic observational
errors. In practice today, the catalog equinox is observationally located from the observations of the stars in a given
catalog, so that the dynamical equinox is not used. For the
reference system, the only use of the dynamical equinox is for

determining the catalog equinox when a new reference system is being established, as for the FK5. This is done so the catalog equinox and the dynamical equinox agree to some accuracy, but in practice they will never be the same and will always have some relative motion.

The implication of this fact for the present is that observational data should be recorded and retained in its original observing reference system to the greatest extent possible. This will permit future improvements in knowledge concerning the relations between the coordinate systems to be used to re-reduce the observational data. There should be no non-reversable processing of data and the specific methods and constants used in the processing should be specified. While there may not be any purely impersonal fundamental observations, nevertheless, that should be the goal of positional observations.

IV. CONCLUSION

The astronomical reference system has developed from the observational and computational capabilities of the times. The IAU (1976) System of Astronomical Constants, which are the most accurate, self-consistent set of values which can be determined from current observational data, has been introduced for the FK5 reference system. The details of conversion from the old to the new reference system reveal some of the conceptual problems with reference systems in general.

As we consider future observations of increased accuracies at a wide range of frequencies and magnitudes from many instrumental platforms, it may be desirable to consider new concepts for reference systems. An ideal reference system would be a space-fixed, nonrotating coordinate system. A principal change is the separation of the origin of right ascension from the dynamical equinox. Observational data should be recorded such that the full inherent accuracy of the observations can be realized with improved knowledge concerning the relationships between different coordinate systems.

REFERENCES

Brandt, V. E. 1974, Astronomicheskii Zhurnal 51, 1100
    (English translation in Soviet Astronomy 18, 649, 1975).
Eichhorn, H. 1979, Prochazka, F. V.,and Tucker, R.H. ed.,
    Institute of Astronomy, Vienna, p. 391-410.
Fricke, W. 1975, IAU Colloq. No. 26, Kolaczek B., and
    Weiffenbach, G. ed., Warsaw, Poland.
Fricke, W. 1980, private communication.

Grafarend, E. W., Mueller, I. I., Papo, H. B., Richter, B.
    1979, Bull. Geod. 53, p. 195-213.
Guinot, B. 1979, IAU Symp. 82, McCarthy, D.D. and Pilkington,
    J. D. H. ed., D. Reidel Publ., Dordrecht, p. 7-18.
Kovalevsky, J. 1979, IAU Symp. 82, McCarthy, D.D. and Pil-
    kington, J. D. H. ed., D. Reidel Publ., Dordrecht, p. 151-
    163.
Lieske, J. H., Lederle, T., Fricke, W., and Morando, B. 1977,
    Astron. Astrophys. 58, 1-16.
Murray, C.A. 1979, IAU Symp. 82, McCarthy, D.D. and Pilkingto
    J. D. H. ed., D. Reidel Publ., Dordrecht, p. 165-167.
Murray, C. A. 1978, Q. J. R. Astr. Soc. 19, 187-193.
Wade, C. M. 1976, VLA Scientific Memorandum 122.
——— 1977, Transactions of the IAU, Volume XVI B, D. Reidel
    Publ., Dordrecht.
——— 1980, Transactions of the IAU, Volume XVII B, D. Reidel
    Publ., Dordrecht.

APPORT DES OBSERVATIONS D'OCCULTATIONS STELLAIRES EN VUE DU RATTACHEMENT DU SYSTEME DYNAMIQUE AU SYSTEME GEOMETRIQUE.

M. Froeschlé et C. Meyer
C.E.R.G.A., Grasse (France)

ABSTRACT. The possibilities of tying the geometrical and dynamical reference frames by use of lunar occultations are reviewed. Actually, the best accuracy available is about 0".4, essentially limited by stellar positions and limb corrections. One can expect an improvement of one order of magnitude with the accomplishment of POLO and HIPPARCOS Projects.

INTRODUCTION

Au mois d'août 1969, a été obtenue la première détermination de la distance Terre-Lune par télémétrie laser. La précision des mesures surpassait celle dont on disposait jusque là par la technique radar. C'est pourquoi il devenait possible d'améliorer le système de référence dynamique par des mesures de ce type. Ce système dynamique planétaire repose sur l'intégration des équations différentielles régissant le mouvement de quelques planètes et de la Lune par rapport à un repère galiléen. Dans ce système, la représentation du mouvement de la Lune à partir d'observations de distances faites depuis la Terre nécessite une modélisation poussée des forces agissantes et des paramètres physiques de la Terre et de la Lune. Cependant, l'utilisation des seules mesures de distance rend le système Terre-Lune aveugle, c'est à dire sans possibilité de rattachement au système géométrique. Ce dernier est défini par des directions fixes matérialisées par les directions d'étoiles repérées par rapport à des objets extrêmement éloignés et dont les mouvements propres sont négligeables (galaxies lointaines).

Notre propos est de passer en revue les techniques de rattachement du système Terre-Lune au système géométrique en analysant plus particulièrement celle du positionnement de la Lune par rapport aux étoiles par l'observation d'occultations.

Lors des vols Apollo 16 et 17, deux émetteurs ont été déposés à la surface de la Lune (ALSEP). Les techniques d'interférométrie à longue base permettent de rattacher leurs directions à des radio-sources

E. M. Gaposchkin and B. Kołaczek (eds.), Reference Coordinate Systems for Earth Dynamics, 317–324.

naturelles avec une précision d'ensemble de l'ordre de 0",001 (Slade,
1977 ; Baudry, 1979). Si l'on sait par ailleurs raccorder ces radio-
sources au système géométrique, ce qui peut être obtenu par des mesures
interférométriques en cascade (Kovalevsky, 1975), le système dynamique
sera rattaché au système géométrique. Notons enfin qu'on ne connait
qu'une quarantaine de sources qui permettent d'accomplir ce processus.

Remarquons dès à présent qu'un grand nombre de techniques de rattachement utilisent le système terrestre comme intermédiaire. La position de la Terre peut être déterminée soit par des procédés astrométriques classiques (astrolabe, PZT), soit par interférométrie à très
longue base (VLBI). Dans tous ces cas, le système terrestre se trouve
rattaché au système géométrique. Les réseaux obtenus par l'un ou l'autre de ces procédés peuvent être étendus par télémétrie laser sur satellite ou par des observations Doppler.

On conçoit donc que les mesures de distance Terre-Lune effectuées
par télémétrie laser dans le système terrestre puissent être elles-mê-
mes rattachées au système géométrique. C'est ainsi que la campagne
internationale de mesure de la rotation de la Terre par télémétrie
laser (EROLD) a été entreprise depuis 1977, conduisant à une précision
de l'ordre de la milliseconde dans la détermination du temps universel
(Valein, 1979).

Il existe d'autres techniques pour effectuer ce rattachement :

- La photographie de la Lune sur fond d'étoiles : compte tenu des
  difficultés pour l'obtention des clichés et pour la réduction,
  des erreurs actuelles des catalogues stellaires, on ne peut espérer faire un rattachement à mieux que 0",3.

- Les observations méridiennes de la Lune ont été utilisées (Klock,
  1970 ; Yasuda, 1971) pour orienter le système FK 4 tout en déter-
  minant de nouveau les paramètres orbitaux de la Lune. Les difficultés de pointage sont telles que la précision d'une mesure est
  de l'ordre de 0",8.

- Les occultations d'étoiles par la Lune : dans ce cas, on lie la
  direction de l'étoile occultée (système géométrique) à la direction Terre-Lune (matérialisant, en l'occurence, le système dynamique). Cette technique a servi successivement à la détermination
  de longitudes terrestres (XVIIIe siècle), à l'étude du mouvement
  orbital de la Lune et, en particulier, à la détermination de son
  accélération séculaire. Avant l'introduction du Temps Atomique,
  elle permettait une lecture du Temps des Ephémérides. Puis, par
  l'analyse de longues séries d'observations, elle a permis un
  calage du système FK 4 (Morrison, 1979). Plus généralement, cette technique permet un positionnement des astres occultés avec
  une précision qui dépend évidemment des éphémérides dont on dispose et de la connaissance de la topographie lunaire. Par exemple, le projet EXOSAT (E.S.A.) a pour but de déterminer les

dimensions, mais aussi les positions de sources de rayonnement X
à partir d'occultations observées depuis un satellite.

Nous allons examiner plus en détail l'état actuel et les perspec-
tives de rattachement des systèmes dynamique et géométrique par occultations stellaires.

1. POSITION DU PROBLEME

A l'instant d'une occultation, la direction observateur-étoile
tangente en (P) le disque apparent de la Lune. Il s'agit de relier entre eux les vecteurs (figure 1) qui définissent :

- la position de l'observateur (O) par rapport à un système lié
  à la Terre,

- la direction ($\vec{\Gamma}$) centre des masses de la Terre - centre des masses de la Lune,

- la direction ($\vec{\Delta}$) centre des masses de la Lune - point où se produit l'occultation,

- la direction ($\vec{\gamma}$) de l'étoile occultée.

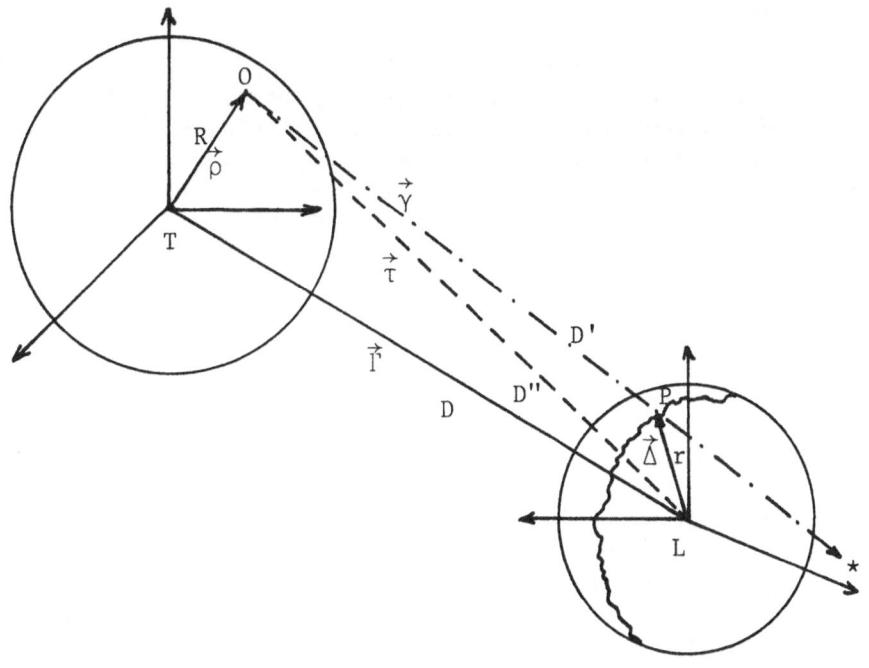

Figure 1.

Soit $\vec{\tau}$ la direction OL. Dans les triangles OTL et OLP, nous pouvons
écrire les relations vectorielles suivantes :

$$D'' \vec{\tau} = - R \vec{\rho} + D \vec{l} \qquad (1)$$

où D'', R et D sont respectivement les distances observateur-centre Lune,
observateur-centre Terre et centre Terre-centre Lune.

$$D'' \vec{\tau} = D' \vec{\gamma} - (r + \varepsilon) \vec{\Delta} \qquad (2)$$

où ε correspond à l'élévation du point (P) au-dessus de la sphère de
référence de rayon r.

Le vecteur $\vec{l}$ est une direction du système dynamique alors que le
vecteur $\vec{\gamma}$ est une direction du système géométrique. Le vecteur $\vec{\tau}$ est
une direction intermédiaire qui sert au rattachement. L'équation (1)
montre que la direction du vecteur $\vec{\tau}$ dépend de l'éphéméride lunaire par
le vecteur $\vec{l}$ et de la position de l'observateur par le vecteur $\vec{\rho}$. Par
ailleurs, dans l'équation (2), $\vec{\tau}$ apparaît comme fonction de la position
de l'étoile occultée et des paramètres qui fixent par rapport au système sélénodésique la direction $\vec{\Delta}$ (librations optiques et physiques).
La précision avec laquelle ces différents paramètres sont connus a évo-
lué rapidement ces dernières années, tant par l'apport de nouvelles
techniques d'observation que par l'usage de moyens de calcul importants.

2.   LE RATTACHEMENT.

La donnée brute d'une observation d'occultation est l'instant du
phénomène. La précision de cette datation varie dans d'importantes limites selon la technique d'observation utilisée. Elle est typiquement
de ± $0^s,5$ pour des observations visuelles, de ± $0^s,1$ par l'utilisation
du micromètre à double image (Meyer, 1974) et de ± $0^s,001$ par les techniques photoélectriques (Nather, 1970 ; de Vegt, 1976). Dans ce dernier
cas, l'incertitude sur la datation peut être considérée comme faible
devant celles qui affectent les autres paramètres du phénomène liés par
les équations (1) et (2). L'équation (1) exprime le passage des coordonnées géocentriques de la Lune à ses coordonnées topocentriques. La
position de la station par rapport à l'axe instantané de rotation et
l'équateur vrai est connue avec une précision plus ou moins grande selon les stations. Par l'analyse d'occultations des Pléiades observées
à partir de plusieurs stations (Morrison, 1971), on a pu évaluer à
± 0",1 l'incertitude sur la position de l'observateur, ce qui correspond, pour la distance moyenne Terre-Lune, à ± 200 mètres environ. Cependant, pour des stations raccordées aux systèmes internationaux, la
précision est actuellement de 5 à 10 mètres, compte tenu des écarts
systématiques entre les différents réseaux disponibles obtenus à partir de techniques telles que Doppler, VLBI, laser. Dans l'avenir, on
peut espérer une nette amélioration sur ce paramètre par une meilleure
modélisation des phénomènes de marées terrestres et du déplacement re-
latif des plaques tectoniques (Kovalevsky, 1979).

Les coordonnées géocentriques de la Lune ($\vec{\Gamma}$) sont données par les éphémérides. On utilise actuellement l'éphéméride j = 2 basée sur la théorie de Brown modifiée. Les réductions effectuées dans ces conditions (Morrison, 1971) font apparaître une incertitude de ± 0",16 sur la position de la Lune, soit ± 320 mètres à la distance moyenne Terre-Lune. Un grand effort a été accompli pour améliorer cette situation. On peut disposer aujourd'hui d'éphémérides obtenues par intégration numérique (LURE 2) et dont la précision peut être estimée à quelques dizaines de mètres sur la période de validité de cette intégration. Les éphémérides les plus récentes (CERGA-TEXAS, JPL) rendent compte des observations laser-Lune à mieux que 0",001, soit ± 2 mètres à la distance Terre-Lune, cela dans le système de référence propre à l'éphéméride utilisée.

L'équation (2) fait intervenir, outre la direction topocentrique de la Lune, la direction de l'étoile observée et celle sélénocentrique, du point où se produit l'occultation. Si l'on multiplie vectoriellement chacun des membres de l'équation (2) par le vecteur $\vec{\gamma}$, il vient :

$$D"\vec{\tau} \wedge \vec{\gamma} + (r + \varepsilon) \vec{\Delta} \wedge \vec{\gamma} = \vec{0} \tag{3}$$

Les directions $\vec{\gamma}$ sont extraites des catalogues stellaires. Les plus usuels (S.A.O., AGK 3, Z.C.) ont aujourd'hui une cohérence interne de l'ordre de 0",4. Pour le S.A.O., le nombre d'étoiles occultables est de l'ordre de 105.000 jusqu'à la magnitude visuelle 9. On gagne évidemment en précision par l'usage des catalogues FK4 et Sup. dont les positions stellaires sont données à ± 0",1 près ; mais ces catalogues ne comportent que des étoiles brillantes dont seulement 1500 sont occultables. Le projet HIPPARCOS (ESA, 1979) accepté par l'ESA, fournira un catalogue de 100.000 étoiles environ avec une erreur probable sur les positions de ± 0",002 à la date d'observation. L'incertitude sur les mouvements propres entraînera une dégradation de 0",002 par an. Parmi les étoiles de ce catalogue, 30.000 environ seront occultables jusqu'à la magnitude visuelle 9. Remarquons que ce catalogue définit un système géométrique "libre" qu'il faudra rattacher à des directions fixes (sources extragalactiques) (Kovalevsky, 1975).

Le vecteur (r + ε) $\vec{\Delta}$ d'origine (L), centre des masses de la Lune, a pour extrémité le point de la surface lunaire où disparaît l'étoile pour l'observateur (O). L'élévation ε de ce point rapportée au contour moyen défini par les librations topocentriques l, b, se lit dans les atlas de profils lunaires. Les profils sont orientés par rapport à la projection de l'axe de rotation de la Lune sur le plan du contour moyen : l'angle de position du point est compté positivement vers l'Est. Le profil apparent se compose de la projection d'accidents du relief qui, selon leur altitude réelle, peuvent se trouver à ± 180 km du contour moyen (Weimer, 1954). Cette projection fait que l'on peut considérer au premier ordre près, les deux vecteurs $\vec{\Delta}$ et $\vec{\gamma}$ perpendiculaires. L'équation (3) s'écrit alors :

$$\left[\cos(\delta)\cos(\delta_\star)(\alpha-\alpha_\star)^2 + (\delta-\delta_\star)^2\right]^{1/2} - (r + \varepsilon) = \sigma \tag{4}$$

où α, δ sont les coordonnées topocentriques de la Lune,
   α_*, δ_* sont les coordonnées de l'étoile.

   L'écart entre les deux systèmes est alors donné à l'instant de
l'occultation par σ . Cet écart est dû pour une grande part à l'impré-
cision des atlas. Actuellement, on dispose de deux atlas de profils :
l'Atlas de profils lunaires (Weimer, 1952) et The marginal zone of the
Moon (Watts, 1963), tous deux obtenus à partir de clichés photographiques. Ces profils se réfèrent dans le cas de Weimer au centre de figure
de la Lune et au centre des masses dans le cas de Watts. Une comparai-
son systématique entre ces deux atlas (Meyer, 1976) a montré des dis-
persions qui peuvent atteindre localement 0",5 et en moyenne, 0",3. La
réduction d'une campagne d'occultations faite avec l'un ou l'autre de
ces documents conduit par ailleurs à une même précision de $\pm$ 0",26.
Remarquons que l'Atlas de Watts est le résultat d'un lissage de 867 profils alors que l'atlas de Weimer est un relevé fidèle de profils tels
qu'ils apparaissent sur les plaques photographiques. Leur nombre réduit
(139) nécessite une densification qui est en cours (Meyer, 1978) pour
permettre une interpolation entre profils compatible avec la précision
de l'Atlas. En toute rigueur, pour tirer le maximum de précision de cet
Atlas, il faudrait recalculer les librations des clichés utilisés par
Weimer à partir des théories modernes de la libration (Migus, 1976 ;
Calame, 1975). En effet, une erreur de 3' dans l'orientation du système
sélénocentrique peut entraîner des variations du profil apparent de
l'ordre de 0",1 et une erreur d'orientation de celui-ci qui peut atteindre 1'.

   On a vu que la précision du rattachement des deux systèmes géométrique et dynamique dépend pour une large part de la précision que l'on
a sur la topographie lunaire et sur la position du centre des masses
par rapport à la surface de la Lune. Par la télémétrie laser, à partir
d'un satellite circumlunaire, on peut apporter à la solution de ce pro-
blème une amélioration importante. Les missions Apollo 15, 16 et 17
avaient déjà procédé à des mesures altimétriques à l'aide de lasers
embarqués. La précision relative sur les dénivellations a été estimée
à $\pm$ 10 mètres, alors que la précision absolue sur l'altitude du satellite n'a été que de $\pm$ 100 mètres après restitution globale de la tra-
jectoire. Les mesures ont été effectuées tous les 30 km, le diamètre
du spot au sol étant de 30 mètres environ (Kaula, 1974). Le projet
européen POLO (Polar Orbiting Lunar Observatory) de l'ESA doit permettre, par une campagne altimétrique systématique, d'obtenir un gain
appréciable sur ces précisions. Ce satellite circulera sur une orbite
inclinée de 85° à 95° sur l'équateur lunaire, à une altitude variant
de 50 à 150 km (ESA, 1979), ce qui est particulièrement favorable à la
description des zones marginales de la Lune. Deux lasers altimétriques
pourront être embarqués, l'un fonctionnant en mode déclenché, l'autre
en mode continu. Le rôle du premier est d'obtenir des mesures absolues
d'altitude servant à caler les mesures de dénivellations relatives réa-
lisées par le second (Gaignebet, 1980). Les spécificités demandées sont:
un contrôle d'attitude du satellite à $\pm$ 5' et une restitution d'orbite
à $\pm$ 10 mètres sur les trois coordonnées spatiales, après modélisation

du potentiel lunaire. Ceci entraîne une erreur possible de ± 0°,005
sélénocentrique sur la position du point mesuré et de ± 30 mètres sur
l'altitude de ce point compte tenu de ce que les pentes moyennes lunaires ne dépassent pas la dizaine de degré (Evans, 1971). Pour obtenir
une bonne couverture des zones marginales, il faudrait 500 orbites. Le
long de chacune de ces orbites, le laser continu donne les variations
significatives d'altitude du relief entre deux mesures du laser déclan-
ché. Avec la réalisation d'un tel projet, les profils apparents seraient
connus avec une précision améliorée d'un facteur 10 par rapport aux
atlas actuels.

Après l'examen des différents paramètres qui conditionnent la précision du rattachement entre le système géométrique et le système dynamique, nous pouvons, en regard des techniques d'observation, les ras-
sembler dans le tableau 1, en indiquant l'incertitude qui les affecte
présentement et dans le futur.

Tableau 1.

Techniques d'observation	visuelle	double image	photoélec- trique	futur
Datation	$0^s,5$	$0^s,1$	$0^s,001$	$0^s,001$
	500 m	100 m	1 m	1 m
Position de la station	0",005	0",005	0",005	0",001
	10 m	10 m	10 m	2 m
Positions stellaires	0",26	0",26	0",26	0",02
	520 m	520 m	520 m	40 m
Ephémérides	(0",16)	(0",16)	(0",16)	0",001
	320 m	320 m	320 m	2 m
Bord lunaire	0",30	0",30	0",30	0",03
	600 m	600 m	600 m	60 m
Précision du rattachement	0",5	0",4	0",4	0",04
	1000 m	800 m	800 m	80 m

3. CONCLUSION

Le rattachement du système dynamique au système géométrique par
l'utilisation des observations d'occultations stellaires ne peut se
faire aujourd'hui qu'avec une précision médiocre due essentiellement
à la mauvaise connaissance de la topographie des zones marginales de
la Lune et à la qualité actuelle des catalogues stellaires. On peut
attendre un gain d'un facteur 10 sur la précision de ces deux paramètres par la réalisation des projets POLO et HIPPARCOS.

L' observation systématique d'occultations stellaires permet-
trait alors un rattachement des deux systèmes a 0",04 près.

REFERENCES

Baudry, A.: 1979, Radio astrométrie; principe; précision actuelle
    et future des mesures de positions, Coll. GS V, Bordeaux.
Calame, O.: 1975, Etude des mouvements libratoires lunaires et
    localisation des stations terrestres à partir de mesures
    laser de distances, Thèses, Paris.
de Vegt, C., Gehlich, U. K.: 1976, Astron. Astroph., 48, p. 245.
ESA: 1979, HIPPARCOS space astrometry, report on phase A study,
    SCI (79) 10.
ESA: 1979, Polar orbiting lunar observatory, assessment study,
    SCI (79) 7.
Evans, D. S.: 1971, in proceedings of the IAU General Assembly,
    Highlights of Astronomy, C.de Jager ed., p. 601.
Gaignebet, J.: 1980, Communication personnelle.
Kaula, W. M., Schubert, G., Lingenfelder, R. E., Sjogren, W. L.,
    Wollenhaupt, W. R.: 1974, Proc. Fifth Lunar Conf., 3, p. 3049.
Klock, B. L., Scott, D. K.: 1970, Astr. J., 75, p. 851.
Kovalevsky, J.: 1975, in Space Astrometry, ESRO, SP 108, p. 67.
Kovalevsky, J.: 1979, IAU Symposium No. 82, p. 151.
Meyer, C.: 1974, Contribution à la détermination des positions
    lunaires par l'observation d'occultations au moyen d'un
    micromètre à double image, Thèses, Paris.
Meyer, C.: 1976, The Moon, 16, p. 27.
Meyer, F.: 1978, Méthode pour l'amélioration et l'extension d'un
    catalogue de profils lunaires, Thèses, Nice.
Migus, A.: 1976, The Moon, 15, p. 165.
Morrison, L. V.: 1971, Proc. IAU Symposium No. 47, p. 395.
Morrison, L. V.: 1979, Mon. Not. R. Astr. Soc., 187, p. 41.
Nather, R. E., Evans, D. S.: 1970, Astr. J., 75, p. 575 and p. 953.
Slade, M. A., Preston, R. A., Harris, A. W., Skjerve, L. J., and
    Spitzmesser, D. J.: 1977, The Moon, 17, p. 133.
Valein, J. L.: 1979, Etude de la rotation de la Terre à partir
    des mesures laser-Lune, Thèses, Paris.
Watts, C. B.: 1963, Astr. Pap. Washington, 17.
Weimer, T.: 1952, Atlas de profils lunaires, Observatoire de Paris.
Weimer, T.: 1954, Recherches sélénographiques: allongement du sélé-
    noide, libration physique, profils lunaires, Thèses, Paris.
Yasuda, H., Miyauchi, N.: 1971, Ann. Tokyo Astr. Obs., 12, p. 188.

DYNAMICAL CONSTANTS AND REFERENCE PARAMETERS FOR MARS

William H. Michael, Jr. and George M. Kelly
NASA-Langley Research Center and Analytical Mechanics
   Associates, Inc.
Hampton, Virginia, 23665, USA

Abstract. Dynamical constants and other fundamental reference parameters for Mars have been derived from analyses of Viking lander ranging
and Doppler tracking data covering a time span of nearly four years.
Precise values have been obtained for the coordinates of the spin axis
and for the rotation rate, suggesting that these Viking-derived values
are definitive and are suitable for adoption by the IAU. Preliminary
results have been obtained for a small seasonal variation in the rotation rate, and progress has been made toward a direct determination of
the precession constant.

INTRODUCTION

Tracking data from the Viking Mars landers and orbiters have provided a wealth of information for analyses of properties of Mars and its
environment (Michael et al., 1977). The lander Doppler tracking data
are uniquely appropriate for determination of parameters of the physical
ephemeris of Mars and two components of lander position. Ranging data
are primarily used in determining the Earth-Mars ephemeris and the
component of lander position parallel to the spin axis. Results of data
analyses for time spans ranging from a few days to 16 months have been
reported previously (Michael et al., 1976; Mayo et al., 1977; Michael,
1979). Reported here are results of the most recent analyses of the
data available to date, covering a time span of 46 months. In general,
these results have been obtained using analytical procedures and the
tracking systems described elsewhere (Mayo et al., 1977; Michael et al.,
1972). The stability and precision of the present and earlier solutions
with continuously increasing data spans indicate that results for the
Mars pole position and rotation rate are now definitive.

MARS POLE POSITION AND ROTATION RATE

Solutions for the coordinates of the north pole of Mars and for the
rotation rate converged rapidly for even the short data spans of the

325

E. M. Gaposchkin and B. Kołaczek (eds.), Reference Coordinate Systems for Earth Dynamics, 325–328.
Copyright © 1981 by D. Reidel Publishing Company.

early analyses of Viking data and have exhibited remarkable stability
for successively increasing data spans. On the basis of this experience
it is not expected that these results will be significantly affected
through incorporation of additional data.

The recommended values for the right ascension α and the declination δ of the Mars spin axis, referred to the Earth mean equator and
equinox of 1950.0, are

$$\alpha = (317.340^\circ \pm 0.003^\circ) - 0.10106^\circ T$$

and $$\delta = (52.711^\circ \pm 0.002^\circ) - 0.05706^\circ T$$

where T is measured in Julian centuries from 0 hours, January 1, 1950
(Julian Day 2433282.5). The secular variations in α and δ are obtained
from the Lowell (1914) predicted precession of -708 arc s per Julian
century. An assumed uncertainty of $\pm$ 50 arc s per century in this value
of the precession constant is the primary contributor to the quoted
uncertainties in α and δ at the 1950 epoch. The uncertainties in the
instantaneous coordinates of the pole at any epoch within the Viking
time frame are approximately $\pm$ 0.001 deg.

The recommended value for the sidereal rotation rate of Mars,
determined as the average rate for essentially two full Martian years,
is 350.891985 $\pm$ 5 x 10^{-6} deg per day. This corresponds to a sidereal
rotation period of 24h 37m 22.6631s $\pm$ 0.0013s. By convention, the
sidereal rotation is referred to the vernal equinox of Mars, which is
assumed to precess at the rate mentioned earlier. The Viking-derived
value for the rotation period is in very close agreement with the
recently modified value of 24h 37m 22.662s $\pm$ 0.002s obtained by de
Vaucouleurs (1980) from albedo station data for the past 300 years,
suggesting that these improved results are definitive.

SEASONAL VARIATION IN THE MARS ROTATION RATE

The rather significant seasonal variation in atmospheric pressure
measured at the Viking landing sites, due to alternate deposit and
release of carbon dioxide at the polar caps, suggests that this effect
could lead to a detectable variation in the rotation rate. The magnitude of such an effect can be estimated by a simple model which assumes
conservation of angular momentum about the polar axis and relates changes
in atmospheric pressure to changes in the polar moment of inertia. A
combination of simple expressions leads to the relation $\Delta \omega = -(5/3)\omega \, \Delta M/M$
where ω is the rotation rate, ΔM is the change in mass of a spherical
shell representing the well-mixed Martian atmosphere (3.9 x 10^{18}g per
mbar change in surface pressure), and M is the mass of Mars (6.4 x 10^{26}g)
Using smoothed lander 1 pressure data from Hess et al. (1980), with
periodic extrapolation beyond the first Martian year, the predicted
seasonal variation in the Mars rotation rate, $\Delta \omega$, is shown in Figure 1.

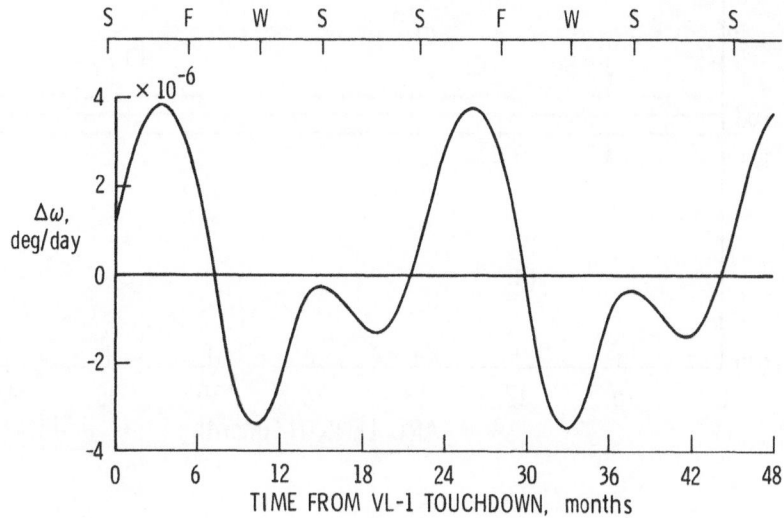

Fig. 1. Predicted Variation in the Mars Rotation Rate.

This model for the rotation rate variation, together with its integrated effect representing displacement, has been incorporated in the data analysis program, with the normalized amplitude as an additional solution parameter. Preliminary estimates for the amplitude are 0.4 that of Figure 1, giving an indication of the existence of a small seasonal variation in the rotation rate but at a lower level than predicted with the limited considerations of the simple model. This experimental result tends to support the analytical work of Philip (1979). Definitive confirmation of this effect would be of considerable interest in geophysical and atmospheric studies of Mars.

THE PRECESSION CONSTANT FOR MARS

With a sufficiently long data span it should be possible to determine the precession constant μ for Mars from the motion of the spin axis, even though the yearly motion is quite small. Figure 2 shows values obtained for μ from solutions with various data spans. The values of μ obtained for data spans longer than three years are just within the limits of the assumed uncertainty and, although not definitive, exhibit an encouraging trend. With additional tracking data from lander 1, which is expected to continue operating for some years, a definitive determination of μ can be anticipated. Such a determination, in turn, could have important applications to the polar moment of inertia of Mars and to analyses of its interior structure.

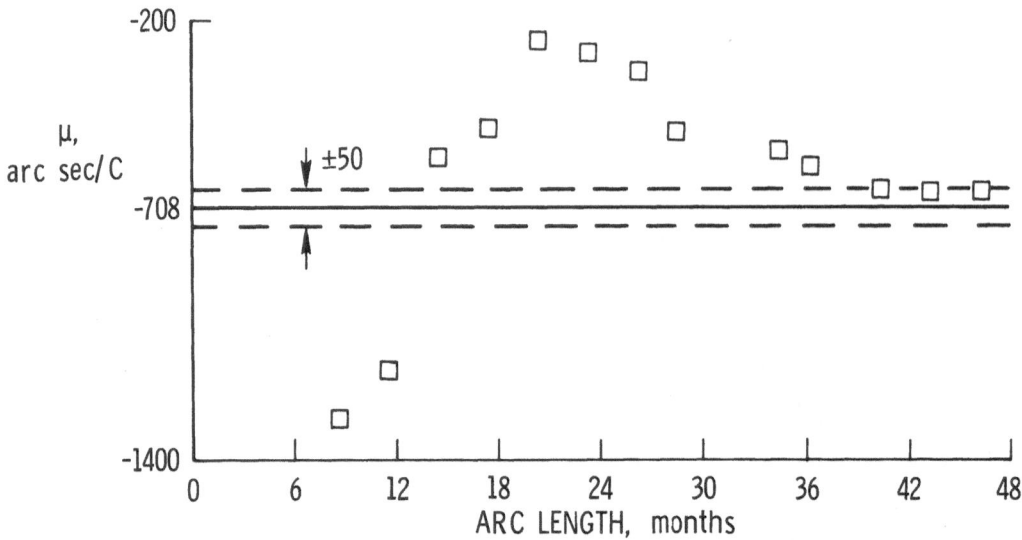

Fig. 2.   Results of Solutions for the Mars Precession Constant.

We thank J. P. Brenkle, T. A. Komarek, T. M. Kaufman, H. N. Royden, J. Breidenthal, D. Kline, and E. Klumpe for experiment management and data calibration; M.S. Keesey and D. L. Cain for providing improved developmental ephemerides; and A. P. Mayo for analytical formulations.

REFERENCES

de Vaucouleurs,G.: 1980, Astron. J. 85, pp. 945-960.
Hess,S.L., Ryan,J.A., Tillman,J.E., Henry,R.M., and Leovy,C.B.: 1980, Geophys. Res. Letters 7, pp. 197-200.
Lowell,P.: 1914, Astron. J. 28(21), pp. 169-171.
Mayo,A.P., Blackshear,W.T., Tolson,R.H., Michael,W.H., Kelly,G.M., Brenkle,J.P., and Komarek,T.A.: 1977, J. Geophys. Res. 82, pp. 4297-4303.
Michael,W.H.,Jr.: 1979, The Moon and the Planets 20, pp. 149-152.
Michael,W.H.,Jr., Cain,D.L., Fjeldbo,G., Levy,G.S., Davies,J.G., Grossi,M.D., Shapiro,I.I., and Tyler,G.L.: 1972, Icarus 16, pp. 57-73.
Michael,W.H.,Jr., Tolson,R.H., Mayo,A.P., Blackshear,W.T., Kelly,G.M., Cain,D.L., Brenkle,J.P., Shapiro,I.I., and Reasenberg,R.D.: 1976, Science 193, p. 803.
Michael,W.H.,Jr., Tolson,R.H., Brenkle,J.P., Cain,D.L., Fjeldbo,G., Stelzried,C.T., Grossi,M.D., Shapiro,I.I., and Tyler,G.L.: 1977, J. Geophys. Res. 82, pp. 4293-4295.
Philip,J.R.: 1979, Geophys. Res. Letters 6, pp. 727-730.

ORIENTATION OF FK4 SYSTEM FROM MERIDIAN OBSERVATIONS OF PLANETS

D.P. Duma, L.N. Kizjun, Yu.I.Safronov
Main Astronomical Observatory, Ukrainian Academy of Sciences
Kiev, U.S.S.R.

ABSTRACT

The orientation of the FK4 coordinate axes was made by two methods: using the classical interpretation of the orientation problem and by the new method suggested and developed at the Main Astronomical Observatory.

The orientation elements of the FK4 were obtained using 11232 meridian observations of Mercury, Venus, Mars and 5987 observations of Ceres, Pallas, Juno and Vesta made at six observatories between 1928 and 1971. The pseudo-perturbation method was used for the adjustment of the ill-conditioned system of the equations.

The computations showed that the new method for orientation of the axes of star catalogues is efficient and the usefulness of the pseudo-perturbation method has been ascertained.

The following values of the equinox and equator corrections of the FK4 have been obtained using:

the new method (14 unknowns)

arithmetical means over the planets	weighted means over the planets
$\Delta A = + 0^S.028 \pm 0^S.014$	$\Delta A = + 0^S.013 \pm 0^S.005$
$\Delta \delta_O = + 0".016 \pm 0".030$	$\Delta \delta_O = + 0".021 \pm 0".024$

the classical method (12 unknowns)

arithmetical means over the planets	weighted means over theplanets
$\Delta A = + 0^S:028 \pm 0^S.014$	$\Delta A = + 0^S.016 \pm 0^S.008$
$\Delta \delta = + 0".024 \pm 0".030$	$\Delta \delta_O = 0".042 \pm 0".023$

E. M. Gaposchkin and B. Kołaczek (eds.), Reference Coordinate Systems for Earth Dynamics, 329.
Copyright © 1981 by D. Reidel Publishing Company.

DEFINITION AND PRACTICAL REALIZATION OF THE REFERENCE FRAME IN THE FK5 — THE ROLE OF PLANETARY DYNAMICS AND STELLAR KINEMATICS IN THE DEFINITION

Walter Fricke
Astronomisches Rechen-Institut
Heidelberg, Federal Republic of Germany

ABSTRACT

The formulation of the fundamental reference system to be represented by the FK5 includes the determination of the equinox and equator on the basis of planetary dynamics and the application of the new expressions for the general precession in longitude adopted in the IAU (1976) System of Astronomical Constants. The role of hypotheses which entered the determination of the lunisolar precession is explained. Results are presented for the equinox and equator of the FK5 which are based on observations of the Sun, planets, and lunar occultations.

CONSTRUCTION OF INERTIAL SYSTEMS AND DYNAMICAL DETERMINATIONS OF THE GENERAL PRECESSION IN LONGITUDE

In assuming that the motions in the solar system are governed by the gravitational forces of its members only — i.e., that no external forces have to be taken into account — the law of gravitation allows one to determine inertial planes and directions. The orbital planes, their inclinations and nodes, and the directions to the perihelia have — aside from secular perturbations which can be computed — fixed (inertial) positions and may be employed for the formulation of an inertial system. Anding (1905) has shown how a purely dynamical reference system can be constructed. Bauschinger (1922) has followed Anding's ideas in order to find out in how far the star catalogue PGC by Boss (1910) represents an inertial system. He applied the comparisons which Newcomb (1895) had made between the secular variations of the elements of the four inner planets derived from observations and those resulting from the planetary masses, and he took into account the relativistic motions of perihelia predicted by Einstein's theory. Bauschinger came to the following remarkable conclusion: "The star catalogue by Boss can be considered an inertial system in the sense of Newtonian mechanics and of the law of gravitation corrected for the Einsteinian motion of perihelia provided that Newcomb's (1898) precession is corrected by + 0$\overset{''}{.}$86 and that a secular motion of 1$\overset{''}{.}$16 of the equinox on the equator with respect to Newcomb's equinox is taken into account" (translated from German).

E. M. Gaposchkin and B. Kołaczek (eds.), Reference Coordinate Systems for Earth Dynamics, 331–340.

It has been known for a long time that precession should be determined from the moments of inertia of the earth's body, but these are not known with sufficient accuracy. Hence, determinations based upon observations of the Sun and planets have been considered of great importance, because they are free of hypotheses. However, the attempts made by Brouwer (1950) and Clemence (1966) on the basis of observations of the Sun, Mercury and Venus failed to yield any reliable corrections; the uncertainty was at least as large as the correction $\Delta p = + 1\overset{''}{.}12$ per tropical century (to Newcomb's general precession) which resulted from these determinations. A considerable improvement was achieved by Laubscher (1976) who applied the dynamical method to the optical observations of Mars made from 1751 to 1969. These observations were reduced to the system of the FK3 which is very near to the FK4 system in the zodiacal zone. They yielded the following values for 1850

$$\Delta p = + 1\overset{''}{.}21 \pm 0\overset{''}{.}44,$$
$$\Delta e = + 1\overset{''}{.}12 \pm 0\overset{''}{.}29, \qquad\qquad (1)$$
$$\Delta \varepsilon = - 0\overset{''}{.}14 \pm 0\overset{''}{.}11;$$

Δp is the correction to Newcomb's value of the general precession per tropical century, Δe, a correction to the centennial proper motions in right ascension of the FK3, and $\Delta \varepsilon$, a correction to Newcomb's value of the obliquity of the ecliptic for 1850. The errors are standard deviations; they are large in Δp and Δe. In discussions on the revision of Newcomb's precession Laubscher's determination was duely considered. It was not taken into account in the new system of constants, because there remained uncertainty about the effect of correlations between the unknowns as explained by Laubscher himself.

HYPOTHESES INCLUDED IN THE DETERMINATION OF THE GENERAL PRECESSION ADOPTED BY THE IAU

In the IAU (1976) System of Astronomical Constants the new value of the general precession in longitude is the result of a revision of Newcomb's precession in two parts. The change consists of the following components: a correction $\Delta p_1 = + 1\overset{''}{.}10$ per tropical century at 1900 to Newcomb's lunisolar precession, and a correction to Newcomb's planetary precession arising from the adoption of new values of planetary masses. The correction to the lunisolar precession was determined by Fricke (1967) from proper motions of FK4 stars, and the expressions for precession quantities based upon the IAU (1976) System were developed as a function of the constants by Lieske et al. (1977). The motivating ideas for changing precession were described by Fricke (1977 a), who also presented the basic material for the determination of precession and a review of methods and results (Fricke, 1977 b).

The last mentioned paper contains full information on the assumptions involved in the determination of the lunisolar precession from fundamental proper motions. Before entering a discussion on kinematical hypotheses it should be pointed out that the fundamental proper motion components of each star consist of the following parts:
(a) a parallactic motion of each star due to the solar motion with respect to the stars under consideration;
(b) a rotation due to precessional errors and the rotation of the assembly of stars considered;
(c) a deformation (shear) of the velocity field of the stars leading to terms produced by non-rigid rotation in which the assembly of stars take part; and
(d) a residual motion due to the peculiar motion of the stars and internal errors of the motions.
This list is complete and free of assumptions. The crucial point is the decomposition of the angular rotation vector (ω_1, ω_2, ω_3) resulting from (b). The decomposition involves the hypothesis that the rotation vector describes the accumulative effect of three parts of different origin:
(a) precessional corrections (Δp_1 to the lunisolar precession, $\Delta\lambda$ to the planetary precession);
(b) a common correction Δe to the fundamental proper motions μ_α due to erroneous determinations of the vernal equinox, and
(c) galactic rotation (the rotation parameter Q).

They allow one to determine from the rotation vector $\vec{\omega}$ the following three unknowns

$$\Delta n, \quad \Delta\lambda + \Delta e, \quad Q,$$

according to the equations

$$
\begin{aligned}
\omega_1 &= -0.868\,Q \,, \\
\omega_2 &= -0.188\,Q - \Delta n \,, \\
\omega_3 &= +0.460\,Q + \Delta n \cot \varepsilon - (\Delta\lambda + \Delta e) \,,
\end{aligned}
\tag{2}
$$

where Δn, the correction to the general precession in declination, is related to the correction Δp_1 to the lunisolar precession by $\Delta n = \Delta p_1 \sin \varepsilon$. One notices that not more than three unknowns can be determined from the components of the velocity vector. The proper motions of distant stars in FK4 and FK4 Sup yield the values (per tropical century)

$$
\begin{aligned}
\Delta p_1 &= +1\overset{''}{.}10 \pm 0\overset{''}{.}15, \\
\Delta\lambda + \Delta e &= +1\overset{''}{.}20 \pm 0\overset{''}{.}16, \\
Q &= -0\overset{''}{.}23 \pm 0\overset{''}{.}06.
\end{aligned}
\tag{3}
$$

The errors are standard deviations. In adopting the correction $\Delta\lambda = -0\overset{''}{.}03$ to planetary precession determined dynamically by Laubscher (1972) and Lieske et al. (1977) with an accuracy better than $0\overset{''}{.}01$, one obtains the centennial corrections

$$\Delta p = + 1\overset{''}{.}13 \pm 0\overset{''}{.}15$$
$$\Delta e = + 1\overset{''}{.}23 \pm 0\overset{''}{.}16$$

(4)

to Newcomb's general precession in longitude and to the proper motions
in right ascensions of the FK4, respectively. The quantity Δe tells
us that all μ_α of the FK4 are too small and that, hence, the equinox
of the FK4 (zero point of right ascensions) is in motion with respect
to the true (dynamical) equinox.

The decomposition of the angular velocity vector and the
formulation of Eqs. (2) is based on the following hypotheses:
(1) no other rotation than those selected in (a) to (c) contributes
to the angular velocity vector;
(2) with the coefficients of B the Oort-Lindblad model of galactic
rotation was adopted;
(3) even if the decomposition is well-founded, one may suspect that
the interpretation of Δe is hypothetical, because Δe may also indicate
either a rotation of the assembly of stars under consideration about
the earth's axis of rotation or a rotation of the planetary system with
respect to the stars;
(4) finally, least-square solutions are based on a Gaussian
distribution of the residuals.

In referring to the hypotheses in the order of its numbering
the following comments can be given:
(1) no other physically comprehensible or spurious rotation
(originating from errors of observation) can be seen;
(2) in an investigation of the same material of proper motions
Du Mont (1977) has replaced the Oort-Lindblad model by the three-
dimensional method proposed by Ogorodnikov (1932) and Milne (1935)
with the result that no additional significant information was found;
(3) the quantity Δe is a correction to the fundamental proper motions
in right ascensions. For clarity, the name "non-precessional motion
of the equinox" previously also used by the author should be replaced
by "a motion of the catalogue equinox (zero point of fundamental
right ascensions) with respect to the dynamical (true) equinox".
Fricke (1979) has shown that we are dealing with a zero point error
of the proper motions. It occured in Newcomb's FC (1899) and was not
eliminated later. The zero point error is due to an error (magnitude
equation) of observations of bright stars made for the determination
of the equinox in the 19th century. An alternative interpretation of
Δe proposed by Balakirev (1980) consists in a real rotation of the
plane of the ecliptic about an axis lying in this plane. He found that
the axis of rotation is nearly identical with the nodal line of the
galactic plane on the ecliptic, and that the period of rotation is
about 3×10^8 years. Balakirev has conceded that no physical
explanation can be given. The arithmetic is correct but the interpretation fantastic in view of the disillusionment by facts that do
not allow us to assume that equinox determinations were free of
systematic errors.

In this connection it deserves mentioning that Van Woerkom (1943) has investigated a precessional motion of the invariable plane of the planetary system with respect to the galactic plane due to the attractive force of the galaxy on massive planets. The effect turned out so small that it need not be considered.

(4) The residuals of the proper motions, which resulted from the determination of precessional corrections etc., were investigated by Fricke (1967). It was found that no systematic effects can be recognized in the residuals. Because the distances of the stars were approximately known, parallax factors were applicable, and the remaining residuals are not much greater than the errors of the fundamental proper motions. The assumption of an arbitrary distribution of the residual proper motions has often been considered a fundamental hypothesis underlying determinations of precession. This is no longer correct.

DYNAMICAL DEFINITION OF THE EQUINOX AND EQUATOR AND ITS REALIZATION IN THE FK5

As mentioned before, the systems of all fundamental catalogues in the series from Auwers' FC to the FK4 are defined by the dynamics of the planetary system as far as the equinox and equator are concerned. Pioneer work was done by Bessel (1830) who determined the equinox from observations of the Sun made by Bradley from 1750 to 1762 and his own observations made at Königsberg from 1822 to 1835. On this basis Bessel determined the mean and apparent places of 36 "Maskelyne Stars" from 1750 to 1850 presented in his famous "Tabulae Regiomontanae", which was widely used as a fundamental reference coordinate system in the 19th century. Newcomb (1872) has extended Bessel's equinox determination in employing all available observations of the Sun from 1756 to 1869, as far as they were reduced by the observers to give fundamental right ascensions of clock stars. The difference between α(obs) and α(Bessel) turned out to be $+ 0.^{s}014$ at 1840, and no significant secular variation was found. Newcomb (1882) made use of his equinox determination in fixing the zero point of right ascensions of equatorial stars in his "Catalogue of 1098 Standard Clock and Zodiacal Stars. This zero point is commonly called Newcomb's equinox N_1.

At the beginning of this century deficiencies of Newcomb's N_1 became apparent from new observations of the Sun and planets carried out with improved transit circles equipped with impersonal micrometers. One noticed that all right ascensions required a negative correction ΔN_1, and Newcomb's error was explained as the consequence of the imperfect methods of observing (eye and ear, key, hand driven travelling threads of micrometers). On the basis of observations of the Sun and planets made with impersonal micrometers a correction $\Delta N = -0.^{s}050$ was adopted for the FK3. Hence, the equinox of the FK3 is $N_1 - 0.^{s}050$, and it was adopted independent of

time. After careful consideration this equinox was maintained in the
FK4, although the equator of the FK4 was determined from the observations of the Sun and planets from 1900 to 1958 which indicated a
correction $\Delta\delta = -0\overset{''}{.}017$ at 1928 to the equator of the FK3. This
correction was applied.

In view of the demand of an improved fundamental catalogue (FK5)
and of a new set of high-precision lunar and planetary ephemerides
it became clear that in the formulation of the FK5 system every
effort has to be made in order to ensure that the FK5 corresponds as
closely as possible to the dynamical reference frame. The FK5 equinox
and equator have to correspond to the dynamical equinox and to the
dynamically determined equator. In the latest status report on the
FK5 Fricke (1980) has given detailed information on the methods
applied for determination of the equinox and equator and on results
so far achieved. Hence, a brief summary may be sufficient here.

(1) Definition of the equinox: The vernal equinox is the average
location of the ascending node of the earth's moving mean orbit on
the equator. This definition is consistent with the common practice
of comparing observations of the Sun with the places computed from the
ephemeris, and it is in accordance with Newcomb's definition.

(2) Motion of the FK4 equinox: Three different sources have provided
evidence that the correction E to the FK4 equinox increases with time
according to

$$E(T) = E(T_0) + \dot{E}(T - T_0) , \tag{5}$$

where $\dot{E}$ is the time derivative of E. The sources are (a) the FK4
proper motions, (b) the observations of lunar occultations, and
(c) the equinox determinations $E(T_\nu)$ for different mean epochs T_ν
from about 1900 to 1970. They yield the following results for $\dot{E}$:

(a) FK4 proper motions; after application of the new value of the
general precession in longitude the centennial proper motions require
a correction Δe such that

$$(\mu_\alpha)_{FK4} + \Delta e = (\mu_\alpha)_{dyn} \tag{6}$$

is fulfilled, where

$$\Delta e = \dot{E} = + 1\overset{''}{.}23 \pm 0\overset{''}{.}16 \tag{7}$$

(b) The lunar occultations from 1820 to 1970 analysed by Van Flandern
(private communication) have resulted in

$$E(T) = + 0\overset{''}{.}65 + 1\overset{''}{.}31 (T - 19.50) , \tag{8}$$

where $\dot{E} = + 1\overset{''}{.}31$ and T is counted in centuries.

(c) Available are 35 equinox determinations which have yielded discrete values for $E(T_\nu)$ from 1900 to 1970 at the mean epochs of the respective observations. The following list gives information on the objects, the number of values E, and the observatories or authorities.

Objects	Determinations	Observatories or Authorities
Sun	22	Washington (8), Cape (6), Greenwich (4), Breslau (1), Herstmonceux (1), Ottawa (1), Pulkovo (1)
Mercury	1	Duma, Fricke
Venus	1	Duma, Fricke
Mars	3	Duma, Glebova, Niimi
Minor Planets	5	Branham (2), Duma, Kristensen, Orelskaya
Lunar Occultations	2	Morrison, Van Flandern
Various Objects: (optical, radar, laser ranging)	1	Standish

A least-squares solution based on all 35 values of E has yielded

$$\dot{E} = + 1\overset{''}{.}28 \pm 0\overset{''}{.}15 \quad (\text{m.e.}) \tag{9}$$

The agreement of the results given in Eqs. (7) to (9) is excellent; it is better than we had expected in view of the independent sources and the differences between the methods of determination of $\dot{E}$. Hence, the mean value of the results (7) to (9), which is

$$\dot{E} = + 1\overset{''}{.}276 = + 0\overset{s}{.}085 \quad \text{per century} \tag{10}$$

appears to be the most likely value of the correction to the motion of FK4 equinox.

(3) Correction to the FK4 equinox: According to Eq. (5) the value $E(T_o)$ is the correction to all right ascensions of the FK4

$$\alpha_{FK4} + E = \alpha_{dyn} , \tag{11}$$

at the equinox and epoch T_o. In adopting the view that $E(T_o)$ is best determined from the most recent observations of different type , we have employed all equinox determinations $E(T_\nu)$ from 1950.0 to 1970.0 leading to the weighted average value

$$E = + 0\overset{s}{.}042 \pm 0\overset{s}{.}003 \quad \text{at} \quad T = 1958.5. \tag{12}$$

From Eqs. (10) and (12) we obtain

$$E = + 0^{s}_{\cdot}035 \pm 0^{s}_{\cdot}003 + (0^{s}_{\cdot}085 \pm 0^{s}_{\cdot}010) \ (T-19.50), \quad (13)$$

where T is counted in centuries. Hence, the zero point correction is $E = + 0^{s}_{\cdot}035$ at 1950, while previously Fricke (1980) reported the preliminary value $+ 0^{s}_{\cdot}031$ which resulted from all determinations from 1900 to 1970. From Eq. (13) we conclude that in the transition from FK4 to FK5 the following operations have to be made in order to achieve that the FK5 equinox is "at all times" as nearly as possible identical to the dynamical equinox

$$\alpha_{FK4} + 0^{s}_{\cdot}035 = \alpha_{FK5} \quad \text{at } 1950.0,$$

$$(\mu_{\alpha})_{FK4} + 0^{s}_{\cdot}085 = (\mu_{\alpha})_{FK5}, \quad (14)$$

where in μ_{α} the change in the general precession in longitude has to be taken into account. These results were reported in March 1980 to the astronomers engaged in the construction of new lunar and planetary ephemerides.

(4) Definition of the equator and its practical realization: The equator is dynamically defined as the plane of symmetry of the ecliptic in declination. Its determination may most easily be explained in considering observations of the sun only. Such observations will indicate differences $\Delta\delta$ between observed declinations of the Sun and those computed from the ephemeris

$$\Delta\delta = (\delta_{\odot})_{obs} - (\delta_{\odot})_{comp}. \quad (15)$$

If the observations are reduced to the FK4 and if the FK4 equator represents the plane of symmetry, the annual mean value of $\Delta\delta$ will be zero. In practice, one will find

$$<\Delta\delta> = - D, \quad (16)$$

where D is the correction which has to be applied to the equatorial declinations of the FK4 in order to find the dynamical equator. This classical method does not take into account that a catalogue equator may be tilted with respect to the dynamical equator as emphasized by Duma (1978). However, a tilt of the FK4 equator must become apparent in the comparisons between absolute observations of equatorial declinations and the FK4. The modern absolute observations (see section 4) don't indicate any significant tilt, a fact, which is plausible, because terms $\Delta\delta_{\alpha}$ have always formed an essential part in the construction of the fundamental catalogues from the FC to the FK4. Furthermore, the equator point corrections D as determined from all suitable recent observations of the Sun and planets do not indicate a significant change of the position of the FK4 equator. No change is therefore intended in the FK5.

In future, one should expect that the equator will be determined by absolute radio-interferometric measurements of the declination of extragalactic radio sources. The position of the vernal
equinox, however, will remain to rest upon the application of the
dynamical method, and the knowledge of the equinox point will be
required as long as planetary ephemerides are needed.

EXTENSION OF THE DYNAMICAL SYSTEM TO THE POLES AND TO FAINTER STARS

The extension of the system from the equator to the poles or
vice versa is established by absolute or quasi-absolute observations
of stars in both coordinates. After the completion of the FK4 about
30 catalogues of observations of this category have become available;
they will contribute to the elimination zonal and regional systematic
errors of the FK4. The analysis of these catalogues is based on the
methods for the determination of systematic differences developed by
Brosche (1966) and Schwan (1977) and refined and tested by Bien et al.
(1978). The latest paper describes precisely the procedure applied
for the FK5. Good progress has been made in the analyses; thus
the completion of the FK5 system can be expected in 1982.

For the determination of differential corrections to the
positions and proper motions of the FK4, there are many modern
catalogues of observation available, and for the extension of the
system to fainter stars we exploit all suitable observations of
about 50,000 stars (including the stars in FK4 and FK4 Sup) that
have become available since about 1900. So far, about 630,000
catalogue positions have entered the data-processing for FK5. Our
criteria for selecting new fundamental stars, however, will not
allow us to make use of all these observations. The criteria are:
(a) reduction to the FK4 must be possible free of errors depending
on the magnitude (this criterion sets the magnitude limit $m_v \approx 9.2$
for the FK5); (b) stars with best observational history;
(c) no nearby stars and no visual binaries or doubles; (d) fit
into a uniform distribution on the sky and in spectral types.

REFERENCES

Anding, E.: 1905, In "Encyklopädie Math. Wiss.", Vol. VI 2.1 p. 1.
    Eds. K. Schwarzschild, S. Oppenheimer. Verlag Teubner,
    Leipzig 1905-1923.
Balakirev, A. N.: 1980, Astron. Tsirk Ak. Nauk. Moskva, No. 1100, 4.
Bauschinger, J.: 1922, Naturwissenschaften 10, 1005.
Bessel, F. W.: 1830, Tabulae Regiomontanae. Regiomonti Prussorum.
Bien, R., Fricke, W., Lederle, T., and Schwan, H.: Veröff. Astron.
    Rechen-Institut, Heidelberg, No. 29.
Boss, L.: 1910, Preliminary General Catalogue. Carnegie Institu-
    tion, Washington, D.C.
Brosche, P.: 1966, Veröff. Astron. Rechen-Institut, Heidelberg,
    No. 17.
Brouwer, D.: 1950, Bull. Astron. Paris 15, 176.
Clemence, G. M.: 1966, Quart. J. Roy. Astron. Soc. 7, 10.
Duma, D. P.: 1978, Astron. Zh. 55, 1103 (Soviet Astron. 22 (5),
    628).
DuMont, B.: 1977, Astron. Astrophys. 61, 127.
Fricke, W.: 1967, Astron. J. 72, 1368.
Fricke, W.: 1977a, Astron. Astrophys. 54, 363.
Fricke, W.: 1977b, Veröff. Astron. Rechen-Institut, Heidelberg,
    No. 28.
Fricke, W.: 1979, In "Dynamics of the Solar System", p. 133.
    IAU Symp. No. 81. Ed. R. L. Duncombe. Reidel Publ. Comp.
    Dordrecht-Holland.
Fricke, W.: 1980, Mitteil. Astron. Gesellschaft 48, 29.
Laubscher, R. E.: 1972, Astron. Astrophys. 20, 407.
Laubscher, R. E.: 1976, Astron. Astrophys. 51, 13.
Lieske, J. H., Lederle, T., Fricke, W., Morando, B.: 1977, Astron.
    Astrophys. 58, 1.
Milne, E. A.: 1935, Monthly Notices R. A. S. 95, 560.
Newcomb, S.: 1872, Washington Observations for 1870. App. III.
    Washington.
Newcomb, S.: 1882, Astron. Pap. Washington Vol. 1, Part 4.
Newcomb, S.: 1895, The Elements of the Four Inner Planets and
    the Fundamental Constants of Astronomy, Washington.
Newcomb, S.: 1898, Astron. Pap. Washington Vol. 8, Part 1.
Newcomb, S.: 1899, Astron. Pap. Washington Vol. 8, Part 2.
Ogorodnikov, K. F.: 1932, Z. Astrophysik 4, 190.
Schwan, H.: 1977, Veröff. Astron. Rechen-Institut, Heidelberg,
    No. 27.
Van Woerkom, A. J. J.: 1943, Bull. Astron. Inst. Neth. 9, 427.

THE STELLAR REFERENCE FRAME FROM SPACE OBSERVATIONS

C. A. Murray
Royal Greenwich Observatory

ABSTRACT: The HIPPARCOS Satellite, to be launched by the European Space Agency, will provide a stellar reference frame over the whole celestial sphere with an average accuracy of ± 0".002 in each coordinate and component of annual proper motion, for some 100,000 stars.

The origin of coordinates will be arbitrary. Absolute rotation of the system of proper motions can be obtained by measuring quasars relative to stars in the HIPPARCOS catalogue, either with the NASA Space Telescope or by conventional ground based astrometric observations.

1. INTRODUCTION

In March 1980, the European Space Agency selected the astrometric satellite, HIPPARCOS, to be the next scientific mission. This satellite will produce astrometric data on some 10^5 stars which will be far superior in accuracy and homogeneity to those generally available from ground based observations.

The satellite will probably be launched in 1986 and the mission duration is to be about 2.5 years. The final results could be available by 1991.

The likely impact of the HIPPARCOS project in many fields of astronomy and astrophysics has been described in the ESA Phase A Study report (ESA 1979) and in numerous papers published in recent years, for example in the proceedings of the Colloquium on European Satellite Astrometry held in Padua in 1978.

A major feature of the scientific case for this mission has been the determination of absolute trigonometric parallaxes and also annual proper motions, with average accuracy of ± 0".002, for many thousands of stars which are of astrophysical interest. However, in this paper I want to confine attention to the implications of the HIPPARCOS project on the stellar reference frame which is used in dynamical and geodetic astronomy.

E. M. Gaposchkin and B. Kołaczek (eds.), Reference Coordinate Systems for Earth Dynamics, 341–348.
Copyright © 1981 by D. Reidel Publishing Company.

Astrometric observations will also be made with the Space
Telescope;   these however will be confined to accurate relative
measurements in small fields and are therefore complementary to data
from HIPPARCOS.   The various possibilities have been discussed by
van Altena (1979).   For the present discussion we need only consider
the use of the fine guidance sensors (FGS) which, at any one time,
will give a field of about 70 square arc minutes available for
astrometry;   an accuracy of ± 0".002 is expected.

2.   HIPPARCOS OBSERVATIONS

The essential feature of HIPPARCOS is simultaneous observation
in two directions, of the order of 70° apart, by means of a beam
splitting mirror, known as the "complex mirror", which is placed in
front of the main telescope.   A schematic layout is shown in Fig 1.

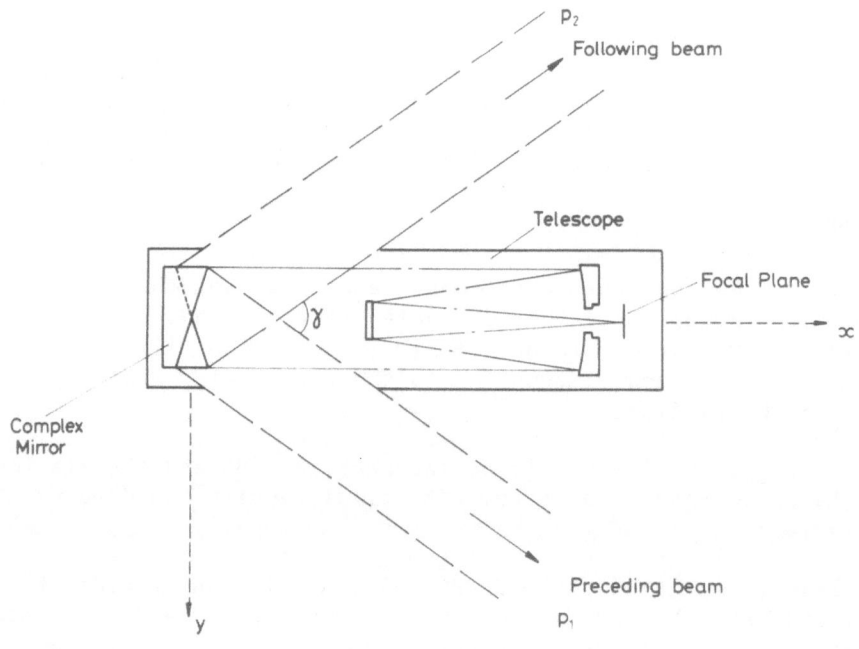

Fig 1

The sky is scanned by rotating the telescope at a rate of ten
revolutions per day about an axis which is nominally perpendicular to
the two beams.   As the instrument rotates, star images from either
beam will transit across the focal plane, and, if the rotation axis
remains fixed, both beams will scan along approximately the same great
circle.   The angle between the centres of the two beams is known as
the "basic angle";   its stability during a complete scan through 360°
is crucial to the success of the mission.

Only preselected stars will be observed.   The instantaneous
field of view (IFOV) is defined by an image dissector tube (IDT) which

can be switched rapidly from one location to another in the focal plane. The elementary observation consists of the detection of a signal from a star within the IFOV, which is modulated by a grid with parallel slits 0".275 wide and 1".1 apart, oriented perpendicular to the direction of image transit. The total area covered by the modulating grid is to be about 0º.9 x 0º.9, and the IFOV has a diameter of 30". Details of the observing technique are given in the Phase A Study report. For the purpose of the present discussion it is sufficient to regard an observation as being the time of transit across some fiducial mark, for example the central slit of the grid.

2.1 Geometry of the instrument

The plane parallel to the normals to the two faces of the complex mirror is known as the principal plane. The instrument can be represented by a triad of unit vectors $Z \equiv [x\ y\ z]$ such that z is normal to the principal plane, x is along the internal bisector of the angle between the normals and $y = z \times x$.

In Fig 1, p_i ($i = 1,2$) denotes the unit vector parallel to the centre of beam i. We can write

$$p_i = \cos\tfrac{1}{2}\gamma\ x - (-1)^i \sin\tfrac{1}{2}\gamma\ y \tag{1}$$

where γ is the basic angle. If the sense of rotation is positive about z, then p_1 is known as the "preceding" direction and p_2 the "following" direction.

Let r denote the unit vector in the direction to a star. This can be expressed symbolically as

$$r = Z \begin{bmatrix} x'r \\ y'r \\ z'r \end{bmatrix} \tag{2}$$

where the prime denotes scalar multiplication.

Now let ε be a small rotation of the triad Z. The corresponding small changes in the directions of x, y, z can be expressed as $\Delta Z = \varepsilon \times Z$ and a general small change in r can be written

$$\Delta r = \varepsilon \times r + Z \begin{bmatrix} \Delta(x'r) \\ \Delta(y'r) \\ \Delta(z'r) \end{bmatrix} \tag{3}$$

Multiplying this equation scalarly by $\underline{z}$ x $\underline{r}$ and rearranging we obtain

$$(\underline{x}'\underline{r})\Delta(\underline{y}'\underline{r})-(\underline{y}'\underline{r})\Delta(\underline{x}'\underline{r})+|\underline{z}x\underline{r}|^2\ \underline{z}'\underline{\varepsilon}\ =(\underline{z}x\underline{r})'\Delta\underline{r}-(\underline{z}'\underline{r})(\underline{z}x\underline{r})'\Delta\underline{z} \tag{4}$$

where we have put $\Delta\ \underline{z} = \underline{\varepsilon}\ x\ \underline{z}$

For a star in beam i we can express $\underline{r}$ in the form

$$\underline{r} = \cos\zeta\ \cos\eta\ \underline{p}_i + \cos\zeta\ \sin\eta\ \underline{z}\ x\ \underline{p}_i + \sin\zeta\ \underline{z} \tag{5}$$

where η,ζ are small angular displacements parallel and perpendicular to the principal plane; these correspond to coordinates of the reflected image in the focal plane of the main telescope. Combining (5) with (1) we derive

$$\underline{x}'\underline{r} = \cos\zeta\ \cos(\eta \pm \tfrac{1}{2}\gamma)$$

$$\underline{y}'\underline{r} = \cos\zeta\ \sin(\eta \pm \tfrac{1}{2}\gamma) \tag{6}$$

$$\underline{z}'\underline{r} = \sin\zeta$$

where the upper and lower signs refer to the preceding and following field respectively. Differentiating the first two of (6) and using (4) we obtain

$$\Delta\eta \pm \tfrac{1}{2}\Delta\gamma + \underline{z}'\underline{\varepsilon} - \tan\zeta\ \underline{s}'\Delta\underline{z} = \sec\zeta\ \underline{s}'\Delta\underline{r} \tag{7}$$

where $\underline{s}$ is the unit vector along $\underline{z}$ x $\underline{r}$. Equation (7) is the fundamental equation for HIPPARCOS data reduction. The displacement $\Delta\underline{z}$ of the $\underline{z}$ axis is multiplied by the small factor $\tan\zeta$ and therefore need not be known with extreme accuracy. The quantity $\underline{z}'\underline{\varepsilon}$ enters as an arbitrary zero point of the "abscissa" coordinate η; errors in $\underline{z}'\underline{\varepsilon}$ are virtually eliminated by making quasi-simultaneous observations of stars in both beams of the instrument. The correction $\Delta\gamma$ to the nominal basic angle can be obtained from each complete scan since all stars will be observed in both beams.

2.2   Attitude determination

The orientation of $\underset{\sim}{Z}$ can be determined from observations of transits of selected bright stars across an auxiliary grid near the edge of the field, which is known as the "star mapper". This grid, which is much coarser than the main modulating grid has some slits parallel to those of the main grid and some inclined at 45°; thus both coordinates can be measured.

Differentiating the third of equations (6) we obtain

$$\Delta\zeta = (\underline{z}'\Delta\underline{r} + \underline{r}'\Delta\underline{z})\sec\zeta \tag{8}$$

Provided that Δr is small, (as it certainly will be after a first iteration of the data), equations (7) and (8) can be used to determine Δz and $z'\epsilon$ with sufficient accuracy. This can be done in principle from simultaneous observations of a star in each beam, but in practice the change in attitude will be monitored continuously by readings of three, or possibly four, gyros.

2.3 Scanning geometry

Complete sky coverage will be achieved by changing the direction of the axis of rotation. Let c be the unit vector along this axis, and h the unit vector toward the mean position of the Sun in the ecliptic. If k is the ecliptic pole we have

$$c = \sin\xi \, \sin\nu \, k + \sin\xi \, \cos\nu \, k \times h + \cos\xi \, h \qquad (9)$$

where ξ, the "revolving angle", is the angle between c and h, and ν is the phase of revolution. The provisional scheme is for ξ to remain constant over extended periods and the rate of revolution of c about h will be chosen so that the locus of c is a series of overlapping loops.

When c crosses the ecliptic, $c'k = 0$ and $\nu = n\pi$ (n integer). In this case $h \times c = (-1)^n \sin\xi \, k$. Now if the longitude of h is $2\pi (t-t_0)$ where t is the time measured in years, and c revolves about h K times per year, the longitude of c at the nth crossing of the ecliptic is given by

$$\lambda_n = \frac{n\pi}{K} - 2\pi t_0 + (-1)^n \xi \qquad (10)$$

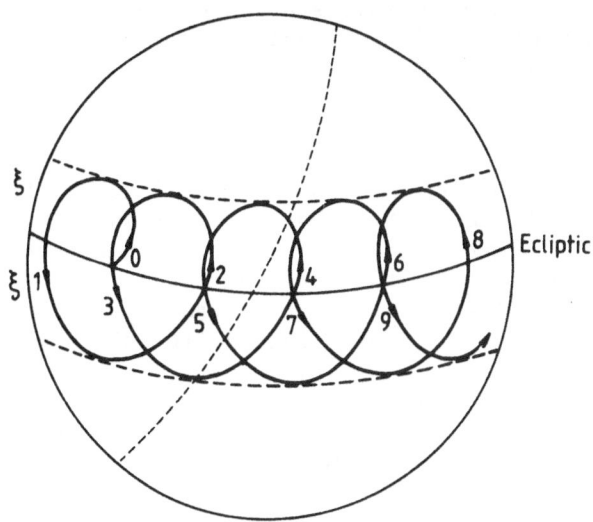

Fig 2

As is shown in Fig 2, a sufficient condition for successive loops to overlap is $\lambda_{2m} \geqslant \lambda_{2m+3}$ which is equivalent to

$$K \geqslant \frac{3\pi}{2\xi} \tag{11}$$

The revolving angle is also the angle of incidence of sunlight on to the solar panels, and therefore cannot be too large. On the other hand, stellar parallactic displacement is proportional to sin ξ, and should be maximised. The constants provisionally chosen are $\xi = 36^{\circ}$, $K = 7.5$. With this pattern, it is seen from Fig 2 that any great circle will intersect the locus of $\underline{c}$ in at least six points so any point in the sky will be scanned at least six times per year.

2.4 Data reduction

The initial stages in processing the data will be to determine the attitude parameters as functions of the time, from the star mapper and gyro readings, and then to derive for each star, the angular coordinate along the scanning direction from the IDT photon counts obtained during a single great circle scan.

The procedure proposed in the Phase A Study report for deriving the final catalogue of astrometric parameters follows a so-called "three step" method. The first step is to combine all data from a "set" of five consecutive scans to provide abscissae for all stars projected on to a common reference great circle; these abscissae would be referred to an arbitrary zero point. In the second step, all abscissae for some 500-1000 primary stars obtained throughout the mission, will be analysed to determine the zero points for each "set". The third and final step will be to apply the zero point corrections to the abscissae for all programme stars and thence calculate the astrometric parameters. It is envisaged that several iterations of this procedure will be needed.

2.5 Observational programme

Taking into account the mission duration (assumed to be 2.5 years) and the need to have sufficient integration time on each star, the main features of a possible observing programme have been given in the Phase A Study report. In this programme the total number of stars would be 10^5; all 6×10^4 stars brighter than $m_{pg} = 9$ would be included and the remainder would be selected from stars with $9 \leqslant m_{pg} \leqslant 13$.

The mean errors of the final astrometric parameters will vary with magnitude and ecliptic latitude. The all sky average error in position for a star brighter than $m_{pg} = 10$ should not exceed $\pm0''0022$ in ecliptic longitude and $\pm0''0010$ in latitude. At $m_{pg} = 13$ the errors will increase by a factor of between two and three. The errors in annual proper motion components will be approximately fifty per cent larger than in the corresponding position coordinates.

3. ABSOLUTE REFERENCE FRAME

The final output from HIPPARCOS will be a fully coherent catalogue of relative positions and proper motions, and absolute parallaxes.

The method of measurement, with the modulating grid, precludes the observation of objects with non-stellar images; it will therefore not be possible to include solar system objects, except perhaps for a few bright minor planets. But it is not expected that any of these will be observed sufficiently frequently during the 2.5 year mission, to derive good orbits or to define a dynamical reference frame independent of ground based work. It will therefore be necessary to relate the HIPPARCOS system empirically to the fundamental system, FK5.

Any residual rotation of the HIPPARCOS proper motion system could also be determined from observations of extra galactic objects. Unfortunately only one quasar, 3C 273, is likely to be bright enough for direct observation. It is just possible that a few bright Seyfert galaxy nuclei could also be observed, but these would be very close to the faint limit for HIPPARCOS; nevertheless they should be observed at least once in order to assess their usefulness.

During the last decade, astrometric observations of extra-galactic sources in both radio and optical wavebands have been used for referring the position system of the radio interferometric observations to the optical fundamental system FK4.

Because many radio sources have significant angular structure even though their optical counterparts appear to be quasi-stellar, it is necessary to select for this purpose only those sources which have small structure at radio frequencies. Accordingly at the IAU Colloquium No 48 on "Modern Astrometry" which was hald in Vienna in 1978, Commission 24 set up a working group of radio and optical astrometrists which was charged with the task of drawing up a list of suitable "bench mark" sources. This working group, under the chairmanship of K. J. Johnston, reporting to the Commission at Montreal, presented a list of 81 primary candidates and a further list of 32 possible candidates; nearly all these sources are quasars. Since these sources are all optically too faint for HIPPARCOS it is necessary that at least one bright star, close to each source should be included in the HIPPARCOS programme. Measurement of the angular separation between each source and its "associate" star, and its rate of change, would be sufficient to refer radio positions to the HIPPARCOS system and to determine the rotation of the HIPPARCOS proper motions relative to an extra-galactic frame.

Initially this might be feasible from ground based photographic astrometry since there is already a considerable body of optical astrometric data on many of the sources extending over a decade or more. A much better alternative would be to use the fine guidance

sensors of the Space Telescope. For this to be feasible, the angular
separation between a source and its associate star must be less than
about 15'. From examination of a sample of northern radio source
fields which have been observed in recent years at the Royal Greenwich
Observatory, it is found that all have associate stars bright enough
to be observed with HIPPARCOS and close enough to be measured with the
Space Telescope; eighty per cent have AGK3 stars satisfying both
these criteria. After one year a single relative displacement between
quasar and star could be measured with the Space Telescope with a
mean error of about ±"004; combining this with ±"002 for the HIPPARCOS
proper motions and averaging over fifty well distributed fields, the
mean error of one component of the rotation of the system would be
only about ±"001.

4.   ACKNOWLEDGEMENTS

This description of the HIPPARCOS mission is based on the work
of the ESA HIPPARCOS Science Team as given in the Phase A Study report
and also the ESTEC technical study (ESA 1979b). Unpublished working
papers by Dr L Lindegren and Dr E Høg have been particularly useful.

I am also grateful to Mr E D Clements who has identified
possible associate stars in many of the radio source fields and also
examined photographs of several bright Seyfert galaxies obtained with
the 26 inch refractor at Herstmonceux.

1980 July 25

References

European Space Agency 1979a   HIPPARCOS.   Space Astrometry.   Report on
                                          Phase A Study (SCl(79)10)

     "   "   "   1979b   HIPPARCOS.   Spacecraft design and
                                          development (PF 616)

van Altena, W. F.   1979   IAU Colloquium No 48.   "Modern
                                          Astrometry" (eds. F. V. Prochazka and
                                          R. H. Tucker, Vienna), pp.561-571

DYNAMICAL EQUINOX AND ANALYTICAL THEORY OF THE SUN

Krasinsky G.A., Sveshnikov M.L.
Institute of Theoretical Astronomy
Academy of Sciences
USSR, Leningrad

ABSTRACT*

A concept of dynamical equinox and its relation to the analytical form of the adopted theory of the Sun is discussed. Connection between the FK4 equinox and the dynamical equinox is determined by comparing two analytical theories of the Sun (the adopted Newcomb's theory and a new one (AT-1) constructed at the Instituted of Theoretical Astronomy) with solar meridian observations made at US Naval Observatory (1911-1971 yr.). Corrections $\Delta\alpha$ to the FK4 right ascensions, $\Delta\delta_O$ to declinations and $\Delta\varepsilon$ to the angle between the equator and the ecliptic are:

$$\Delta\alpha = \begin{cases} 0^S.024 \pm 0^S.003 - (0^S.003 \pm 0^S.023)*(T-19.40) & \text{Newcomb} \\[1em] 0^S.032 \pm 0^S.003 - (0^S.012 \pm 0^S.022)*(T-19.40) & \text{AT-1} \end{cases}$$

$$\Delta\delta_O = \begin{cases} 0".25 \pm 0".02 + (0".67 \pm 0".07)*(T-19.40) & \text{Newcomb} \\[1em] 0".24 \pm 0".02 + (0".68 \pm 0".10)*(T-19.40) & \text{AT-1} \end{cases}$$

$$\Delta\varepsilon = \begin{cases} 0".26 \pm 0".01 + (0".12 \pm 0".09)*(T-19.40) & \text{Newcomb} \\[1em] 0".30 \pm 0".01 - (0".17 \pm 0".08)*(T-19-40) & \text{AT-1} \end{cases}$$

Secular variations in $\Delta\alpha$ and $\Delta\varepsilon$ may be explained by improvement of observational conditions after reconstruction of the telescope pavilion in the forties.

---

*The full paper will be published in Celestial Mechanics.

E. M. Gaposchkin and B. Kołaczek (eds.), Reference Coordinate Systems for Earth Dynamics, 349.

DEVELOPMENT OF A RADIO-ASTROMETRIC CATALOG BY MEANS OF VERY LONG BASELINE INTERFEROMETRY OBSERVATIONS*

J. L. Fanselow, O. J. Sovers, J. B. Thomas, F. R. Bletzacker, T. J. Kearns, E. J. Cohen, G. H. Purcell, Jr., D. H. Rogstad, L. J. Skjerve, L. E. Young

Jet Propulsion Laboratory, California Institute of Technology, Pasadena, CA 91103.

ABSTRACT

The Jet Propulsion Laboratory of the California Institute of Technology has been developing a radio-astrometric catalogue for use in the application of radio interferometry to interplanetary navigation and geodesy. The catalogue consists of approximately 100 compact extragalactic radio sources whose relative positions have formal uncertainties of the order of $0\overset{''}{.}01$. The sources cover nearly all of the celestial sphere above $-40°$ declination. By using the optical counterparts of many of these radio sources, we have tied this radio reference frame to the FK4 optical system with a global accuracy of approximately $0\overset{''}{.}1$. This paper describes the status of this work.

INTRODUCTION

Development of a radio-astrometric catalog is an essential element in the application of radio interferometry to both spacecraft navigation and geodesy. For this reason, the Jet Propulsion Laboratory of the California Institute of Technology has been developing a catalog of precise positions for compact extragalactic radio sources. Our goal has been a catalog of approximately 100 sources, uniformly distributed over the celestial sphere. In order to support the navigation of the Voyager mission, an accuracy in these positions of approximately $0\overset{''}{.}01$ is required in 1980. Further, it is required that this catalog have negligible ($0\overset{''}{.}1$) mean offset in right ascension relative to the FK4 system, since all interplanetary navigation to date has been based on that system.

*This paper presents the results of one phase of research carried out at the Jet Propulsion Laboratory, California Institute of Technology, under Contract No. NAS 7-100, sponsored by the National Aeronautics and Space Administration.

E. M. Gaposchkin and B. Kołaczek (eds.), Reference Coordinate Systems for Earth Dynamics, 351–357.

INSTRUMENTATION AND DATA REDUCTION

The observations on which the results of this paper are based were obtained over the period from 1971 to March 1980. Throughout the nine years of this development, the interferometry instrumentation has been steadily improved, so that the nine most recent measurements employ dual frequency observations at S- and X-band (13 and 3.6 cm wavelengths, respectively), hydrogen maser frequency standards, and a 4 Mbs data acquisition system. The data were obtained during observing sessions of 8 to 24 hours in duration, utilizing antennas of the Deep Space Network. These DSN facilities provided an 8400 km baseline from Califonia to Spain, and a 10,600 km baseline from Califonia to Australia. A total of 44 such observing sessions are included in the results presented here. Altogether the observations include 3941 independent measurements, of which 1844 are measurements of delay and 2097 are of delay rate.

For convenience in processing this large amount of data, we have separated the data into two time sequences, with each sequence containing about half the data. Sequence #1 contains all data obtained prior to January 1, 1979, while sequence #2 is all data collected after that date. In the last step of this processing, we fit all observations of a given sequence with an analytic model, using a conventional least squares technique to adjust selected parameters of that model. In sequence #1 we solved for the values of 431 parameters, including 136 parameters for source positions. In sequence #2, 148 of the 273 solve-for parameters pertained to source positions. The list of sources observed in sequence #2 was not identical to that observed in sequence #1, as we were attempting to expand the number of sources in our catalog. However, in sequence #2 we have reobserved 33 of the sources from sequence #1 so as to provide overlap between the two parts of the catalog. In each sequence, the sources observed were fairly well distributed over the entire celestial sphere.

The delay model used in processing consisted of geometric, clock, ionospheric and tropospheric components. In the geometric component, the adjusted parameters included baseline, source position, UT1 and polar motion. Precession, nutation, solid earth tides and gravitational bending were all modeled but no associated parameters were adjusted. One of the major deficiencies in our model was the use of the standard nutation series, which has known errors as large as approximately 0.''02 in magnitude. We also had to "patch in" an improved precession rate in order to fit the data. Both of these deficiencies will be corrected in the near future as we incorporate better Earth models. With regard to the clock model, we typically had to assign only one epoch offset and one rate offset to each baseline for each observing session. On occasion, however, we had to introduce discontinuities in epoch and rate within a session. In the case of the ionosphere, a simple diurnal model was used whenever we were observing at only one radio frequency. For those observations involving dual frequencies, the effect of the ionosphere, as well as all other charged particle contributions to the measurements, were removed by exploiting the dispersive character of a plasma at these frequencies. All of the data in sequence #2 were obtained on the basis of this dual frequency technique. For the troposphere, a monthly-mean model was used as a priori, but the zenith troposphere delay was adjusted for each station under the constraint of that a priori.

One of our goals is to provide a catalog with the smallest possible rotation in right ascension relative to the FK4 system. Thus, we employed the following two-step process in the final reduction of our data:

(1)   A preliminary multiparameter adjustment was performed. In this adjustment the right ascensions of those sources that had suitably measured optical counterparts were statistically constrained to the FK4 system on the basis of the apriori errors in the right ascensions of these counterparts. This procedure is mathematically equivalent to adding to our observations a set of measurements of right ascension specific to the subset of sources with optical counterparts. This parameter adjustment step resulted in an uncertainty in right ascension alignment given approximately by:

$$\sigma_a \approx \left(\sum_{i=1}^{N} \frac{1}{\sigma_1^2} \right)^{-\frac{1}{2}}$$

where σ_i is the uncertainty in right ascension of the ith optical counterpart, and where the summation is over the N optical counterparts.

(2)   A final multiparameter adjustment was then made. In this step all constraints on the source positions were removed except for the constraint on the right ascension of a "mean reference" source. The reference source was tightly ($0.\!''0000002$) constrained to the right ascension obtained for that particular source in the previous estimation step. The selection of this source was relatively arbitrary, although it appears that a source at about 30° declination was best for the particular baselines involved in these experiments.

This procedure produces a global minimization of the right ascension offset between the FK4 system and the radio reference system. Currently, we believe the accuracy of this alignment is approximately 0."1. Another advantage of this procedure is that the intrinsic precision (i.e. relative position error) is directly printed out in the final fit as the right ascension error of each source. At this point in the analysis of the data, we have executed this procedure only for sequence #1 of the data, and have chosen NRAO 140 as the "mean reference" source. In the subsequent processing of the data in sequence #2, we adopted this reference position without resorting to another preliminary fit. However, for the final analysis of this data, the procedure outlined above eventually will be performed for the entire data set as a single unit.

RESULTS

The position catalog we have obtained has been designated JPL 1980-1 and is listed in Table I. In presenting this catalog, we have excluded all sources that were observed fewer than 3 times. One source was observed 67 times, though more typically each source was observed 10-40 times. The source positions are given in 1950.0 solar-system-barycentric coordinates while the position errors are the formal uncertainties obtained by adjusting chi-square for the fit residuals to 1.0. For convenience, we have listed the "elliptical aberration" terms that must be added to our results to obtain the coordinates conventionally used in optical catalogues. In all, 109 sources are listed, with most of the positions having formal uncertainties less than 0."01. One check on the quality of the data is to compare common source positions between the two sequences of data. When the 33 common sources were compared, almost all of the differences were less than about 0."03, with the larger differences resulting primarily from inadequate observations in one of the two sequences. As a test of the formal uncertainties, these position differences were compared with the errors obtained from the formal uncertainties. We found that an additional error of about 0."01 had to be root-sum-squared with the formal uncertainties in order to make the total errors statistically consistent with the position differences.

SUMMARY AND PLANS FOR THE FUTURE

Radio-astrometric positions have been obtained for 109 extragalatic radio sources. The formal uncertainties in these positions fall primarily in the range 0."003-0."02 while the accuracy is presently estimated to be approximately 0."01 - 0."02. This work is part of an ongoing effort to develop an astrometric catalog of extragalactic radio sources distributed over the entire celestial sphere with positional accuracies of 0."01 or better. Improvements in the hardware and modeling scheduled for 1981 should allow us to improve the accuracy of the current catalog to the level of 0."003-0."005 within the next year or two.

T A B L E 1

JPL SOURCE POSITION CATALOG 1980-1 (1950.0 SOLAR-SYSTEM-BARYCENTRIC COORDINATES)

SOURCE	\multicolumn{8}{c}{1950.0 SSB POSITION}	\multicolumn{2}{c}{E TERMS}
	\multicolumn{3}{c}{RIGHT ASCENSION}	ERROR	\multicolumn{3}{c}{DECLINATION}	ERROR	R.A.	DECL.
	H	M	SEC	SEC	D	M	ARC SEC	ARC SEC	SEC	ARC SEC
P 0008-264	00	08	28.89062	0.00074	-26	29	14.7068	0.0098	-0.00398	-0.1761
P 0104-408	01	04	27.57593	0.00053	-40	50	21.4313	0.0066	0.00263	-0.2442
P 0106+01	01	06	04.51802	0.00028	01	19	01.0517	0.0051	0.00215	-0.0207
P 0111+021	01	11	08.57066	0.00204	02	06	24.7017	0.0313	0.00265	-0.0161
P 0113-118	01	13	43.21948	0.00074	-11	52	04.5578	0.0118	0.00297	-0.0977
DA 55	01	33	55.09519	0.00028	47	36	12.5292	0.0028	0.00724	0.2273
P 0202+14	02	02	07.38976	0.00027	14	59	50.7876	0.0056	0.00784	0.0558
DW 0224+67	02	24	41.13548	0.00050	67	07	39.3413	0.0031	0.02483	0.2741
CTD 20	02	34	55.57744	0.00029	28	35	11.1996	0.0050	0.01203	0.1198
GC 0235+16	02	35	02.62012	0.00027	16	24	03.8667	0.0066	0.01110	0.0579
P 0237-23	02	37	52.77814	0.00115	-23	22	06.2887	0.0157	0.01179	-0.1455
QE 400	03	00	10.08884	0.00060	47	04	33.4153	0.0055	0.01868	0.1882
3C 84	03	16	29.54633	0.00033	41	19	51.6859	0.0040	0.01869	0.1564
P 0332-403	03	32	25.21858	0.00091	-40	18	23.9515	0.0099	0.01999	-0.1861
NRAO 140	03	33	22.38543	0.00000	32	08	36.4807	0.0036	0.01809	0.1104
CTA 26	03	36	58.93966	0.00034	-01	56	16.9227	0.0068	0.01559	-0.0370
NRAO 150	03	55	45.22857	0.00058	50	49	20.0966	0.0038	0.02673	0.1598
P 0402-362	04	02	02.57565	0.00094	-36	13	11.7743	0.0107	0.02144	-0.1545
GC 0406+12	04	06	35.45953	0.00029	12	09	49.2242	0.0084	0.01799	0.0178
VRO 41.04.01	04	20	27.90873	0.00035	41	43	07.8786	0.0053	0.02470	0.1123
P 0420-01	04	20	43.52420	0.00059	-01	27	28.7187	0.0085	0.01846	-0.0337
3C 120	04	30	31.58567	0.00064	05	14	59.5673	0.0091	0.01908	-0.0112
P 0434-188	04	34	48.94761	0.00091	-18	50	48.1149	0.0126	0.02033	-0.0862
P 0438-43	04	38	16.14141	0.00092	-43	38	53.4625	0.0098	0.02687	-0.1437
NRAO 190	04	40	05.27323	0.00071	-00	23	20.5961	0.0126	0.01951	-0.0298
P 0451-28	04	51	15.10887	0.00081	-28	12	29.3160	0.0100	0.02276	-0.1017
P 0528+134	05	28	06.73759	0.00027	13	29	42.6034	0.0053	0.02214	0.0017
P 0537-441	05	37	21.05095	0.00063	-44	06	44.6034	0.0065	0.03038	-0.0891
DA 193	05	52	21.73457	0.00058	39	48	21.9227	0.0048	0.02889	-0.0277
P 0605-08	06	05	36.01023	0.00164	-08	34	20.2896	0.0216	0.02271	-0.0368
P 0607-15	06	07	25.96313	0.00052	-15	42	03.2793	0.0079	0.02336	-0.0424
DW 0723-00	07	23	17.81774	0.00042	-00	48	55.0426	0.0076	0.02246	-0.0277
P 0727-11	07	27	58.07964	0.00074	-11	34	52.5369	0.0106	0.02284	-0.0150
P 0735+17	07	35	14.10394	0.00061	17	49	09.3584	0.0082	0.02335	-0.0502
OI 363	07	38	00.15143	0.00029	31	19	02.1767	0.0087	0.02594	-0.0655
DW 0742+10	07	42	44.43300	0.00043	10	18	32.7510	0.0043	0.02241	-0.0435
P 0748+126	07	48	05.03818	0.00035	12	38	45.5747	0.0061	0.02246	-0.0484
OJ 425	08	14	51.63812	0.00057	42	32	07.9250	0.0043	0.02854	-0.1099

T A B L E 1 - CONT'D

JPL SOURCE POSITION CATALOG 1980-1 (1950.0 SOLAR-SYSTEM-BARYCENTRIC COORDINATES)

SOURCE	\multicolumn{4}{}{1950.0 SSB POSITION}					\multicolumn{2}{}{E TERMS}	
	\multicolumn{3}{}{RIGHT ASCENSION}	ERROR	\multicolumn{3}{}{DECLINATION}	ERROR	R.A.	DECL.
	H	M	SEC	SEC	D	M	ARC SEC	ARC SEC	SEC	ARC SEC
P 0823+033	08	23	13.52100	0.00038	03	19	15.5035	0.0065	0.02073	-0.0368
B2 0827+24	08	27	54.37577	0.00051	24	21	07.8035	0.0068	0.02250	-0.0875
4C 55.16	08	31	04.33624	0.00147	55	44	41.5632	0.0287	0.03617	-0.1428
4C 71.07	08	36	21.48266	0.00066	71	04	22.6662	0.0032	0.06203	-0.1609
OJ 287	08	51	57.23118	0.00048	20	17	58.5891	0.0046	0.02063	-0.0894
OJ 499	08	59	39.95317	0.00093	47	02	57.0712	0.0139	0.02778	-0.1586
P 0859-14	08	59	54.93147	0.00100	-14	03	38.8076	0.0143	0.01950	-0.0186
4C 39.25	09	23	55.29490	0.00024	39	15	23.8158	0.0024	0.02258	-0.1608
AO 0952+17	09	52	11.78721	0.00092	17	57	44.7993	0.0149	0.01635	-0.1042
GC 1004+14	10	04	59.77021	0.00068	14	11	11.0960	0.0092	0.01506	-0.0920
GC 1034-293	10	34	55.81427	0.00065	-29	18	26.9805	0.0084	0.01399	-0.1164
OL 064.5	10	38	10.62252	0.00041	06	25	58.6886	0.0575	0.01196	-0.0610
3C 245	10	40	05.98969	0.00056	12	19	15.1695	0.0077	0.01204	-0.0904
P 1055+01	10	55	55.30470	0.00033	01	50	03.6896	0.0057	0.01039	-0.0383
P 1104-445	11	04	50.35912	0.00091	-44	32	53.0777	0.0093	0.01346	-0.1971
GC 1111+14	11	11	21.30050	0.00069	14	58	47.8907	0.0085	0.00932	-0.1087
P 1116+12	11	16	20.76764	0.00050	12	51	06.9046	0.0118	0.00876	-0.0983
P 1123+26	11	23	14.86160	0.00034	26	26	50.2673	0.0042	0.00883	-0.1683
P 1127-14	11	27	35.66118	0.00037	-14	32	54.3755	0.0057	0.00775	-0.0534
GC 1128+38	11	28	12.50587	0.00074	38	31	51.9262	0.0252	0.00951	-0.2236
P 1144-379	11	44	30.85450	0.00046	-37	55	30.7109	0.0058	0.00747	-0.1804
P 1148-00	11	48	10.12252	0.00065	-00	07	13.0260	0.0089	0.00554	-0.0279
P 1222+037	12	22	19.09700	0.00042	03	47	27.2259	0.0070	0.00220	-0.0510
3C 273	12	26	33.24591	0.00030	02	19	43.4662	0.0045	0.00178	-0.0424
3C 274	12	28	17.56987	0.00054	12	40	01.9607	0.0072	0.00164	-0.1026
P 1244-255	12	44	06.71495	0.00056	-25	31	26.6764	0.0076	0.00004	-0.1215
3C 279	12	53	35.83516	0.00080	-05	31	07.8978	0.0119	-0.00091	-0.0044
B2 1308+32	13	08	07.56619	0.00076	32	36	40.5509	0.0037	-0.00279	-0.2073
OP -322	13	13	29.81113	0.00065	-33	23	09.7439	0.0087	-0.00343	-0.1627
DW 1335-12	13	34	59.81113	0.00117	-12	42	09.6906	0.0091	-0.00511	-0.0455
GC 1342+663	13	42	41.06705	0.00132	66	21	13.5572	0.0054	-0.01428	-0.3145
P 1349-439	13	49	59.25598	0.00118	-43	57	54.2530	0.0127	-0.00892	0.2071
P 1354+19	13	54	42.09628	0.00035	19	33	44.1912	0.0051	-0.00730	-0.1360
GC 1418+54	14	18	06.21090	0.00053	54	36	58.4161	0.0056	-0.01564	-0.2722
OQ-151	14	30	10.65791	0.00089	-17	48	24.2834	0.0131	-0.01065	-0.0664
OR 103	15	01	08.17377	0.00030	-10	41	17.9005	0.0048	-0.01310	-0.0804
P 1510-08	15	10	08.91683	0.00072	-08	54	47.5743	0.0103	-0.01369	0.0144
P 1519-273	15	19	37.25585	0.00057	-27	19	30.3141	0.0079	-0.01607	0.0969

T A B L E   1 - CONT'D

JPL SOURCE POSITION CATALOG 1980-1 (1950.0 SOLAR-SYSTEM—BARYCENTRIC COORDINATES)

| SOURCE | 1950.0 SSB POSITION | | | | | | E TERMS | |
| | RIGHT ASCENSION | | ERROR | DECLINATION | | | ERROR | R.A. | DECL. |
	H	M	SEC	SEC	D	M	ARC SEC	ARC SEC	SEC	ARC SEC
DW 1555+00	15	55	17.71303	0.00061	00	06	43.6138	0.0087	-0.01686	-0.0290
DA 406	16	11	47.94478	0.00057	34	20	20.0350	0.0060	-0.02170	-0.1427
GC 1633+38	16	33	30.65680	0.00027	38	14	10.2645	0.0033	-0.02440	-0.1368
NRAO 512	16	38	48.20296	0.00026	39	52	30.2680	0.0032	-0.02534	-0.1361
3C 345	16	41	17.63965	0.00024	39	54	10.9820	0.0026	-0.02551	-0.1341
DW 1656+05	16	56	05.64376	0.00132	05	19	47.1145	0.0208	-0.02037	-0.0429
GC 1717+17	17	17	00.35082	0.00043	17	48	08.6088	0.0068	-0.02221	-0.0661
NRAO 530	17	30	13.55741	0.00045	-13	02	45.8174	0.0077	-0.02217	-0.0033
DT 465	17	38	36.35562	0.00118	-47	39	28.8671	0.0083	-0.03244	-0.0908
P 1741-038	17	41	20.64125	0.00067	-03	48	48.8811	0.0100	-0.02197	-0.0223
1749+701	17	49	03.47807	0.00087	70	06	39.7254	0.0100	-0.06503	-0.0866
3C 371	18	07	18.63145	0.00043	69	48	57.1646	0.0045	-0.06517	-0.0616
P 1821+10	18	21	41.68538	0.00033	10	42	43.9245	0.0103	-0.02307	-0.0344
OV-236	19	21	42.25872	0.00121	-29	20	26.3558	0.0132	-0.02579	-0.0520
P 1933-400	19	33	51.14815	0.00099	-40	04	47.4408	0.0107	-0.02909	-0.0689
OV-198	19	58	04.63144	0.00065	-17	57	16.9631	0.0095	-0.02272	-0.0604
OW-637	20	21	13.36085	0.00104	61	27	17.9739	0.0071	-0.04349	-0.1094
P 2029+121	20	29	32.70544	0.00038	12	09	28.6943	0.0067	-0.02090	-0.0039
OW 551	20	30	29.14186	0.00057	54	44	49.0654	0.0056	-0.03532	-0.1081
B2 2113+29B	21	13	20.60309	0.00029	29	21	06.5346	0.0046	-0.02081	-0.0765
P 2134+004	21	34	05.22698	0.00052	00	28	24.9999	0.0084	-0.01682	-0.0267
P 2145+06	21	45	36.09733	0.00028	06	43	40.8124	0.0049	-0.01614	-0.0001
OX 082	21	49	07.71970	0.00036	05	38	06.7847	0.0069	-0.01585	-0.0042
OX-192	21	55	23.25930	0.00081	-15	15	30.1166	0.0130	-0.01588	-0.0941
VRO 42.22.01	22	00	39.38807	0.00026	42	02	08.3514	0.0029	-0.02010	-0.1516
CTA 102	22	30	07.82046	0.00079	11	28	22.6653	0.0042	-0.01286	-0.0286
GC 2234+28	22	34	01.74630	0.00027	28	13	23.0082	0.0042	-0.01393	-0.1110
OY-172.6	22	43	39.80604	0.00036	-12	22	40.3085	0.0064	-0.01173	-0.0912
P 2245-328	22	45	51.51834	0.00056	-32	51	44.3630	0.0072	-0.01342	-0.1852
3C 454.3	22	51	29.53403	0.00047	15	52	54.1807	0.0061	-0.01121	-0.0549
GC 2253+41	22	53	19.86196	0.00036	41	46	50.9891	0.0065	-0.01424	-0.1802
P 2320-035	23	20	57.53518	0.00050	-03	33	33.6885	0.0091	-0.00813	-0.0483
P 2345-16	23	45	27.69119	0.00068	-16	47	52.6153	0.0102	-0.00606	-0.1229

SYSTEMATIC DIFFERENCES BETWEEN RADIO ASTROMETRIC SURVEYS

H.G. Walter
Astronomisches Rechen-Institut
Heidelberg, Fed.Rep. of Germany

ABSTRACT

A radio astronomical reference frame has been established from four interferometric position surveys of extragalactic objects. The resulting systematic differences between surveys were studied with different weighting schemes and averaged systematic corrections were derived.

1. INTRODUCTION

Four radio astrometric surveys of extragalactic objects accurate to a few hundredths of seconds of arc in both coordinates formed the basis for a reference frame with respect to which systematic differences of the independent surveys have been determined. The four lists of positions are due to Clark et al. (1976), Elsmore and Ryle (1976), Wade and Johnston (1977), and Fanselow et al. (1979); below they are called, in turn, C_1, C_2, C_3, C_4. Before solving simultaneously for corrections of the source positions and the zero points, allowance was made for the zero points adopted by the different authors (Elsmore, 1979). On adding 2 ms, 8 ms, 3 ms and 6 ms to the positions in C_1, C_2, C_3 and C_4, respectively, the right ascensions of the four surveys were adjusted to RA = $12^h26^m33^s250$ as recently derived by de Vegt and Gehlich (1980) for 3C 273 B.

2. RESULTS AND DISCUSSION

Having made allowance for the different zero points, two weighting schemes (w_{ik}) were applied when constraining the sum of systematic differences to zero which is common astrometric practice in setting up the reference frame of a general catalogue, i.e.

$$\sum_i w_{ik} \, \Delta C_{ik} = 0, \quad k = 1, \ldots, N,$$

E. M. Gaposchkin and B. Kołaczek (eds.), Reference Coordinate Systems for Earth Dynamics, 359–362.
Copyright © 1981 by D. Reidel Publishing Company.

where i is associated with the surveys and k with the observations comprised within each survey. In Model I, uniform weights are chosen while the weighting factors in Model II are related to the standard deviations of the observations quoted in the surveys, thus taking more directly account of the influences of baseline geometry and source positions on the precision of observations. Table 1 shows the averaged systematic differences of each survey.

Table 1. Systematic differences and errors obtained by averaging the differences (survey position – reference frame position).

Survey	Systematic differences				Number of contributing objects
	Model I		Model II		
	$\Delta\alpha\cos\delta$	$\Delta\delta$	$\Delta\alpha\cos\delta$	$\Delta\delta$	
	[10^{-3} arcsec]		[10^{-3} arcsec]		
C_1	−10 ± 6	1 ±4	−2 ±4	−1 ±7	13
C_2	−4 ±8	14 ±9	−13 ±12	6 ±12	15
C_3	8 ±7	−19 ±6	0 ±9	−35 ±7	30
C_4	7 ±6	18 ±6	1 ±1	2 ±1	28

The individual systematic differences resulting from Model II are depicted in Figs. 1 and 2 versus RA and Dec for the more abundant surveys C_3 and C_4. Of course, C_4 dominates the results owing to observation accuracies superior to those of C_3. Nevertheless, the plots illustrate common features in as much as the residuals versus RA are notably noisy in the region $16^h < RA < 21^h$. It is not obvious how to explain the fluctuations. On the other hand, the residuals versus Dec diminish with increasing declination apart from a few exceptions. This behaviour may be interpreted as being due to the reduced sensitivity of radio interferometric measurements at declinations near zero. Another issue of the comparison is the discovery of a systematic difference in declination of about 0."03 between C_3 and C_4.

On the whole, the adoption of a zero point in RA common to all surveys and the introduction of an observation-dependent weighting scheme (Model II) in the conditional equations for the systematic differences led to a significant reduction of the noise in the residuals. This is underlined by comparing with an analogous study (Walter, 1980) which omitted introducing a common zero point and which omitted discriminating between weights of the surveys. It produced graphs corresponding to

Figs. 1 and 2 of basically the same pattern but with larger amplitudes.

The source coordinates ensuing from the different models vary by about ±0."03 and stay comfortably within the 3σ limits of the positions defining the reference frame.

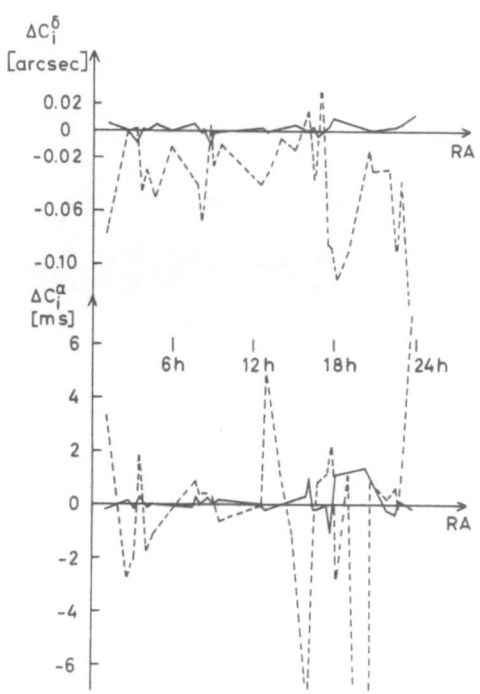
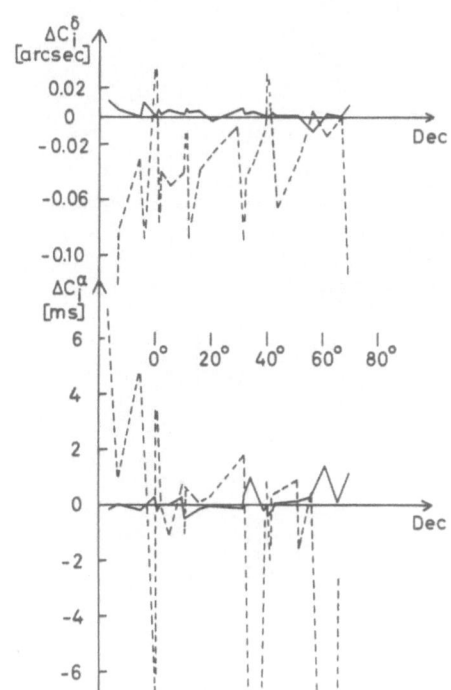

Figure 1. Systematic differences versus right ascension between C_3, C_4 and the reference frame.
C_3: ---- C_4: ———

Figure 2. Systematic differences versus declination between C_3, C_4 and the reference frame.
C_3: ---- C_4 ———

REFERENCES

Clark,T.A.,Hutton,L.K;Marandino,G.E.,Counselman,C.C.,Robertson,D.S.,
    Shapiro,I.I.,Wittels,J.J.,Hinteregger,H.F.,Knight,C.A.,
    Rogers,A.E.E.,Whitney,A.R.,Niell,A.E.,Rönnäng,B.O.,Rydbeck,O.E.H.:
    1976,Astron.J.81,pp.599-603.
de Vegt,Chr.,and Gehlich,U.K.: 1980,Mitt.Astron.Gesellschaft,in press.
Elsmore,B.: 1979,private communication.
Elsmore,B.,and Ryle,M.: 1976,Monthly Notices Roy.Astron.Soc. 174,pp.
    411-423.
Fanselow,J.L.,Thomas,J.B.,Cohen,E.J.,Purcell,Jr.,G.H.,Rogstad,D.H.,
    Sovers,O.J.,Skjerve,L.J.: 1979,"Measurements of Earth Orientation

    Using Very Long Baseline Interferometry",paper presented at the
    IAU General Assembly,Montreal,Canada.
Wade,C.M.,and Johnston,K.J.: 1977,Astron.J.82,pp.791-795.
Walter,H.G.: 1980,Astron.Astrophys., in press.

CONTRIBUTION DES ASTROLABES AU RACCORDEMENT DES SYSTEMES DE REFERENCE
"OPTIQUE" ET "RADIO".

Suzanne Débarbat
Observatoire de Paris, Paris, France

ABSTRACT

Twenty-two radio sources, being bright objects of the Galaxy are
observable with astro labe (plus six for those having photoelectric de-
vices). It is about half of the total number (forty-six, figure 2) of
objects of the list of galactic objects given by the working group of
Commission six of the IAU. Seventeen astro labe stations (figure 2)
are interested in a cooperation which final objective is to link the
radio and optical reference systems.

1. INTRODUCTION

La Commission 24 de l'Union Astronomique Internationale a constitué un
Groupe de travail, dont l'intitulé "Working Group on the Identification
of Radio/Optical Astrometric Sources", est significatif des objectifs
développés, notamment, lors de la XVIIème Assemblée Générale de l'UAI
(Montréal 1979). La recherche de radiosources à réplique optique et
l'identification de ces objets, conséquence directe du Colloque UAI n°
48 (Vienne 1978), sont fondées sur les bases jetées, dans le domaine de
ce qu'il est maintenant convenu d'appeler radioastrométrie, dès le Symposium UAI n° 61 (Perth 1973).

2. LES SOURCES RADIOASTROMETRIQUES

Ce groupe de travail a établi un rapport comprenant, outre des recom-
mandations aux observateurs potentiels, plusieurs listes d'objets. Deux
premières Tables (Ia et Ib) concernent des sources extragalactiques
primaires et secondaires ; il s'agit d'objets faibles (magnitudes, pour
la plupart, de l'ordre de 18) dont quelques-uns (magnitude de 12 à 15)
devraient pouvoir être observés, optiquement, par un assez grand nombre
d'instruments différents. La Table II comprend 46 objets galactiques,
donc brillants, dont la magnitude s'échelonne entre 1.1 et 15.5 ou 16.

E. M. Gaposchkin and B. Kołaczek (eds.), Reference Coordinate Systems for Earth Dynamics, 363–367.
Copyright © 1981 by D. Reidel Publishing Company.

L'ensemble des sources ainsi sélectionnées doit permettre de constituer, d'une part, le repère nécessaire pour établir un système de référence inertiel rapporté à des objets extragalactiques et, d'autre part, l'intermédiaire indispensable pour raccorder les systèmes de référence "optique" et "radio". Les objets des Tables I (a et b) permettent de prendre en compte les deux objectifs considérés ; ceux de la Table II serviront au seul rattachement "optique-radio".

Pour les objets extragalactiques, des études approfondies ont été faites (voir par exemple, de Vegt et Gehlich 1978, Walter et West 1980, Johnston et al. 1980). Les positions, dans le domaine "radio" sont déterminées à la précision de 0".05 (Wade et Johnston 1977). Cette précision n'est guère atteinte dans le domaine "optique" que pour les étoiles du Catalogue Fondamental de Référence (actuellement FK4, bientôt FK5).

A Paris, notamment, et après des campagnes conjointes à d'autres instruments (Tucker et al. 1973), des observations de β Perseï ont été systématiquement entreprises (Débarbat 1978) dès 1975. L'intérêt d'observations astrométriques précises de cette "radiosource-étoile fondamentale" résidait dans son emploi comme "zéro" pour les mesures radio-astrométriques (Elsmore et Ryle 1976).

Il apparaît maintenant souhaitable de définir ce "zéro", non plus par un seul objet mais par un ensemble de sources réparties sur tout le

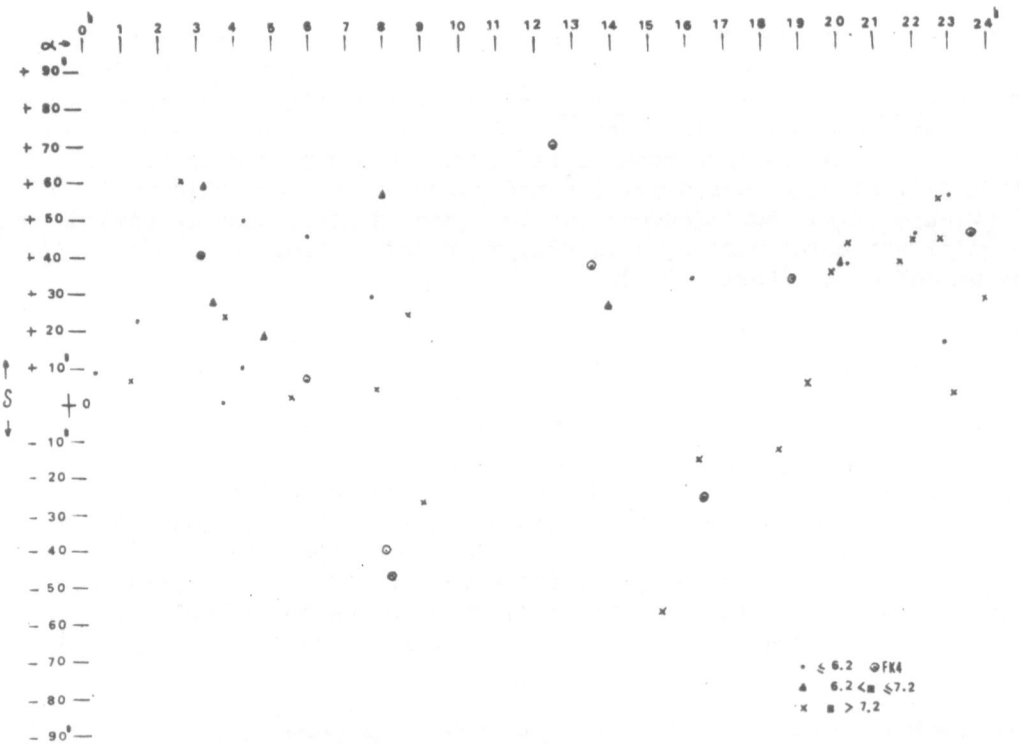

Figure 1 - Les radiosources.

ciel. C'est la raison pour laquelle le Département d'Astronomie Fondamentale de l'Observatoire de Paris a proposé aux observatoires dispo-
sant d'un astrolabe de participer à des campagnes systématiques d'observations des objets de la Table II établie par le groupe de travail
de la Commission 24.

3.   CONTRIBUTION DES ASTROLABES

Les radiosources à réplique optique de cette Table, ayant une magnitude
inférieure ou égale à 6.2, ont été selectionnées. La position de ces
21 objets est reportée sur la figure 1 (représentation par des points).
On observe une très faible présence de tels objets dans l'hémisphere
sud (3) et un trou entre 8h 07m et 12h 31m. En ajoutant les 6 objets
accessibles aux astrolabes photoélectriques (triangles) et ceux dont la
magnitude dépasse 7.2 (croix), on représente l'ensemble des objets de
la Table II. On remarque que l'hémisphere sud n'est guère enrichi, et
qu'on ne comble pas le vide horaire, seulement diminué et ramené a l'in-
tervalle 8h 57m - 12h 31m.

Ainsi la moitié des objets au moins est accessible aux astrolabes, et
de nombreux observatoires ont fait connaître leur intérêt pour cette
proposition : certains d'entre eux avaient déjà contribué à ce travail
en ayant, par hasard, introduit dans leurs programmes des étoiles fond-
amentales qui devaient se révéler être des radiosources. D'autres observatoires avaient déjà entrepris des campagnes systématiques pour de
tels objets. Enfin, parmi les observatoires intéressés, certains ont
une longue expérience, d'autres viennent seulement de débuter ; d'autres
- encore - sont en cours d'installation.

Sur la figure 2 sont représentés les observatoires qui ont, à cette

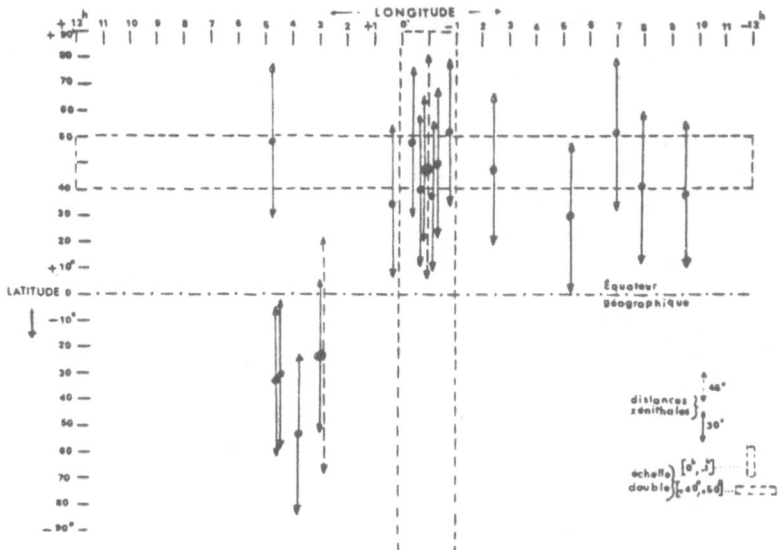

Figure 2 - Les stations "astrolabe".

date, fait connaître leur intérêt pour participer à ces campagnes systématiques d'observations de radiosources à réplique optique brillante.
Chacun d'entre eux est figuré par sa latitude, puisque c'est elle qui
conditionne les limites (en déclinaison) pour chaque station. La com-
paraison des figures 1 et 2 montre que toutes les radiosources observables à des astrolabes sont accessibles à l'une et/ou l'autre des
stations. L'absence d'astrolabe à basse latitude sera compensée par
l'installation, en cours au Brésil, de l'instrument transféré – depuis
le Royal Greenwich Observatory – aussi près que possible de l'équateur.
L'équerre optique à 45° dont dispose l'instrument du Cerga (France) per-
mettra d'accroître le recouvrement des observations effectuées en des
stations de latitude différente, de même celle qui sera, prochainement,
mise en place à Sao Paulo.

Parmi les radiosources de la Table II, 9 sont des étoiles du Catalogue
Fondamental de Référence (représentation, figure 1, par un point cer-
clé) ; elles sont toutes observables à l'un ou l'autre des astrolabes
représentés figure 2.

4.   CONCLUSIONS

Le raccordement des systèmes de référence "optique" et "radio", rendu
nécessaire par l'introduction de l'interférométrie à très longue base
(VLBI) passe nécessairement par des mesures conjointes à différents
instruments, parallèlement à celles des radiotélescopes. Pour les objets très faibles, seuls les astrographes pourront apporter leur con-
cours. Pour les objets plus brillants, les instruments méridiens seront les outils nécessaires ; sous leur forme moderne de méridiens
automatiques, ils permettront d'atteindre la précision interne de 0ʺ04
après six observations (Réquième, 1979). Leur nombre est limité à quelques unités seulement ; les astrolabes représentent, parallèlement, un
potentiel instrumental existant important.

Une coopération de tous les instruments, de tout type, doit permettre
de raccorder les systèmes de référence. L'avantage des méridiens et des
astrolabes est de fournir un rattachement direct au Système Fondamental.
Les erreurs de zone seront fortement atténuées aux astrolabes : les
groupes d'étoiles de rattachement s'étendent sur 6 h en ascension droite
et 60° en déclinaison et, de plus, leur composition sera différente
puisque les stations auront des latitudes variées.

Les aléas des conditions météorologiques seront palliés grâce à la répartition géographique des différents astrolabes. Il convient de remar-
quer que pour tirer le parti maximal de leur coopération il est nécessaire, en outre, que soit assurée une bonne répartition en latitude ;
on atteint ainsi au maximum de précision pour chacune des deux coordonnées (ascension droite et déclinaison). La participation des instruments de l'hémisphère sud (y compris en zone équatoriale) est fondamentale.

La précision des catalogues d'astrolabe est généralement estimée à
0."10. Des campagnes systématiques et assez denses (10 à 12 observations
à chaque passage) devraient permettre d'obtenir une solution globale
dont l'erreur interne ne devrait pas dépasser 0."05 , puisque cette va-
leur est très souvent atteinte dans les différents catalogues primaires.
Pour l'erreur externe on peut se reporter à ce que donne le Catalogue
Général d'Astrolabe (CGA) pour lequel les erreurs estimées (Billaud
1978) sont de 0.004 s en ascension droite et de 0.06" en déclinaison.

La contribution des astrolabes s'inscrit dans le cadre du raccordement
des systèmes de référence "optique" et "radio" au niveau de précision
actuel des observations radioastrométriques. Cette contribution ser-
vira également à déterminer les radiosources qu'il faudra effective-
ment utiliser lors de la synthèse définitive.

REFERENCES

Billaud, G. : 1978, Thèse, Contribution de l'astrolabe à l'améliora-
　　　　　　tion du système fondamental de référence : le Catalogue
　　　　　　Général.
Débarbat , S. : 1978, Colloque UAI n° 48.
Elsmore, B., Ryle, M. : 1976, MNRAS, 174, 411.
Johnston, K.J., Spencer, J.H., Kaplan, G.M., Klepczynski, W.J. and
McCarthy, D.D. : 1980, Cel. Mech., 22,　　.
Réquième, Y. : 1979, Colloque du GS V, Publication de l'Observatoire
　　　　　　de Bordeaux.
Tucker, R.H., Yallop, B.D., Argue, A.N., Kenworthy, C.M., Elsmore, B.
and Ryle, M. : 1973, MNRAS, 164, 27.
Vegt (de), Chr. and Gehlich, U.K. : 1978, Colloque UAI n° 48.
Wade, C.M., and Johnston, K.J. : 1977, Astron. J., 82, 791.
Walter, H.G. and West, R.M. : 1980, Astron. Astrophys. 86, 1.

AN ATTEMPT TO COMPARE THE RADIO ASTRONOMICAL SYSTEM OF
COORDINATES OF QUASARS WITH FK4

I. I. Kumkova
Pulkovo Observatory
USSR

ABSTRACT

A method of comparison of the FK4 system to a system based on
extragalactic sources is tested on actual data for 140 sources
observed by radio-interferometric and whose optical counterparts
have positions referred to FK4. The difference of coordinates
is analyzed in spherical harmonics series. The computation show
that the method gives satisfactory results. However, the precision of the data prevents from any interpretation in terms of
systematic errors of the FK4.

Positional observations of compact extragalactic radio sources by
radio interferometers have achieved high accuracy. There exist
more than two hundred sources with $0\overset{''}{.}02-0\overset{''}{.}05$ coordinate accuracy
and their number increases continuously (1,2,3). This permits
the hope that a high accuracy inertial coordinate system depending
upon a large number of extragalactic radio sources will be
established in the future.

The investigation of using such a system for improvement of a
fundamental system is also of interest. One of the possible ways
is to compare precise positions of radio sources derived from
radio interferometric observations with corresponding optical
data. The evaluation of this method was described by Gubanov and
Kumkova (1978). In the present paper, the suggested technique is
applied to real data. So, for this purpose, 140 identified radio
sources, mostly quasars, which have known precise radio and
optical positions were chosen. A list of above sources is given
in Table I. Unfortunately radio measurements deal mainly with
objects on the north hemisphere which is caused by the location
of radio antennae.

E. M. Gaposchkin and B. Kołaczek (eds.), Reference Coordinate Systems for Earth Dynamics, 369–373.
Copyright © 1981 by D. Reidel Publishing Company.

TABLE I

0106+013	0953+254	2005+403	0836+711	1705+456	1143-245
0123+257	0954+253	2134+004	0859+470	1730-130	1335-127
0133+476	0955+328	2200+420	0859-140	1738+447	1349-439
0134+329	0959-443	2201+315	0952+179	1743+173	145 -375
0229+132	1148-001	2251+158	0954+658	1749+096	1519-273
0235+164	1155+251	0056-002	0954+554	1928+738	1555+001
0237-233	1215+303	0115+027	1034-293	1933-400	1629+680
0300+471	1219+285	0122-004	1055+018	1954+513	1636+680
0316+413	1226+023	0153+744	1104-445	2021+614	1749+700
0332-403	1253-055	0202+319	1127-145	2134+141	1823+568
0333+321	1313-334	0212+735	1144-379	2227-399	1849+670
0338-214	1328+307	0237-027	1245-197	2230+114	1909+309
0420-015	1328+254	0336-019	1252+119	2345-168	2008-158
0518+165	1404+286	0430+052	1430-178	2352+495	2223-052
0642+449	1442+101	0438-436	1430+052	0016+731	2254+074
0735+178	1458+718	0440-004	1510-089	0200+045	
0736+017	1502+106	0454+844	1517+204	0224+671	
0738+318	1514+197	0454-234	1611+343	0402-362	
0827+243	1641+399	0537-441	1616+063	0422-380	
0831+557	1645+174	0538+498	1634+628	0531+194	
0839+187	1716+686	0552+398	1637+626	0607-157	
0851+202	1739+522	0605-085	1638+398	0615+820	
0912+297	1741-038	0749+096	1642+690	0636+680	
0923+392	1807+698	0814+425	1656+571	0829+187	
0945+408	1821+107	0828+494	1656+571	1030+415	

Radio coordinates of the sources are considered to have accuracy of about 0".1 (2,3,5,6). It is suggested that the list of radio sources may be considered as a catalogue with its system. The base of this assumption is a uniform technique of observations and reductions of data. Corresponding optical coordinates have inherent accuracy 0".1-0".3 (7,8,9,10,11,12). Photographic coordinate of northern objects were determined with reference to stars of AGK3 by means of intermediate stars. The system of AGK3 is FK4 system. For the southern objects, the catalogues SAO and Perth-70 were used, their systems being FK4 too, but to a less extent. Thus, the comparison of radio and optical coordinates of the sources under consideration may be regarded as the comparison of FK4 with the radio system.

As shown (4), the simultaneous determination of mutual orientation and of systematic errors of these systems does not give a satisfactory result and, therefore, the calculations of the parameters have been made independently. The mutual orientation was determined according to the formulae:

$$\Delta\alpha \ \text{Cos}\delta = \Delta w \ \text{Cos}\delta + \Delta i \ \text{Sin}\delta \ \text{Cos}(w-\alpha) - \Delta\Omega \left[\text{Cos}\delta \ \text{Cos}i + \text{Sin}\delta \ \text{Sin}i \ \text{Sin}(w-\alpha)\right] \ (1)$$

$$\Delta\delta = \Delta i \ \text{Sin}(w-\alpha) + \Delta\Omega \ \text{Sin}i \ \text{Cos}(w-\alpha)$$

where α, δ - equatorial coordinates of radio sources;
        i, Ω, w - Eulerian angles of mutual orientation of the
                coordinate systems;
            i - angle of mutual inclination of the main planes of
                the systems;
            Ω - longitude of ascending node in radio coordinate
                system;

w – right ascension of ascending node in radio system;

$\Delta\alpha, \alpha\delta$ – differences of radio and photographic positions which include inherent measuring errors in photographic reference process, possible discrepancy of radio and optical emission centroids, star coordinate errors in fundamental catalogue.

As we are comparing equatorial coordinates, we have to consider the case i = 0; then equations of errors (1) take from:

$$\Delta\alpha \; Cos\delta = \Delta i \; Sin\delta \; Cos(w-\alpha) + (\Delta w - \Delta\Omega) \; Cos\delta$$

(2)

$$\Delta\delta = \Delta i \; Sin(w-\alpha)$$

The systematic errors are expanded in normalized spherical functions:

$$\Delta\alpha \; Cos\delta = \sum_{m=o} \; \sum_{k=o} \; (A_{mk} \; Cosk\alpha + C_{mk} \; Sink\alpha) \; P_{mk} \; (Sin\delta)$$

$$\Delta\delta = \sum_{m=o} \; \sum_{k=o} \; (C_{mk} \; Cosk\alpha + d_{mk} \; Sink\alpha) \; P_{mk} \; (Sin\delta)$$

(3)

Two lists were chosen from the initial list of 140 objects for computation of the corrections to the orientation parameters. The first one consisted of 55 objects (which are contained in Table I) with differences of radio and optical coordinates less $0\overset{..}{.}6$. The calculations were fulfilled for the initial list and two variants. The results of the calculations of corrections Δi, $(\Delta\omega - \Delta\Omega)$ in the case i = 0, $\omega = 0$ are presented in Table 2. Analysis of the results of Table 2 shows that the evaluations of the same corrections are satisfactorily close and values of the correction Δi derived from right ascensions and from declinations essentially differ. The above results shows us that the systematic part of the differences $\Delta\alpha \; Cos\delta$, $\Delta\delta$ are stable and differ from each other. As it was shown earlier, the correction Δi_{α} correlates with member of expansions (3) A_{20}, the correction $(\Delta\omega - \Delta\Omega)$ correlates with A_{00}, A_{20}, B_{21} and $\Delta i\delta$ correlates with A_{11}. Thus the corrections presented in Table 2 are linear combinations of the corrections themselves and of the mentioned members of the expansions. The correlations between Δi_{α} and $(\Delta\omega - \Delta\Omega)_{\alpha}$ do not exceed 0.07 in all cases.

TABLE 2

Corrections	For 55 objects	For 115 objects	For 140 objects
Δi_{α}	$-0\overset{..}{.}03 \pm 0\overset{..}{.}04$	$-0\overset{..}{.}07 \pm 0\overset{..}{.}05$	$-0\overset{..}{.}06 \pm 0\overset{..}{.}06$
$(\Delta\omega - \Delta\Omega)_{\alpha}$	$-0\overset{..}{.}03 \pm \quad 2$	$-0\overset{..}{.}05 \pm \quad 2$	$-0\overset{..}{.}03 \pm \quad 3$
$\Delta i\delta$	$+0\overset{..}{.}04 \pm \quad 3$	$+0\overset{..}{.}3 \pm \quad 3$	$+0\overset{..}{.}01 \pm \quad 5$

The evaluation of systematic errors by formulae (3) has been made
for α and δ separately because their expansions are identical.
Sixteen members of expansions (3) were calculated, which corresponds to m,k $\leq$ 3. The calculations were fulfilled for three
lists of objects as it was done during the calculations of the
orientation. The values of factors of expansions (3) A_{mk},B_{mk} for
$\Delta\alpha$ Cosδ and for $\Delta\delta$ for variant of 115 objects are given in
Table 3. In the same Table 3, the rms errors of coefficients are
presented. The system of coefficients is not complete, the
factor A_{20} includes, in fact, A_{00} because these members do not
separate for this case of distribution of the objects on the
sphere. The systematic differences $\Delta\alpha$ Cosδ in terms of $0^s.001$
and $\Delta\delta$ in terms of $0''.01$ computed in accordance with data of the
Table 3 are given in Tables 4 and 5, respectively. The values for
$\delta < -45^0$ and $\delta > 60^0$ are not presented for the lack of sources in
these areas. The system of conditional equations for 55 objects
is difficult to solve because of high correlations between the
members of expansions (3). The results for the list of 140
objects are practically in accordance with the results presented
in Tables 4 and 5, but the corresponding errors are larger. In
addition, the list of 115 objects was subdivided in an arbitrary
way in two lists of equal volume. The systematic errors have
been computed for both lists and calculations and gave the same
results for both variants. It proves the nonrandom nature of the
results of the Tables 4 and 5. Though the final purpose of these
calculations is to estimate the systematic errors of FK4, the
quality of initial data does not yet allow us to interpret the
results in this way. The data of the Tables 4 and 5 should be regarded as systematic differences of the two coordinates systems,
viz., radio and optical.

TABLE 3

mk	Expansion $\Delta\alpha$ Cos δ A_{mk}, B_{mk} $0^s.001$		Expansion $\Delta\delta$ C_{mk},d_{mk} $0''.01$	
10	-8	$\pm$ 6	-22	$\pm$ 13
11	4	6	-9	11
11	3	5	0	10
20	12	7	22	13
21	4	4	5	7
21	-4	3	-9	7
22	14	5	5	10
22	2	4	6	9
30	5	7	8	14
31	-11	5	-17	11
31	6	5	1	10
32	3	5	-1	10
32	3	6	6	11
33	-2	4	5	7
33	-12	4	0	9

TABLE 4

δ\α	0ʰ	1	2	3	4	5	6	7	8	9	10	11	12	13	14	15	16	17	18	19	20	21	22	23
60°	4	3	2	2	1	1	1	1	1	0	-1	-0	0	2	4	5	6	6	5	4	4	4	4	4
45°	0	0	-1	-2	-2	-1	-0	1	0	-2	-4	-5	-4	-2	0	3	3	2	-1	-4	-6	-5	-3	-1
30°	-2	-1	-2	-3	-3	-1	1	2	-2	-1	-5	-8	-8	-5	-1	2	1	-3	-9	-15	-17	-15	-11	-6
15°	-1	1	-0	-2	-2	-0	3	4	4	-0	-6	-9	-9	-5	-0	3	1	-5	-14	-22	-25	-22	-15	-7
0°	-3	6	4	1	-0	1	3	4	3	-0	-6	-10	-9	-4	2	6	4	-3	-13	-22	-25	-21	-13	-4
-15°	-9	11	9	5	2	1	1	1	-1	-4	-8	-10	-7	-1	6	10	8	2	-7	-15	-16	-14	-6	3
-30°	15	16	13	8	3	-0	-2	-4	-6	-8	-10	-10	-6	1	8	12	12	8	2	-3	-4	-1	5	11
-45°	18	18	14	9	4	-1	-4	-7	-9	-10	-10	-8	-4	2	8	13	14	13	11	9	8	-10	14	17

TABLE 5

δ\α	0ʰ	1	2	3	4	5	6	7	8	9	10	11	12	13	14	15	16	17	18	19	20	21	22	23
60°	-3	-5	-7	-9	-11	-12	-11	-8	-5	-0	4	8	11	13	14	13	12	10	8	6	4	2	0	-1
45°	-9	-10	-12	-14	-17	-18	-18	-15	-10	-3	1	5	8	8	8	6	4	1	-2	-4	-6	-8	-9	-9
30°	-12	-11	-11	-13	-16	-19	-19	-17	-12	-6	-1	2	3	2	-0	-3	-6	-9	-12	-14	-16	-17	-16	-14
15°	-10	-7	-6	-8	-11	-14	-15	-14	-10	-6	-2	-0	-1	-3	-6	-9	-12	14	-16	-18	-20	-20	-18	-15
0°	-5	-1	1	-0	-3	-7	-10	-10	-8	-5	-3	-2	-3	-5	-7	-9	-9	-10	-11	-12	-14	-14	-13	-9
-15°	4	6	7	6	2	-3	-6	-7	-7	-4	-4	-4	-4	-3	-2	0	2	3	2	0	-1	-1	1	
30°	12	12	11	8	4	-0	-4	-7	-7	-7	-5	-4	-2	1	5	9	14	17	19	18	17	15	13	12
45°	19	17	13	9	5	0	-3	-6	-7	-6	-4	-1	3	7	13	19	24	29	31	31	30	27	25	22

The results obtained for the chosen list show obviously that the improvement of the fundamental system of star coordinates by means of precise positional radio observations may be obtained satisfactorily, if highly accurate uniform catalogue of not less than 200 extragalactic radio sources uniformly distributed on whole sphere and surely identified optically will be established by radio interferometric observations in future.

REFERENCES

Arque, A. N., et al.: 1978, IAU Col. N 48, Vienne, 155.
Bridle, A. H., Goodson, R. E.: 1977, I. Roy. Astron. Soc. Can 71, N 3, 240.
Clark, I. A., et al.: 1978, Astron. J. 81, 599.
Couper, H. A.: 1972, Astrophys. Lett. 10, 121.
de Vegt, C., Gehlich, E. D.: 1978, Astron. Astrophys. 67, 65.
Elsmore, B.: 1978, IAU Col. N 48, Vienne, 93.
Gubanov, V. S., Kumkova, I. I.: 1978, IAU Col. N 48, Vienne, 135.
Johnston, K. J.: 1978, IAU Col. N 48, Vienne 175.

Wade, C. M.: 1974, IAU Col. N 61, 133.
Wade, C. M., Johnston, K. J.: 1977, Astron. J. 82, 791.
Walter, H. A., West, R. S.: 1979, ESO Messenger, 18.

COMMENTS ON CONVENTIONAL TERRESTRIAL AND QUASI-INERTIAL REFERENCE SYSTEMS

Jean Kovalevsky
Centre d'Etudes et de Recherches
Geodynamiques et Astronomiques
06130 Grasse, France

Ivan I. Mueller
Department of Geodetic Science
The Ohio State University
Columbus, Ohio 43210 USA

ABSTRACT. After commenting on certain definitions related to both the terrestrial and celestial (quasi-inertial reference systems (CTS and CIS) to clarify terminology, the geodynamic requirements for such systems are reviewed. This is followed by a discussion of certain problematic aspects of the two systems. The article concludes with a list of required actions aimed to assure that the reference system issue is resolved early. The list is proposed for the joint IAG/IAU working group to be established in accordance with the Colloquium resolution printed elsewhere in this volume.

PREFACE

The authors were asked to review the Colloquium at the closing session and summarize the major conclusions. In this article, in order to save space, only those ideas are elaborated on which either are not discussed in the review articles of the authors elsewhere in this volume, or which require further elaboration. For this reason the reader should consult these articles first, otherwise this paper may appear to be a collection of somewhat disjointed thoughts.

1. COMMENTS ON TERMINOLOGY: IDEAL AND CONVENTIONAL REFERENCE SYSTEMS AND FRAMES

In order to clarify some of the conceptual aspects of various reference systems and frames, we propose to assign specific meanings to terms that have been used somewhat inconsistently in the past.

The purpose of a reference frame is to provide the means to materialize a reference system so that it can be used for the quantitative description of positions and motions on the earth (terrestrial frames), or of celestial bodies, including the earth, in space (celestial frames). In both cases the definition is based on a general statement giving the rationale for an ideal case, i.e., for an *ideal reference system*. For

E. M. Gaposchkin and B. Kołaczek (eds.), Reference Coordinate Systems for Earth Dynamics, 375–384.
Copyright © 1981 by D. Reidel Publishing Company.

example, one would have the concept of an ideal terrestrial system,
through the statement that with respect to such a system the crust
should have only deformations (i.e., no rotations or translations). The
ideal concept for a celestial system is that of an inertial system so
defined that in it the differential equations of motion may be written
without including any rotational term. In both cases the term "ideal"
indicates the conceptual definition only and that no means are proposed
to actually construct the system.

The actual construction implies the choice of a physical structure
whose motions in the ideal reference system can be described by physical
theories. This implies that the environment that acts upon the structure
is modeled by a chosen set of parameters. Such a choice is not unique:
there are many ways to model the motions or the deformations of the
earth; there are also many celestial bodies that may be the basis of a
dynamical definition of an inertial system (moon, planets, or artificial
satellites). Even if the choice is based on sound scientific principles
there remains a part of imperfection or arbitrariness. This is one of
the reasons why it is suggested to use the term "conventional" to characterize this choice. The other reason is related to the means, usually
conventional, by which the reference frames are defined in practice.

At this stage, there are still two steps that are necessary to
achieve the final materialization of the reference system so that one
can refer coordinates of objects to them. First, one has to define in
detail the model that is used in the relationship between the configuration of the basic structure and its coordinates. At this point, the coordinates are fully defined, but not necessarily accessible. We propose
to call such a model *conventional reference system*. The term "system"
thus includes the description of the physical environment as well as the
theories used in the definition of the coordinates. For example, the
FK4 (conventional) reference system is defined by the ecliptic as given
by Newcomb's theory of the sun, the values of precession and obliquity,
also given by Newcomb, and the Woolard theory of nutation. Once a reference system is chosen, it is still necessary to make it available to
the users. The system usually is materialized for this purpose by a
number of points, objects or coordinates to be used for referencing any
other point, object or coordinate. Thus, in addition to the conventional
choice of a system, it is necessary to construct a set of conventionally
chosen (or arrived at) parameters (e.g., star positions or pole coordinates). The set of such parameters, materializing the system, define a
conventional reference frame. For example, the FK4 catalogue of over
1500 star coordinates define the FK4 frame, materializing the FK4 system.
Another example is the BIH Conventional Terrestrial Frame, whose pole is
the origin of the polar motion derived (and published) by the BIH, and
whose longitude origin is the point on the equator of the above pole,
used by the BIH for deriving UT1. This frame materializes the BIH Conventional Terrestrial System (CTS), which itself is defined by the FK4
frame, Newcomb's constants of precession and obliquity, Woolard's series
of nutation, and by all the assumptions made regarding the reference coordinates of the participating observatories and their relative weights,
etc.

Another way of defining the CTS for the deformable earth is through the time varying positions of a number of terrestrial observatories whose coordinates are periodically reobserved by some international service. The frame of this CTS could then be derived from the changing coordinates through transformations containing rotational (and possible translational) parameters. These transformation parameters computed and published by the service would then define the frame of the system. The service, as part of the system definition, thus would have to make the assumption that the progressive changes of the reference coordinates of the observatories do not represent rotations (and translations) in the statistically significant sense. This mode seems to be the consensus for the establishment of the future CTS frame.

It is also necessary to point out that celestial reference systems may be defined *kinematically* (through the positions of extragalactic radio sources), or *dynamically* (through the geocentric or heliocentric motions of artificial satellites, moon, planets). Stellar systems, such as the FK5, are hybrid. Furthermore, approximations must be introduced in the model so that it is not true to say that these systems are realizations of an ideal inertial system. This is why it is appropriate to use the term conventional "quasi" inertial system (CIS) as a common term for all such celestial systems. The corresponding frames would be defined by either the adopted positions of a set of radio sources (kinematic frame) or the adopted geocentric or heliocentric ephemerides (dynamic frames), all serving the materialization of the CIS with greater or lesser success (accuracy).

2.   COMMENTS ON THE CONVENTIONAL TERRESTRIAL SYSTEM (CTS)

2.1 Geodynamic Requirements

Geodynamic requirements for reference systems may be discussed in terms of global or regional problems. The former are required for monitoring the earth's rotation, while the latter are mainly associated with crustal motion studies in which one is predominantly interested in strain or strain rate, quantities which are directly related to stress and rheology. Thus for these studies, global reference systems are not particularly important although it is desirable to relate regional studies to a global frame.

For the rotation studies one is interested in the variations of the earth's rotational rate and in the motions of the rotation axis both with respect to space (CIS) and the crust or the CTS. The problem therefore is threefold: (1) to establish a geometric description of the crust, either through the coordinates of a number of points fixed to the crust, or through polyhedron(s) connecting these points whose side lengths and angles are directly estimable from observations using the new space techniques (laser ranging or VLBI). The latter is preferred because of its geometric clarity. (2) To establish the time-dependent behavior of the polyhedron due to, for example, crustal motion, surface loading or

tides. (3) To relate the polyhedron to both the CIS and the CTS. For
the global tectonic problems only the first two points are relevant although these may also be resolved through point (3).

In the absence of deformation, the definition of the CTS is arbitrary. Its only requirement is that it rotates with the rigid earth,
but common sense suggests that the third axis should be close to the
mean position of the rotation axis and the first axis be near the origin
of longitudes. An arbitrary choice, such as the one presently defined
by the BIH-published polar coordinates and UT1, free from the complica-
tions introduced by the CIO definition, is appropriate.

In the presence of deformations, particularly long periodic or secular ones, the definition is more problematical, because of the inability
to separate rotational (and translational) crustal motions of the crust
from those of the CTS. For example, a westward drift of all observatories cannot be distinguished from a secular change in the rotation,and
neither can the secular motion of the pole be separated from plate tectonic motions. This is why the consensus seems to be the CTS described
in Section 1. If such a system is adapted, the secular type motions
mentioned above will be absorbed in the future CTS, by definition. Residuals with respect to such a CTS will provide estimates of relative
motions between stations, i.e., of the deformations.

One geophysical requirement of the reference system is that other
geophysical measurements can be related to it. One example is the gravity field. The reference frame generally used when giving values of the
spherical harmonic coefficients is tied to the axes of figure of the
earth. This frame should be simply related with sufficient accuracy to
the CTS as well as to the CIS in which, for example, satellite orbits
are calculated. Another example is height measurements with respect to
the geoid.

The vertical motions may require some special attention, because
absolute motions with respect to the center of mass have an immediate
geophysical interest and are realizable. Again, if the center of mass
has significant motions with respect to the crust, such a motion will be
absorbed in the future CTS, if defined as suggested above. At present
there is no compelling evidence that the center of mass is displaced
significantly at least at the decade time scale.

Apart from the geometrical considerations the configuration of observatories should be such that (1) there are stations on most of the
major tectonic plates in sufficient number to provide the necessary statistical strength, (2) the stations lie on relatively stable parts of the
plate so as to reduce the possibility that tectonic shifts in some stations will not overly influence, at least initially, the parameters defining the CTS frame.

2.2 The Future CTS

There is little doubt that the terrestrial reference frame presently adopted and tied to the CIO is of very little practical use because of its insufficient accessibility. Further, the astronomical observations currently used to maintain this and other frames (i.e., those of the BIH and the IPMS) should be replaced by methods which are not tied to the direction of the vertical but rather to directions tied to the crust. Such methods are the laser observations to satellites and to the moon, and VLBI. Portable systems can establish the polyhedron(s) discussed earlier, while permanent stations at suitably chosen locations would become the observatories for the maintenance of the CTS.

The repeatedly determined coordinates of the observatories by means of the above-mentioned techniques, suitably corrected for those variations which are due to well-established (especially periodic) deformations, will serve as the basis of the future CTS.

The definition of the CTS frame could have a similar form proposed by Guinot in this volume: The pole of the conventional terrestrial frame (CTP) is the origin of the polar motion derived by a future international service. The first axis of the frame (CTO) is the point on the equator of the CTP used by such a service for deriving UT1. In these derivations the assumption is made that the progressive changes of the reference coordinates of the observatories contributing to the determination of the earth rotation (position of the instantaneous rotation axis and UT1) do not represent statistically a rotation (and a translation).

Until the new system becomes operational, the above definition could be adopted for a specified existing service (BIH or IPMS), even if the coordinates of some of the contributing observatories are the astronomical latitudes and longitudes. An early adoption of such a definition would reduce the present confusion about the CTS described by Mueller in this volume. Possible alternative computational schemes for the determination of such future CTS parameters are also described there and also by Richter in this volume.

3. COMMENTS ON CONVENTIONAL QUASI-INERTIAL SYSTEMS (CIS)

3.1 Conceptual Considerations

Since the definition of such systems may be based on dynamical properties of the solar system as well as on the kinematics of extragalactic sources, we are led to distinguish between two kinds of quasi-inertial systems:

a) *Conventional kinematical systems*, based on the assumption that the proper motions of some celestial bodies have known statistical properties. In the case of extragalactic sources, it is postulated that remote galaxies have no rotational component in their motions.

  b) *Conventional dynamical systems*, based on the theory of the motion of some bodies in the solar system constructed in such a way that
there remains no rotational term in the equations of motion.

  If, in the framework of Newtonian mechanics, both definitions are
equivalent, this is no more true in the theory of general relativity. A
dynamical system of coordinates is a local reference that is locally
tangent to the general space-time manifold. In contrast, the celestial
system defined by the apparent directions of remote objects is a coordinate system that is subject to relativistic effects such as the geodetic
precession. Even if this is being suitably corrected for, there remains
a basic difference between the concept, and this is another good reason
to use the terminology *"quasi-inertial"* to characterize both celestial
and dynamical systems.

  It is now well agreed that the best future CIS will be based on the
position of extragalactic radio sources. But even if such a system is
due to play a major role among conventional quasi-inertial systems,
there may be great advantages, in some cases, to use a dynamical system.
This is the case, for instance, when artificial satellites are used to
monitor the earth rotation. This is why we are led to propose a certain
hierarchy among these systems and give to the CIS based on extragalactic
radio sources a role of a *primary system*, a role which is presently
played by the FK4 System and will be played, during the interim, by the
FK5 System before the VLBI based system is really set up and made available. Other systems, and in particular all the conventional dynamical
systems, will have to be connected to the primary system in order to
give consistent results.

3.2 Conventional Quasi-inertial Reference Frames

  The actual availability of the systems is obtained through their
realization in the form of reference frames. This materialization can
be done in two different ways so that one can distinguish between two
kinds of reference frames:

  a) *Stellar reference frames.* The fiducial points are presently
stars. Even if it is expected that they will be extragalactic radio
sources in the future, it will still be necessary to provide connections
to stellar catalogues, so that the celestial system be made available to
optical instrumentation.

  b) *Ephemeris reference frames.* In such frames, one or several moving objects are used as the materialization of the system (e.g., the GPS)
The theory supporting the corresponding reference system provides the
apparent ephemeris of the objects as a function of time and the observed
successive positions are the fiducial points needed to refer the observations to the system.

  It is to be noted that there is not a bi-univocal correspondence
between both types of frames and the two sorts of quasi-inertial systems.

For instance, the FK4 System is dynamical, while the FK4 Frame is
stellar.

3.3 Origin of Quasi-inertial Frames

Astronomers have always used the equator as a fundamental plane of
the coordinate system, and the origin was the equinox, although in studies of celestial mechanics, ecliptic coordinates are preferred. But none
of these fundamental planes appears necessary in a purely celestial reference system. Since the point of the origin has been hotly debated, let
us analyze the problem.

If a dynamical system is based on the motion of planets, the ecliptic plays a privileged role and, naturally, the ecliptic is used in the
definition of coordinates. Since equatorial coordinates are preferred
to ecliptic for obvious instrumental reasons, the ecliptic becomes the
natural origin of right ascensions. When the dynamical system is geocentric, the natural reference plane is the Laplace plane whose position depends upon the relative magnitude of the perturbations. For the moon,
the solar effects are dominant and, practically, the Laplace plane is
the ecliptic and, again, the equinox is the natural origin of equatorial
coordinates. In the case of artificial satellites presently used for
earth dynamics, the perturbations due to the earth flattening are pre-
dominant so that the Laplace plane is the equator. The equator is, therefore, the natural fundamental plane, but the origin may be arbitrary.
This explains why the mean equinox at a given epoch is used and not the
true equinox.

Similarly, the choice of the equinox in the FK4 series is justified
by the fact that they are dynamical systems based upon planetary theories.
However, in the construction of the corresponding stellar frame, the difficulty of maintaining the theoretical origin is so serious that one is
led to distinguish the dynamical equinox which defines the origin of the
system and the catalogue equinox which is the origin of the frame. In
practice, the actual origins of the FKn reference frames are purely conventional and are not the dynamical equinox.

This situation will become even more conspicuous for frames derived
from conventional celestial systems. Even if, for the sake of continuity,
the origin and the fundamental plane of a celestial system should be
close to the equinox and the equator, they should be conventional points
defined only by the realization of the corresponding frame. Otherwise,
it would be necessary to introduce a complex dynamical model to define
the origin at the expense of introducing inaccuracies in the system and
an uncertainty in its realization by the frame. In practice, the solution might be analogous to the present situation for the terrestrial longitude system. One would establish an international organization that
would provide the coordinates of radio sources in the conventional celestial frame, taking into account eventual changes in the number and position of the reference sources, due, for instance, to the disappearance
or motion of quasars.

3.4 Origin of Terrestrial and Geocentric Ephemeris Frames

Finally one should realize that the problem of the geometric origin
of the CTS frame is linked with this topic of its comparison with a geocentric ephemeris frame. The center of mass of the earth is directly
accessible to dynamical methods and is the natural origin of a geocentric
satellite-based dynamical system. But, as such, it is model dependent.
And, unless the terrestrial reference frame is also constructed from the
same satellites (as is the case in various earth models such as GEM, SAO,
GRIM), there may be inconsistencies between the assumed origin of a kinematically obtained terrestrial system and the center of mass. A time-
dependent error in the position of the center of mass, considered as the
origin of a terrestrial frame, may introduce spurious apparent shifts in
the position of stations that may then be interpreted as erroneous plate
motions. To avoid this problem the parameters defining the CTS frame
should include translational terms as suggested in Section 2.2.

4. CONCLUDING COMMENTS

From the discussion it is obvious that a number of actions are required to assure that the reference system issue is resolved early and
that uniformity is assured by means of international agreements. These,
not necessarily in order of importance, are the following:

Re CTS:
1. Selection of observatories whose catalogue will define the CTS.
2. Initiation of measurements at these observatories.
3. Recommendation on the observational and computational maintenance of
    the CTS (e.g., permanent versus temporary and repeated station occu-
    pations, constraints to be used).
4. Decision on how far and which way the earth deformation should be
    modeled initially.
5. Plans and recommendations for the establishment of new international
    service(s) to provide users with the appropriate information regard-
    ing the use of the CTS frame.

Re CIS:
6. Selection of extragalactic radio sources whose catalogue will define
    the CIS.
7. Improvement of the positions of these sources to a few milliseconds
    (arc).
8. Final decision on the IAU series of nutation and to assure that it
    describes the motion of the Celestial Ephemeris Pole.
9. Early completion of the FK5 and revision of astronomical equations
    due to the changed equinox (e.g., transformation between sidereal and
    Universal times).
10. Extension of the stellar catalogues (FK5 and later Hipparcos) to
    higher magnitudes.
11. Connection of the FK5, and later Hipparcos, reference frames to the
    CIS frame.

12. Assure that all dynamical (planetary, lunar and satellite) ephemer-
    ides are referenced to the CIS.
13. Plans and recommendations for the establishment of a service that
    would be in charge of the maintenance of the CIS frame.

It is hoped that the proposed joint IAG/IAU Working Group will consider these items during their deliberations.

As a summary, the following figure shows the hierarchy of the conventional and quasi-inertial reference systems discussed, including examples of possible connections between them. The heavy boxes indicate
the CTS and those CIS-s which are the most important ones from the points
of view of orientation and origin definition. The heavy lines indicate
the connections between these systems. It should be noted that though
in some cases it may be possible to theoretically derive the transformations between two systems (e.g., those based on the motion of the moon
and the planets), the result would not be of high accuracy because of
the dependence on the model used, i.e., due to the degradation of the
materialization of the systems. The connections through the observations
shown actually result in transformations between the various frames, but
this is exactly what the users need. For more details, see the authors'
articles elsewhere in this volume.

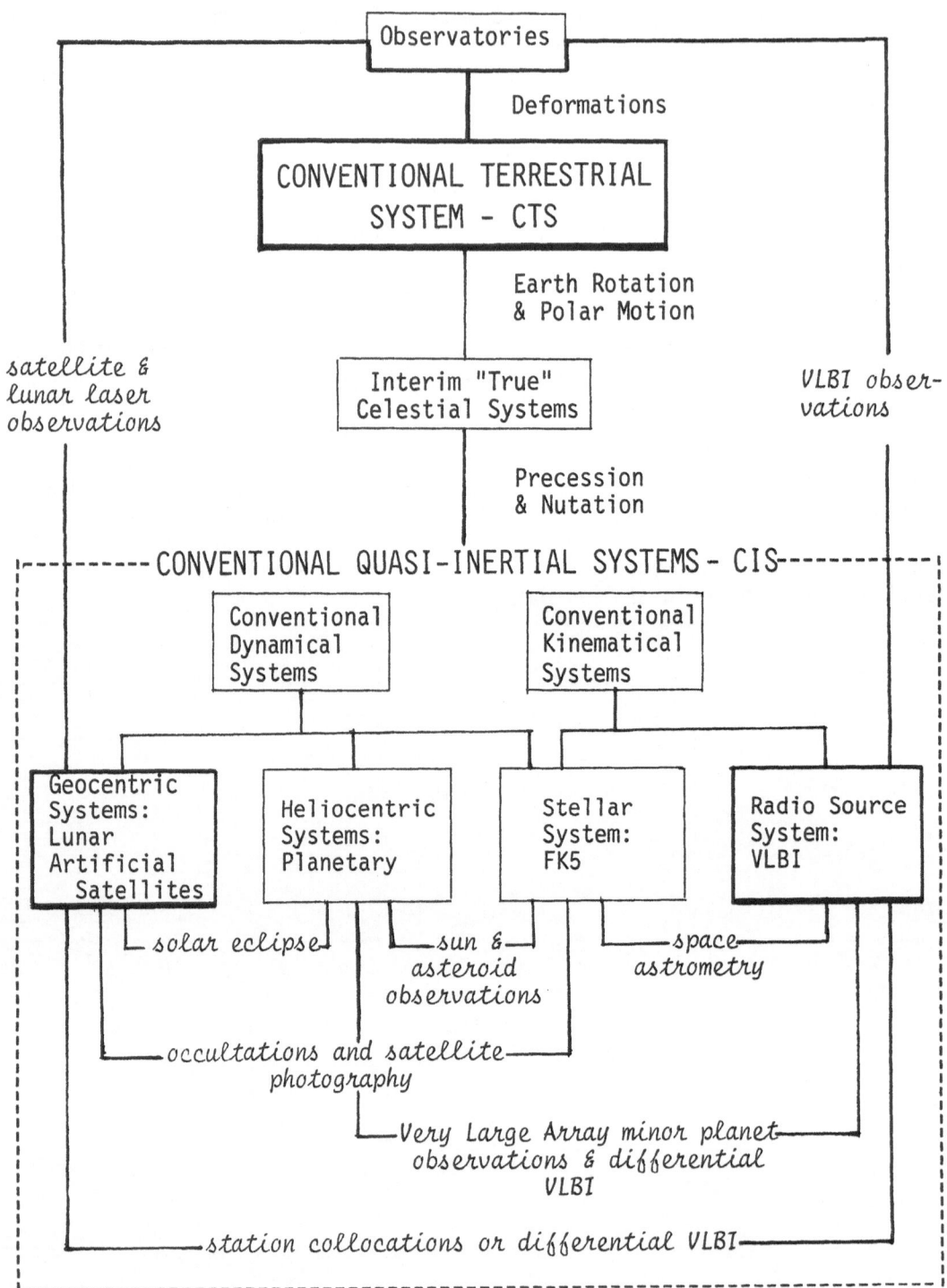

Conventional Terrestrial and Quasi-inertial Reference Systems
with Examples of Possible Connections.

RESOLUTION I

IAU Colloquium No. 56, "Reference Coordinate Systems for Earth Dynamics,"

Recognizing:

1) That geodynamics has become the subject of intensive international research during the last decade,

2) That a common requirement for all investigations is the necessity of a well-defined terrestrial coordinate system not available at present,

3) That the use of new techniques, such as Lageos laser ranging, LLR, VLBI, astrometric satellites, when used in a well-coordinated manner can determine and maintain such a system,

Recommends:

That a working group be established by the Presidents of IAU Commissions 4, 19, 31, and the President of IAG to prepare a proposal for the establishment and maintenance of a Conventional Terrestrial Reference System. This system is to include provisions for the replacement of the presently used terrestrial reference frame (such as the one partially defined by the CIO and the BIH zero meridian), providing continuity, and conformance with the IAU 1976 and 1979 resolutions regarding astronomical constants and the theory of nutation, or its possible modification, as well as with the IUGG Geodetic Reference System 1980.

The Colloquium further recommends that the working group should report its findings for discussion and possible adoption at the 1982/83 assemblies of the IAU and the IUGG/IAG.

E. M. Gaposchkin and B. Kołaczek (eds.), Reference Coordinate Systems for Earth Dynamics, 385.
Copyright © 1981 by D. Reidel Publishing Company.

INDEX

aberration 6, 306, 308
  E-terms 306, 308, 310-312

accelerometers 50

adjustment of data 29-31, 150-151, 229-231, 280-281, 335, 346, 353-356, 359-362

apparent forces 272-274

astrogravimetric technique to compare zero meridian of BIH and NWLD 197-199

astrolabe 84, 318, 363-367

astrometric satellite 5
  Hipparchos 8, 83, 282, 321, 323, 341-348, 382
  space telescope 213, 282, 342, 348

astrometry (optical) 136, 210, 263, 341-348, 384

atmospheric effects on observations 25, 27, 192-194, 225, 354

axis of angular momentum 14, 16, 68-69, 107

axis of figure 13-15, 106-107, 233-234, 240
  zero excitation figure axis 15-16
  mean figure axis 235

axis of rotation (instantaneous) 4, 9, 12, 14, 125, 131, 156, 233, 378-379

baseline distance measurements 64-65, 353
  accuracies 27, 113, 192, 207, 269

Bureau Internationale de l'Heure (BIH) 10, 148, 165-173, 379
  BIH pole 10, 127-129, 138-140, 376, 378
  BIH 1968 10, 157, 166
  BIH 1979 10, 158
  BIH UT 139, 141, 297, 378
  BIH zero meridian 10, 60, 126, 138, 146-147, 152, 167, 196-199, 263, 376

ASTROPHYSICS AND SPACE SCIENCE LIBRARY

Edited by

J. E. Blamont, R. L. F. Boyd, L. Goldberg, C. de Jager, Z. Kopal, G. H. Ludwig, R. Lüst,
B. M. McCormac, H. E. Newell, L. I. Sedov, Z. Švestka, and W. de Graaff

23. A. Muller (ed.), *The Magellanic Clouds. A European Southern Observatory Presentation: Principal Prospects, Current Observational and Theoretical Approaches, and Prospects for Future Research, Based on the Symposium on the Magellanic Clouds, held in Santiago de Chile, March 1969, on the Occasion of the Dedication of the European Southern Observatory*. 1971, XII + 189 pp.

24. B. M. McCormac (ed.), *The Radiating Atmosphere. Proceedings of a Symposium Organized by the Summer Advanced Study Institute, held at Queen's University, Kingston, Ontario, August 3–14, 1970*. 1971, XI + 455 pp.

25. G. Fiocco (ed.), *Mesospheric Models and Related Experiments. Proceedings of the 4th ESRIN-ESLAB Symposium, held at Frascati, Italy, July 6–10, 1970*. 1971, VIII + 298 pp.

26. I. Atanasijević, *Selected Exercises in Galactic Astronomy*. 1971, XII + 144 pp.

27. C. J. Macris (ed.), *Physics of the Solar Corona. Proceedings of the NATO Advanced Study Institute on Physics of the Solar Corona, held at Cavouri-Vouliagmeni, Athens, Greece, 6–17 September 1970*. 1971, XII + 345 pp.

28. F. Delobeau, *The Environment of the Earth*. 1971, IX + 113 pp.

29. E. R. Dyer (general ed.), *Solar-Terrestrial Physics/1970. Proceedings of the International Symposium on Solar-Terrestrial Physics, held in Leningrad, U.S.S.R., 12–19 May 1970*. 1972, VIII + 938 pp.

30. V. Manno and J. Ring (eds.), *Infrared Detection Techniques for Space Research. Proceedings of the 5th ESLAB-ESRIN Symposium, held in Noordwijk, The Netherlands, June 8–11, 1971*. 1972, XII + 344 pp.

31. M. Lecar (ed.), *Gravitational N-Body Problem. Proceedings of IAU Colloquium No. 10, held in Cambridge, England, August 12–15, 1970*. 1972, XI + 441 pp.

32. B. M. McCormac (ed.), *Earth's Magnetospheric Processes. Proceedings of a Symposium Organized by the Summer Advanced Study Institute and Ninth ESRO Summer School, held in Cortina, Italy, August 30–September 10, 1971*. 1972, VIII + 417 pp.

33. Antonin Rükl, *Maps of Lunar Hemispheres*. 1972, V + 24 pp.

34. V. Kourganoff, *Introduction to the Physics of Stellar Interiors*. 1973, XI + 115 pp.

35. B. M. McCormac (ed.), *Physics and Chemistry of Upper Atmospheres. Proceedings of a Symposium Organized by the Summer Advanced Study Institute, held at the University of Orléans, France, July 31–August 11, 1972*. 1973, VIII + 389 pp.

36. J. D. Fernie (ed.), *Variable Stars in Globular Clusters and in Related Systems. Proceedings of the IAU Colloquium No. 21, held at the University of Toronto, Toronto, Canada, August 29–31, 1972*. 1973, IX + 234 pp.

37. R. J. L. Grard (ed.), *Photon and Particle Interaction with Surfaces in Space. Proceedings of the 6th ESLAB Symposium, held at Noordwijk, The Netherlands, 26–29 September, 1972*. 1973, XV + 577 pp.

38. Werner Israel (ed.), *Relativity, Astrophysics and Cosmology. Proceedings of the Summer School, held 14–26 August, 1972, at the BANFF Centre, BANFF, Alberta, Canada*. 1973, IX + 323 pp.

39. B. D. Tapley and V. Szebehely (eds.), *Recent Advances in Dynamical Astronomy. Proceedings of the NATO Advanced Study Institute in Dynamical Astronomy, held in Cortina d'Ampezzo, Italy, August 9–12, 1972*. 1973, XIII + 468 pp.

40. A. G. W. Cameron (ed.), *Cosmochemistry. Proceedings of the Symposium on Cosmochemistry, held at the Smithsonian Astrophysical Observatory, Cambridge, Mass., August 14–16, 1972*. 1973, X + 173 pp.

41. M. Golay, *Introduction to Astronomical Photometry*. 1974, IX + 364 pp.

42. D. E. Page (ed.), *Correlated Interplanetary and Magnetospheric Observations. Proceedings of the 7th ESLAB Symposium, held at Saulgau, W. Germany, 22–25 May, 1973*. 1974, XIV + 662 pp.

43. Riccardo Giacconi and Herbert Gursky (eds.), *X-Ray Astronomy*. 1974, X + 450 pp.

44. B. M. McCormac (ed.), *Magnetospheric Physics. Proceedings of the Advanced Summer Institute, held in Sheffield, U.K., August 1973*. 1974, VII + 399 pp.

45. C. B. Cosmovici (ed.), *Supernovae and Supernova Remnants. Proceedings of the International Conference on Supernovae, held in Lecce, Italy, May 7–11, 1973*. 1974, XVII + 387 pp.

46. A. P. Mitra, *Ionospheric Effects of Solar Flares*. 1974, XI + 294 pp.

47. S.-I. Akasofu, *Physics of Magnetospheric Substorms*. 1977, XVIII + 599 pp.

48. H. Gursky and R. Ruffini (eds.), *Neutron Stars, Black Holes and Binary X-Ray Sources*. 1975, XII + 441 pp.

49. Z. Švestka and P. Simon (eds.), *Catalog of Solar Particle Events 1955–1969. Prepared under the Auspices of Working Group 2 of the Inter-Union Commission on Solar-Terrestrial Physics*. 1975, IX + 428 pp.

50. Zdeněk Kopal and Robert W. Carder, *Mapping of the Moon*. 1974, VIII + 237 pp.

51. B. M. McCormac (ed.), *Atmospheres of Earth and the Planets. Proceedings of the Summer Advanced Study Institute, held at the University of Liège, Belgium, July 29–August 8, 1974*. 1975, VII + 454 pp.

52. V. Formisano (ed.), *The Magnetospheres of the Earth and Jupiter. Proceedings of the Neil Brice Memorial Symposium, held in Frascati, May 28–June 1, 1974*. 1975, XI + 485 pp.

53. R. Grant Athay, *The Solar Chromosphere and Corona: Quiet Sun*. 1976, XI + 504 pp.

54. C. de Jager and H. Nieuwenhuijzen (eds.), *Image Processing Techniques in Astronomy. Proceedings of a Conference, held in Utrecht on March 25–27, 1975*. XI + 418 pp.

55. N. C. Wickramasinghe and D. J. Morgan (eds.), *Solid State Astrophysics. Proceedings of a Symposium, held at the University College, Cardiff, Wales, 9–12 July 1974*. 1976, XII + 314 pp.

56. John Meaburn, *Detection and Spectrometry of Faint Light*. 1976, IX + 270 pp.

57. K. Knott and B. Battrick (eds.), *The Scientific Satellite Programme during the International Magnetospheric Study. Proceedings of the 10th ESLAB Symposium, held at Vienna, Austria, 10–13 June 1975*. 1976, XV + 464 pp.

58. B. M. McCormac (ed.), *Magnetospheric Particles and Fields. Proceedings of the Summer Advanced Study School, held in Graz, Austria, August 4–15, 1975*. 1976, VII + 331 pp.

59. B. S. P. Shen and M. Merker (eds.), *Spallation Nuclear Reactions and Their Applications*. 1976, VIII + 235 pp.

60. Walter S. Fitch (ed.), *Multiple Periodic Variable Stars. Proceedings of the International Astronomical Union Colloquium No. 29, held at Budapest, Hungary, 1–5 September 1976*. 1976, XIV + 348 pp.

61. J. J. Burger, A. Pedersen, and B. Battrick (eds.), *Atmospheric Physics from Spacelab. Proceedings of the 11th ESLAB Symposium, Organized by the Space Science Department of the European Space Agency, held at Frascati, Italy, 11–14 May 1976*. 1976, XX + 409 pp.

62. J. Derral Mulholland (ed.), *Scientific Applications of Lunar Laser Ranging. Proceedings of a Symposium held in Austin, Tex., U.S.A., 8–10 June, 1976*. 1977, XVII + 302 pp.

63. Giovanni G. Fazio (ed.), *Infrared and Submillimeter Astronomy. Proceedings of a Symposium held in Philadelphia, Penn., U.S.A., 8–10 June, 1976*. 1977, X + 226 pp.

64. C. Jaschek and G. A. Wilkins (eds.), *Compilation, Critical Evaluation and Distribution of Stellar Data. Proceedings of the International Astronomical Union Colloquium No. 35, held at Strasbourg, France, 19–21 August, 1976*. 1977, XIV + 316 pp.

65. M. Friedjung (ed.), *Novae and Related Stars. Proceedings of an International Conference held by the Institut d'Astrophysique, Paris, France, 7–9 September, 1976*. 1977, XIV + 228 pp.

66. David N. Schramm (ed.), *Supernovae. Proceedings of a Special IAU-Session on Supernovae held in Grenoble, France, 1 September, 1976*. 1977, X + 192 pp.

67. Jean Audouze (ed.), *CNO Isotopes in Astrophysics. Proceedings of a Special IAU Session held in Grenoble, France, 30 August, 1976*. 1977, XIII + 195 pp.

68. Z. Kopal, *Dynamics of Close Binary Systems*, XIII + 510 pp.

69. A. Bruzek and C. J. Durrant (eds.), *Illustrated Glossary for Solar and Solar-Terrestrial Physics*. 1977, XVIII + 204 pp.

70. H. van Woerden (ed.), *Topics in Interstellar Matter*. 1977, VIII + 295 pp.

71. M. A. Shea, D. F. Smart, and T. S. Wu (eds.), *Study of Travelling Interplanetary Phenomena*. 1977, XII + 439 pp.

72. V. Szebehely (ed.), *Dynamics of Planets and Satellites and Theories of Their Motion. Proceedings of IAU Colloquium No. 41, held in Cambridge, England, 17–19 August 1976*. 1978, XII + 375 pp.

73. James R. Wertz (ed.), *Spacecraft Attitude Determination and Control*. 1978, XVI + 858 pp.

74. Peter J. Palmadesso and K. Papadopoulos (eds.), *Wave Instabilities in Space Plasmas. Proceedings of a Symposium Organized Within the XIX URSI General Assembly held in Helsinki, Finland, July 31–August 8, 1978.* 1979, VII + 309 pp.

75. Bengt E. Westerlund (ed.), *Stars and Star Systems. Proceedings of the Fourth European Regional Meeting in Astronomy held in Uppsala, Sweden, 7–12 August, 1978.* 1979, XVIII + 264 pp.

76. Cornelis van Schooneveld (ed.), *Image Formation from Coherence Functions in Astronomy. Proceedings of IAU Colloquium No. 49 on the Formation of Images from Spatial Coherence Functions in Astronomy, held at Groningen, The Netherlands, 10–12 August 1978.* 1979, XII + 338 pp.

77. Zdeněk Kopal, *Language of the Stars. A Discourse on the Theory of the Light Changes of Eclipsing Variables.* 1979, VIII + 280 pp.

78. S.-I. Akasofu (ed.), *Dynamics of the Magnetosphere. Proceedings of the A.G.U. Chapman Conference 'Magnetospheric Substorms and Related Plasma Processes' held at Los Alamos Scientific Laboratory, N.M., U.S.A., October 9–13, 1978.* 1980, XII + 658 pp.

79. Paul S. Wesson, *Gravity, Particles, and Astrophysics. A Review of Modern Theories of Gravity and G-variability, and their Relation to Elementary Particle Physics and Astrophysics.* 1980, VIII + 188 pp.

80. Peter A. Shaver (ed.), *Radio Recombination Lines. Proceedings of a Workshop held in Ottawa, Ontario, Canada, August 24-25, 1979.* 1980, X + 284 pp.

81. Pier Luigi Bernacca and Remo Ruffini (eds.), *Astrophysics from Spacelab*, 1980, XI + 664 pp.

82. Hannes Alfvén, *Cosmic Plasma*, 1981, X + 160 pp.

83. Michael D. Papagiannis (ed.), *Strategies for the Search for Life in the Universe*, 1980, XVI + 254 pp.

84. H. Kikuchi (ed.), *Relation between Laboratory and Space Plasmas*, 1981, XII + 386 pp.

85. Peter van der Kamp, *Stellar Paths*, 1981, forthcoming.

86. E. M. Gaposchkin and B. Kołaczek (eds.), *Reference Coordinate Systems for Earth Dynamics*, 1981, XIV + 396 pp.